Multivariate Nonparametric Regression and Visualization

WILEY SERIES IN COMPUTATIONAL STATISTICS

Wiley Series in Computational Statistics is comprised of practical guides and cutting edge research books on new developments in computational statistics. It features quality authors with a strong applications focus. The texts in the series provide detailed coverage of statistical concepts, methods and case studies in areas at the interface of statistics, computing, and numerics.

With sound motivation and a wealth of practical examples, the books show in concrete terms how to select and to use appropriate ranges of statistical computing techniques in particular fields of study. Readers are assumed to have a basic understanding of introductory terminology.

The series concentrates on applications of computational methods in statistics to fields of bioinformatics, genomics, epidemiology, business, engineering, finance and applied statistics.

Billard and Diday · *Symbolic Data Analysis: Conceptual Statistics and Data Mining*
Bolstad · *Understanding Computational Bayesian Statistics*
Dunne · *A Statistical Approach to Neural Networks for Pattern Recognition*
Ntzoufras · *Bayesian Modeling Using WinBUGS*
Klemela · *Multivariate Nonparametric Regression and Visualization: With R and Applications to Finance*

Multivariate Nonparametric Regression and Visualization

With R and Applications to Finance

JUSSI KLEMELÄ

Department of Mathematical Sciences
University of Oulu
Oulu, Finland

Published by John Wiley & Sons, Inc., Hoboken, New Jersey.
Published simultaneously in Canada.

Library of Congress Cataloging-in-Publication Data:

Klemelä, Jussi, 1965–
Multivariate nonparametric regression and visualization : with R and applications to finance / Jussi Klemelä.
pages cm. — (Wiley series in computational statistics ; 699)
Includes bibliographical references and index.
ISBN 978-0-470-38442-8 (hardback)
1. Finance—Mathematical models. 2. Visualization. 3. Regression analysis. I. Title.
HG176.5.K55 2014
519.5'36—dc23 2013042095

Printed in Singapore.

10 9 8 7 6 5 4 3 2 1

To my parents

CONTENTS IN BRIEF

CONTENTS

PREFACE

The book is intended for students and researchers who want to learn to apply nonparametric and semiparametric methods and to use visualization tools related to these estimation methods. In particular, the book is intended for students and researchers in quantitative finance who want to apply statistical methods and for students and researchers of statistics who want to learn to apply statistical methods in quantitative finance. The book continues the themes of Klemelä (2009), which studied density estimation. The current book focuses on regression function estimation.

The book was written at the University of Oulu, Department of Mathematical Sciences. I wish to acknowledge the support provided by the University of Oulu and the Department of Mathematical Sciences.

The web page of the book is http://cc.oulu.fi/~jklemela/regstruct/.

JUSSI KLEMELÄ

Oulu, Finland
October 2013

INTRODUCTION

We study regression analysis and classification, as well as estimation of conditional variances, quantiles, densities, and distribution functions. The focus of the book is on nonparametric methods. Nonparametric methods are flexible and able to adapt to various kinds of data, but they can suffer from the curse of dimensionality and from the lack of interpretability. Semiparametric methods are often able to cope with quite high-dimensional data and they are often easier to interpret, but they are less flexible and their use may lead to modeling errors. In addition to terms "nonparametric estimator" and "semiparametric estimator", we can use the term "structured estimator" to denote such estimators that arise, for example, in additive models. These estimators obey a structural restriction, whereas the term "semiparametric estimator" is used for estimators that have a parametric and a nonparametric component.

Nonparametric, semiparametric, and structured methods are well established and widely applied. There are, nevertheless, areas where a further work is useful. We have included three such areas in this book:

1. Estimation of several functionals of a conditional distribution; not only estimation of the conditional expectation but also estimation of the conditional variance and conditional quantiles.

2. Quantitative finance as an area of application for nonparametric and semiparametric methods.

3. Visualization tools in statistical learning.

I.1 ESTIMATION OF FUNCTIONALS OF CONDITIONAL DISTRIBUTIONS

One of the main topics of the book are the kernel methods. Kernel methods are easy to implement and computationally feasible, and their definition is intuitive. For example, a kernel regression estimator is a local average of the values of the response variable. Local averaging is a general regression method. In addition to the kernel estimator, examples of local averaging include the nearest-neighbor estimator, the regressogram, and the orthogonal series estimator.

We cover linear regression and generalized linear models. These models can be seen as starting points to many semiparametric and structured regression models. For example, the single index model, the additive model, and the varying coefficient linear regression model can be seen as generalizations of the linear regression model or the generalized linear model.

Empirical risk minimization is a general approach to statistical estimation. The methods of empirical risk minimization can be used in regression function estimation, in classification, in quantile regression, and in the estimation of other functionals of the conditional distribution. The method of local empirical risk minimization is a method which can be seen as a generalization of the kernel regression.

A regular regressogram is a special case of local averaging, but the empirical choice of the partition leads to a rich class of estimators. The choice of the partition is made using empirical risk minimization. In the one- and two-dimensional cases a regressogram is usually less efficient than the kernel estimator, but in high-dimensional cases a regressogram can be useful. For example, a method to select the partition of a regressogram can be seen as a method of variable selection, if the chosen partition is such that it can be defined using only a subset of the variables. The estimators that are defined as a solution of an optimization problem, like the minimizers of an empirical risk, need typically be calculated with numerical methods. Stagewise algorithms can also be taken as a definition of an estimator, even without giving an explicit minimization problem which they solve.

A regression function is defined as the conditional expectation of the distribution of a response variable. The conditional expectation is useful in making predictions as well as in finding causal relationships. We cover also the estimation of the conditional variance and conditional quantiles. These are needed to give a more complete view of the conditional distribution. Also, the estimation of the conditional variance and conditional quantiles is needed in risk management, which is an important area of quantitative finance. The conditional variance can be estimated by estimating the conditional expectation of the squared random variable, whereas a conditional quantile is a special case of the conditional median. In the time series setting the standard approaches for estimating the conditional variance are the ARCH and GARCH modeling, but we discuss nonparametric alternatives. The GARCH estimator is close to a moving average, whereas the ARCH estimator is related to linear state space modeling.

In classification we are not interested in the estimation of functionals of a distribution, but the aim is to construct classification rules. However, most of the regression function estimation methods have a counterpart in classification.

I.2 QUANTITATIVE FINANCE

Risk management, portfolio selection, and option pricing can be identified as three important areas of quantitative finance. Parametric statistical methods have been dominating the statistical research in quantitative finance. In risk management, probability distributions have been modeled with the Pareto distribution or with distributions derived from the extreme value theory. In portfolio selection the multivariate normal model has been used together with the Markowitz theory of portfolio selection. In option pricing the Black-Scholes model of stock prices has been widely applied. The Black-Scholes model has also been extended to more general parametric models for the process of stock prices.

In risk management the p-quantile of a loss distribution has a direct interpretation as such threshold that the probability of the loss exceeding the threshold is less than p. Thus estimation of conditional quantiles is directly relevant for risk management. Unconditional quantile estimators do not take into account all available information, and thus in risk management it is useful to estimate conditional quantiles. The estimation of the conditional variance can be applied in the estimation of a conditional quantile, because in location-scale families the variance determines the quantiles. The estimation of conditional variance can be extended to the estimation of the conditional covariance or the conditional correlation.

We apply nonparametric regression function estimation in portfolio selection. The portfolio is selected either with the maximization of a conditional expected utility or with the maximization of a Markowitz criterion. When the collection of allowed portfolio weights is a finite set, then also classification can be used in portfolio selection. The squared returns are much easier to predict than the returns themselves, and thus in quantitative finance the focus has been in the prediction of volatility. However, it can be shown that despite the weak predictability of the returns, portfolio selection can profit from statistical prediction.

Option pricing can be formulated as a problem of stochastic control. We do not study the statistics of option pricing in detail, but give a basic framework for solving some option pricing problems nonparametrically.

I.3 VISUALIZATION

Statistical visualization is often considered as a visualization of the raw data. The visualization of the raw data can be a part of the exploratory data analysis, a first step to model building, and a tool to generate hypotheses about the data-generating mechanism. However, we put emphasis on a different approach to visualization.

In this approach, visualization tools are associated with statistical estimators or inference procedures. For example, we estimate first a regression function and then try to visualize and describe the properties of this regression function estimate. The distinction between the visualization of the raw data and the visualization of the estimator is not clear when nonparametric function estimation is used. In fact, nonparametric function estimation can be seen as a part of exploratory data analysis.

The SiZer is an example of a tool that combines visualization and inference, see Chaudhuri & Marron (1999). This methodology combines formal testing for the existence of modes with the SiZer maps to find out whether a mode of a density estimate of a regression function estimate is really there.

Semiparametric function estimates are often easier to visualize than nonparametric function estimates. For example, in a single index model the regression function estimate is a composition of a linear function and a univariate function. Thus in a single index model we need only to visualize the coefficients of the linear function and a one-dimensional function. The ease of visualization gives motivation to study semiparametric methods.

CART, as presented in Breiman, Friedman, Olshen & Stone (1984), is an example of an estimation method whose popularity is not only due to its statistical properties but also because it is defined in terms of a binary tree that gives directly a visualization of the estimator. Even when it is possible to find estimators with better statistical properties than CART, the possibility to visualization gives motivation to use CART.

Visualization of nonparametric function estimates, such as kernel estimates, is challenging. For the visualization of completely nonparametric estimates, we can use level set tree-based methods, as presented in Klemelä (2009). Level set tree-based methods have found interest also in topological data analysis and in scientific visualization, and these methods have their origin in the concept of a Reeb graph, defined originally in Reeb (1946).

In density estimation we are often interested in the mode structure of the density, defined as the number of local extremes, the largeness of the local extremes, and the location of the local extremes. The local extremes of a density function are related to the areas of concentration of the probability mass. In regression function estimation we are also interested in the mode structure. The local maxima of a regression function are related to the regions of the space of the explanatory variables where the response variable takes the largest values. The antimode structure is equally important to describe. The antimode structure means the number of local minima, the size of the local minima, and the location of the local minima. The local minima of a regression function are related to the areas of the space of the explanatory variables where the response variable takes the smallest values.

The mode structure of a regression function does not give complete information about the properties of the regression function. In regression analysis we are interested in the effects of the explanatory variables on the response variable and in the interaction between the explanatory variables. The effect of an explanatory variable can be formalized with the concept of a partial effect. The partial effect of an explanatory variable is the partial derivative of the regression function with respect to this variable. Nearly constant partial effects indicate that the regression function is close to a linear function, since the partial derivatives of a linear function are constants. The local maxima of a partial effect correspond to the areas in the space of the explanatory variables where the increase of the expected value of the response variable, resulting from an increase of the value of the explanatory variable, is the largest. We can use level set trees of partial effects to visualize the mode structure and the antimode structure of the partial effects, and thus to visualize the effects and the interactions of the explanatory variables.

I.4 LITERATURE

We mention some of the books that have been used in the preparation of this book. Härdle (1990) covers nonparametric regression with an emphasis on kernel regression, discussing smoothing parameter selection, giving confidence bands, and providing various econometric examples. Hastie, Tibshirani & Friedman (2001) describe high-dimensional linear and nonlinear classification and regression methods, giving many examples from biometry and machine learning. Györfi, Kohler, Krzyzak & Walk (2002) cover asymptotic theory of kernel regression, nearest-neighbor regression, empirical risk minimization, and orthogonal series methods, and they also include a treatment of time series prediction. Ruppert, Wand & Carroll (2003) view nonparametric regression as an extension of parametric regression and treat them together. Härdle, Müller, Sperlich & Werwatz (2004) explain single index models, generalized partial linear models, additive models, and several nonparametric regression function estimators, giving econometric examples. Wooldridge (2005) provides an asymptotic theory of linear regression, including instrumental variables and panel data. Fan & Yao (2005) study nonlinear time series and use nonparametric function estimation in time series prediction and explanation. Wasserman (2005) provides information on nonparametric regression and density estimation with confidence intervals and bootstrap confidence intervals. Horowitz (2009) covers semiparametric models and discusses the identifiability and asymptotic distributions. Spokoiny (2010) introduces local parametric methods into nonparametric estimation.

Bouchaud & Potters (2003) have developed nonparametric techniques for financial analysis. Franke, Härdle & Hafner (2004) discuss statistical analysis of financial markets, with emphasis being on the parametric methods. Ruppert (2004) is a textbook suitable for statistics students interested in quantitative finance, and this book discusses statistical tools related to classical financial models. Malevergne & Sornette (2005) have analyzed financial data with nonparametric methods. Li & Racine (2007) consider various non- and semiparametric regression models presenting asymptotic distribution theory and the theory of smoothing parameter selection, directing towards econometric applications.

PART I

METHODS OF REGRESSION AND CLASSIFICATION

CHAPTER 1

OVERVIEW OF REGRESSION AND CLASSIFICATION

1.1 REGRESSION

In regression analysis we are interested in prediction or in inferring causal relationships. We try to predict the value of a response variable given the values of explanatory variables or try to deduce the causal influence of the explanatory variables to the response variable. The inference of a causal relationship is important when we want to to change the values of an explanatory variable in order to get an optimal value for the response variable. For example, we want to know the influence of education to the employment status of a worker in order to choose the best education. On the other hand, prediction is applied also in the cases when we are not able to, or do not wish to, change the values of the response variable. For example, in volatility prediction it is reasonable to use any variables that have a predictive relevance even if these variables do not have any causal relationship to volatility.

Both in prediction and in estimation of causal influence, it is useful to estimate the conditional expectation

$$E(Y \mid X = x)$$

of the response variable $Y \in \mathbf{R}$ given the explanatory variables $X \in \mathbf{R}^d$. The choice of the explanatory variables and the method of estimation can depend on the purpose of the research. In prediction an explanatory variable can be any variable that has predictive relevance whereas in the estimation of a causal influence the explanatory variables are determined by the scientific theory about the causal relationship. For the purpose of causal inference, it is reasonable to choose an estimation method that can help to find the partial effect of a given explanatory variable to the response variable. The partial effect is defined in Section 1.1.3.

In linear regression the regression function estimate is a linear function:

$$\hat{f}(x) = \hat{\alpha} + \hat{\beta}_1 x_1 + \cdots + \hat{\beta}_d x_d. \tag{1.1}$$

A different type of linearity occurs, if the estimator can be written as

$$\hat{f}(x) = \sum_{i=1}^{n} l_i(x)\, Y_i, \tag{1.2}$$

for some sequence of weights $l_1(x), \ldots, l_n(x)$. In fact, for the linear regression estimate, representations (1.1) and (1.2) hold; see (2.11). In the case of local averaging estimators, like regressogram, kernel estimator, and nearest-neighbor estimator, we use the notation $\hat{f}(x) = \sum_{i=1}^{n} p_i(x)\, Y_i$. In the case of local averaging estimators the weights $p_i(x)$ satisfy the properties that $p_i(x)$ is close to zero when X_i is distant from x and that $p_i(x)$ is large when X_i is near x. Local averaging is discussed in Section 3. There exists regression function estimates that cannot be written as in (1.2), like the orthogonal series estimators with hard thresholding; see (2.72).

In addition to the estimation of the conditional expectation of the response variable given the explanatory variables, we can consider also the estimation of the conditional median of the response variable given the explanatory variables, or the estimation of other conditional quantiles of the response variable given the explanatory variables, which is called quantile regression. Furthermore, we will consider estimation of the conditional variance, as well as estimation of the conditional density and the conditional distribution function of the response variable given the explanatory variables.

In regression analysis the response variable can take any real value or any value in a given interval, but we consider also classification. In classification the response variable can take only a finite number of distinct values and the interest lies in the prediction of the values of the response variable.

1.1.1 Random Design and Fixed Design

Random Design Regression In random design regression the data are a sequence of n pairs

$$(x_1, y_1), \ldots, (x_n, y_n), \tag{1.3}$$

where $x_i \in \mathbf{R}^d$ and $y_i \in \mathbf{R}$ for $i = 1, \ldots, n$. Data are modeled as a realization of a sequence of n random vectors

$$(X_1, Y_1), \ldots, (X_n, Y_n). \tag{1.4}$$

However, sometimes we do not distinguish notationally a random variable and its realization, and the notation of (1.4) is used also in the place of notation (1.3) to denote a realization of the random vectors and not the random vectors themselves.

In regression analysis we typically want to estimate the conditional expectation

$$f(x) = E(Y \mid X = x),$$

and now we assume that the sequence $(X_1, Y_1), \ldots, (X_n, Y_n)$ consists of identically distributed random variables, and (X, Y) has the same distribution as (X_i, Y_i), $i = 1, \ldots, n$. Besides conditional expectation we could estimate conditional mode, conditional variance, conditional quantile, and so on. Estimation of the conditional centers of distribution are discussed in Section 1.1.2 and estimation of conditional risk measures such as variance and quantiles are discussed in Section 1.1.4 and in Section 1.1.6.

Fixed Design Regression In fixed design regression the data are a sequence

$$y_1, \ldots, y_n,$$

where $y_i \in \mathbf{R}$, $i = 1, \ldots, n$. We assume that every observation y_i is associated with a fixed design point $x_i \in \mathbf{R}^d$.

Now the design points are not chosen by a random mechanism, but they are chosen by the conducter of the experiment. Typical examples could be time series data, where x_i is the time when the observation y_i is recorded, and spatial data, where x_i is the location where the observation y_i is made. Time series data are discussed in Section 1.1.9.

We model the data as a sequence of random variables

$$Y_1, \ldots, Y_n.$$

In the fixed design regression we typically do not assume that the data are identically distributed. For example, we may assume that

$$Y_i = f(x_i) + \epsilon_i, \qquad i = 1, \ldots, n,$$

where $x_i = i/n$, $f : [0, 1] \to \mathbf{R}$ is the function we want to estimate, and $E\epsilon_i = 0$. Now the data $Y_1, \ldots, Y_n$ are not identically distributed, since the observations Y_i have different expectations.

1.1.2 Mean Regression

The regression function is typically defined as a conditional expectation. Besides expectation and conditional expectation also median and conditional median can be used to characterize the center of a distribution and thus to predict and explain with the help of explanatory variables. We mention also the mode (maximum of the density function) as a third characterization of the center of a distribution, although the mode is typically not used in regression analysis.

Expectation and Conditional Expectation When the data

$$(X_1, Y_1), \ldots, (X_n, Y_n)$$

are a sequence of identically distributed random variables, we can use the data to estimate the regression function, defined as the conditional expectation of Y given X:

$$f(x) = E(Y \mid X = x), \qquad x \in \mathbf{R}^d, \tag{1.5}$$

where (X, Y) has the same distribution as (X_i, Y_i), $i = 1, \ldots, n$, and $X \in \mathbf{R}^d$, $Y \in \mathbf{R}$. The random variable Y is called the response variable, and the elements of random vector X are called the explanatory variables.

The mean of random variable $Y \in \mathbf{R}$ with a continuous distribution can be defined by

$$EY = \int_{-\infty}^{\infty} y f_Y(y)\, dy, \tag{1.6}$$

where $f_Y : \mathbf{R} \to \mathbf{R}$ is the density function of Y. The regression function has been defined in (1.5) as the conditional mean of Y, and the conditional expectation can be defined in terms of the conditional density as

$$E(Y \mid X = x) = \int_{-\infty}^{\infty} y\, f_{Y \mid X=x}(y)\, dy,$$

where the conditional density can be defined as

$$f_{Y \mid X=x}(y) = \frac{f_{X,Y}(x, y)}{f_X(x)}, \qquad y \in \mathbf{R}, \tag{1.7}$$

when $f_X(x) > 0$ and $f_{Y \mid X=x}(y) = 0$ otherwise, where $f_{X,Y} : \mathbf{R}^{d+1} \to \mathbf{R}$ is the joint density of (X, Y) and $f_X : \mathbf{R}^d \to \mathbf{R}$ is the density of X:

$$f_X(x) = \int_{\mathbf{R}} f_{X,Y}(x, y)\, dy, \qquad x \in \mathbf{R}^d.$$

Figure 1.1 illustrates mean regression. Our data consist of the daily S&P 500 returns $R_t = (S_t - S_{t-1})/S_{t-1}$, where S_t is the price of the index. There are about 16,000 observations. The S&P 500 index data are described more precisely in Section 1.6.1. We define the explanatory and the response variables as

$$X_t = \log_e \sqrt{\frac{1}{k} \sum_{i=1}^{k} R_{t-i}^2}, \qquad Y_t = \log_e |R_t|.$$

Panel (a) shows the scatter plot of (X_t, Y_t), and panel (b) shows the estimated density of (X_t, Y_t) together with the estimated regression functions. The red line shows the linear regression function estimate, and the blue line shows a kernel regression estimate with smoothing parameter $h = 0.4$. The density is estimated using kernel

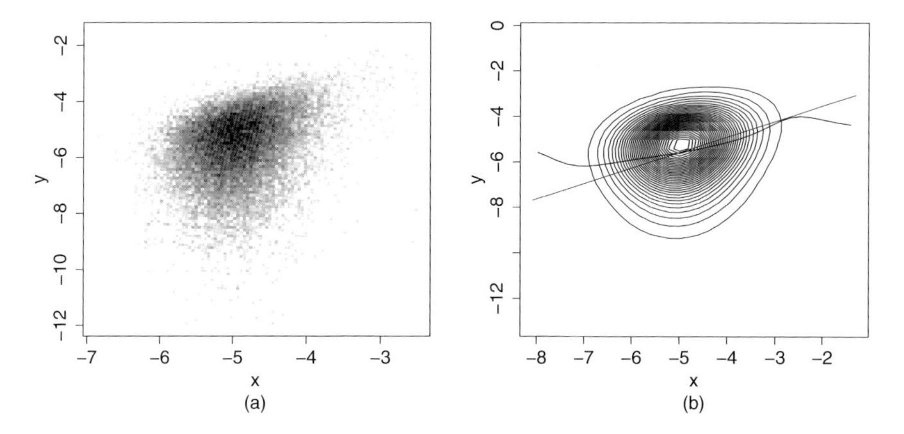

Figure 1.1 *Mean regression.* (a) A scatter plot of regression data. (b) A contour plot of the estimated joint density of the explanatory variable and the response variable. The linear regression function estimate is shown with red and the kernel regression estimate is shown with blue.

density estimation with smoothing parameter $h = 0.6$. Linear regression is discussed in Section 2.1, and kernel methods are discussed in Section 3.2. In the scatter plot we have used histogram smoothing with 100^2 bins, as explained in Section 6.1.1. This example indicates that the daily returns are dependent random variables, although it can be shown that they are nearly uncorrelated.

Median and Conditional Median The median can be defined in the case of continuous distribution function of a random variable $Y \in \mathbf{R}$ as the number $\mathrm{median}(Y) \in \mathbf{R}$ satisfying

$$P(Y \le \mathrm{median}(Y)) = 0.5.$$

In general, covering also the case of discrete distributions, we can define the median uniquely as the generalized inverse of the distribution function:

$$\mathrm{median}(Y) = \inf\{y : P(Y \le y) \ge 0.5\}. \tag{1.8}$$

The conditional median is defined using the conditional distribution of Y given X:

$$\mathrm{median}(Y \mid X = x) = \inf\{y : P(Y \le y \mid X = x) \ge 0.5\}, \qquad x \in \mathbf{R}^d. \tag{1.9}$$

The sample median of observations $Y_1, \ldots, Y_n \in \mathbf{R}$ can be defined as the median of the empirical distribution. The empirical distribution is the discrete distribution with the probability mass function $P(\{Y_i\}) = 1/n$ for $i = 1, \ldots, n$. Then,

$$\mathrm{median}(Y_1, \ldots, Y_n) = Y_{[n/2]+1}, \tag{1.10}$$

where $Y_{(1)} \le \cdots \le Y_{(n)}$ is the ordered sample and $[x]$ is the largest integer smaller or equal to x.

Mode and Conditional Mode The mode is defined as an argument maximizing the density function of a random variable:

$$\text{mode}(Y) = \text{argmax}_{y \in \mathbf{R}} f_Y(y), \tag{1.11}$$

where $f_Y : \mathbf{R} \to \mathbf{R}$ is the density function of Y. The density f_Y can have several local maxima, and the use of the mode seems to be interesting only in cases where the density function is unimodal (has one local maximum). The conditional mode is defined as an argument maximizing the conditional density:

$$\text{mode}(Y \mid X = x) = \text{argmax}_{y \in \mathbf{R}} f_{Y \mid X = x}(y).$$

1.1.3 Partial Effects and Derivative Estimation

Let us consider mean regression, where we are estimating the conditional expectation $E(Y \mid X = x)$, where $X = (X_1, \ldots, X_d)$ is the vector of explanatory variables and we denote $x = (x_1, \ldots, x_d)$. The partial effect of the variable X_1 is defined as the partial derivative

$$p(x_1; x_2, \ldots, x_d) = \frac{\partial}{\partial x_1} E(Y \mid X = x).$$

The partial effect describes how the conditional expectation of Y changes when the value of X_1 is changed, when the values of the other variables are fixed. In general, the partial effect is a function of x_1 that is different for each $x_2, \ldots, x_d$. However, for the linear model $E(Y \mid X = x) = \alpha + \beta' x$ we have

$$p(x_1; x_2, \ldots, x_d) = \beta_1,$$

so that the partial effect is a constant which is the same for all $x_2, \ldots, x_d$. Linear models are studied in Section 2.1. For the additive model $E(Y \mid X = x) = f_1(x_1) + \cdots + f_d(x_d)$ we have

$$p(x_1; x_2, \ldots, x_d) = f'(x_1),$$

so that the partial effect is a function of x_1 which is the same for all $x_2, \ldots, x_d$. Thus additive models provide easily interpretable partial effects. Additive models are studied in Section 4.2. For the single index model $E(Y \mid X = x) = g(\beta' x)$ we have

$$p(x_1; x_2, \ldots, x_d) = g'(\beta' x)\, \beta_1,$$

so that the partial effect is a function of x_1 which is different for each $x_2, \ldots, x_d$. Single index models are studied in Section 4.1.

The partial elasticity of X_1 is defined as

$$\begin{aligned} e(x_1; x_2, \ldots, x_d) &= \frac{\partial}{\partial \log x_1} \log E(Y \mid X = x) \\ &= \frac{\partial}{\partial x_1} E(Y \mid X = x) \cdot \frac{x_1}{E(Y \mid X = x)}, \end{aligned}$$

when $x_1 > 0$ and $E(Y \mid X = x) > 0$. The partial elasticity describes the approximate percentage change of conditional expectation of Y when the value of X_1 is changed by one percent, when the values of the other variables are fixed.[1] The partial semielasticity of X_1 is defined as

$$\begin{aligned} s(x_1; x_2, \ldots, x_d) &= \frac{\partial}{\partial x_1} \log E(Y \mid X = x) \\ &= \frac{\partial}{\partial x_1} E(Y \mid X = x) \cdot \frac{1}{E(Y \mid X = x)}, \end{aligned}$$

when $E(Y \mid X = x) > 0$. The partial semielasticity describes the approximate percentage change of conditional expectation of Y when the value of X_1 is changed by 1 unit, when the values of the other variables are fixed.

We can use the visualization of partial effects as a tool to visualize regression functions. In Section 7.4 we show how level set trees can be used to visualize the mode structure of functions. The mode structure of a function means the number, the largeness, and the location of the local maxima of a function. Analogously, level set trees can be used to visualize the antimode structure of a function, where the antimode structure means the number, the largeness, and the location of the local minima of a function. Local maxima and minima are important characteristics of a regression function. However, we need to know more about a regression function than just the mode structure or antimode structure. Partial effects are a useful tool to convey additional important information about a regression function. If the partial effect is flat for each variable, then we know that the regression function is close to a linear function. When we visualize the mode structure of the partial effect of variable X_1, then we get information about whether a variable X_1 is causing the expected value of the response variable to increase in several locations (the number of local maxima of the partial effect), how much an increase of the value of the variable X_1 increases the expected value of the response variable Y (the largeness of the local maxima of the partial effect), and where the influence of the response variable X_1 is the largest (the location of the local maxima of the partial effect). Analogous conclusions can be made by visualizing the antimode structure of the partial effect.

We present two methods for the estimation of partial effects. The first method is to use the partial derivatives of a kernel regression function estimator, and this method is presented in Section 3.2.9. The second method is to use a local linear estimator, and this method is presented in Section 5.2.1.

1.1.4 Variance Regression

The mean regression gives information about the center of the conditional distribution, and with the variance regression we get information about the dispersion and on the

[1] This interpretation follows from the approximation

$$\log f(x+h) - \log f(x) \approx [f(x+h) - f(x)]/f(x),$$

which follows from the approximation $\log(x) \approx x - 1$, when $x \approx 1$.

heaviness of the tails of the conditional distribution. Variance is a classical measure of dispersion and risk which is used for example in the Markowitz theory of portfolio selection. Partial moments are risk measures that generalize the variance.

Variance and Conditional Variance The variance of random variable Y is defined by

$$\mathrm{Var}(Y) = E(Y - EY)^2 = EY^2 - (EY)^2. \tag{1.12}$$

The standard deviation of Y is the square root of the variance of Y. The conditional variance of random variable Y is equal to

$$\begin{aligned} \mathrm{Var}(Y \mid X = x) &= E\left\{[Y - E(Y \mid X = x)]^2 \mid X = x\right\} & (1.13) \\ &= E(Y^2 \mid X = x) - [E(Y \mid X = x)]^2. & (1.14) \end{aligned}$$

The conditional standard deviation of Y is the square root of the conditional variance. The sample variance is defined by

$$\widehat{\mathrm{Var}}(Y) = \frac{1}{n}\sum_{i=1}^{n}(Y_i - \bar{Y})^2 = \frac{1}{n}\sum_{i=1}^{n} Y_i^2 - \bar{Y}^2,$$

where $Y_1, \ldots, Y_n$ is a sample of random variables having identical distribution with Y.

Conditional Variance Estimation Conditional variance $\mathrm{Var}(Y \mid X = x)$ can be constant not depending on x. Let us write

$$Y = f(X) + \epsilon,$$

where $f(x) = E(Y \mid X = x)$ and $\epsilon = Y - f(X)$, so that $E(\epsilon \mid X = x) = 0$. If $\mathrm{Var}(Y \mid X = x) = E(\epsilon^2)$ is a constant not depending on x, we say that the noise is homoskedastic. Otherwise the noise is heteroskedastic. If the noise is heteroskedastic, it is of interest to estimate the conditional variance

$$\mathrm{Var}(Y \mid X = x) = E(\epsilon^2 \mid X = x).$$

Estimation of the conditional variance can be reduced to the estimation of the conditional expectation by using (1.13). First we estimate the conditional expectation $f(x) = E(Y \mid X = x)$ by $\hat{f}(x)$. Second we calculate the residuals

$$\hat{\epsilon}_i = Y_i - \hat{f}(X_i),$$

and estimate the conditional variance from the data $(X_1, \hat{\epsilon}_1^2), \ldots, (X_n, \hat{\epsilon}_n^2)$.

Estimation of the conditional variance can be reduced to the estimation of the conditional expectation by using (1.14). First we estimate the conditional expectation $E(Y^2 \mid X = x)$ using the regression data $(X_1, Y_1^2), \ldots, (X_n, Y_n^2)$. Second we estimate the conditional expectation $f(x) = E(Y \mid X = x)$ using data $(X_1, Y_1), \ldots, (X_n, Y_n)$.

Theory of variance estimation is often given in the fixed design case, but the results can be extended to the random design regression by conditioning on the design variables. Let us write a heteroskedastic fixed design regression model

$$Y_i = f(x_i) + \sigma(x_i)\,\epsilon_i, \qquad i = 1, \ldots, n, \tag{1.15}$$

where $x_i \in \mathbf{R}^d$, $f : \mathbf{R}^d \to \mathbf{R}$ is the mean function, $\sigma : \mathbf{R}^d \to \mathbf{R}$ is the standard deviation function, and ϵ_i are identically distributed with $E\epsilon_i = 0$. Now we want to estimate both the function f and the function σ. Wasserman (2005, Section 5.6) has proposed making the following transformation. Let $Z_i = \log(Y_i - f(x_i))^2$. Then we have

$$Z_i = \log(\sigma^2(x_i)) + \log \epsilon_i^2.$$

Let $\hat{f}$ be an estimate of f and define $\hat{Z}_i = \log(Y_i - \hat{f}(x_i))^2$. Let $\hat{g}(x)$ be an estimate of $\log \sigma^2(x)$, obtained using regression data $(x_1, \hat{Z}_1), \ldots, (x_n, \hat{Z}_n)$, and define $\hat{\sigma}^2(x) = \exp\{\hat{g}(x)\}$.

A difference-based method for conditional variance estimation has been proposed. Let $x_1 < \cdots < x_n$ be univariate fixed design points. Now $\sigma^2(x)$ is estimated with $2^{-1}\hat{g}(x)$, where $\hat{g}$ is a regression function estimate obtained with the regression data $(x_i, (Y_i - Y_{i-1})^2)$, $i = 2, \ldots, n$. This approach has been used in Wang, Brown, Cai & Levine (2008).

Variance Estimation with Homoskedastic Noise Let us consider the fixed design regression model

$$Y_i = f(x_i) + \epsilon_i, \qquad i = 1, \ldots, n,$$

where $x_i \in \mathbf{R}^d$, $f : \mathbf{R}^d \to \mathbf{R}$ is the mean function, and $E\epsilon_i = 0$. In the case of homoskedastic noise we should estimate

$$\sigma^2 \stackrel{def}{=} E(\epsilon^2).$$

Spokoiny (2002) showed that for twice differentiable regression functions f, the optimal rate for the estimation of σ^2 is $n^{-1/2}$ for $d \leq 8$ and otherwise the optimal rate is $n^{-4/d}$. We can first estimate the mean function f by $\hat{f}$ and then use

$$\widehat{\sigma^2} = \frac{1}{n} \sum_{i=1}^{n} \left(Y_i - \hat{f}(x_i) \right)^2.$$

These types of estimators were studied by Müller & Stadtmüller (1987), Hall & Carroll (1989), Hall & Marron (1990), and Neumann (1994). Local polynomial estimators were studied by Ruppert, Wand, Holst & Hössjer (1997), and Fan & Yao (1998). A difference-based estimator was studied by von Neumann (1941). He used the estimator

$$\widehat{\sigma^2} = \frac{1}{2(n-1)} \sum_{i=2}^{n} (Y_i - Y_{i-1})^2,$$

where it is assumed that $x_1, \ldots, x_n \in \mathbf{R}$, and $x_1 < \cdots < x_n$. The estimator was studied and modified in various ways in Rice (1984), Gasser, Sroka & Jennen-Steinmetz (1986), Hall, Kay & Titterington (1990), Hall, Kay & Titterington (1991), Thompson, Kay & Titterington (1991), and Munk, Bissantz, Wagner & Freitag (2005).

Conditional Variance in a Time Series Setting In a time series setting, when we observe Y_t, $t = 1, 2, \ldots$, the *conditional heteroskedasticity* assumption is that

$$Y_t = \sigma_t \epsilon_t, \qquad t = 0, \pm 1, \pm 2, \ldots, \tag{1.16}$$

where ϵ_t is an i.i.d. sequence, $E\epsilon_t = 0$, $E\epsilon_t^2 = 1$, and σ_t is the volatility process. The volatility process is a predictable random process, that is, σ_t is measurable with respect to the sigma-field generated by the variables $Y_{t-1}, Y_{t-2}, \ldots$. When we assume that ϵ_t is independent from $Y_{t-1}, Y_{t-2}, \ldots$, then under the conditional heteroskedasticity model,

$$\text{Var}(Y_t \mid \mathcal{F}_{t-1}) = \text{Var}(\sigma_t \epsilon_t \mid \mathcal{F}_{t-1}) = \sigma_t^2 \text{Var}(\epsilon_t \mid \mathcal{F}_{t-1}) = \sigma_t^2 \text{Var}(\epsilon_t) = \sigma_t^2, \tag{1.17}$$

where $\mathcal{F}_{t-1}$ is the sigma-algebra generated by variables $Y_{t-1}, Y_{t-2}, \ldots$. In a conditional heteroskedasticity model the main interest is in predicting the value of the random variable σ_t^2, which is thus related to estimating the conditional variance. The statistical problem is to predict σ_t^2 using a finite number of past observations $Y_1, \ldots, Y_{t-1}$. Special cases of conditional heteroskedasticity models are the ARCH model discussed in Section 2.5.2 and the GARCH model discussed in Section 3.9.2.

Partial Moments The variance of random variable $Y \in \mathbf{R}$ is defined as $\text{Var}(Y) = E(Y - EY)^2$. The variance can be generalized to other centered moments

$$E|Y - EY|^k,$$

for $k = 1, 2, \ldots$. The centered moments take a contribution both from the left and the right tails of the distribution. When we are interested only in the left tail or in the right tail (losses or gains), then we can use the lower partial moments or the upper partial moments. The upper partial moment is defined as

$$\text{UPM}_{\tau,k}(Y) = E\left[(Y - \tau)^k I_{[\tau,\infty)}(Y)\right]$$

and the lower partial moment is defined as

$$\text{LPM}_{\tau,k}(Y) = E\left[(\tau - Y)^k I_{(-\infty,\tau]}(Y)\right],$$

where $k = 0, 1, 2, \ldots$, and $\tau \in \mathbf{R}$. In risk management τ could be the target rate. When Y has density f_Y, we can write

$$\text{UPM}_{\tau,k}(Y) = \int_\tau^\infty (y - \tau)^k f_Y(y)\, dy, \qquad \text{LPM}_{\tau,k}(Y) = \int_{-\infty}^\tau (\tau - y)^k f_Y(y)\, dy.$$

For example, when $k = 0$, then

$$\text{UPM}_{\tau,0}(Y) = P(Y \geq \tau), \qquad \text{LPM}_{\tau,0}(Y) = P(Y \leq \tau),$$

so that the upper partial moment is equal to the probability that Y is greater or equal to τ and the lower partial moment is equal to the probability that Y is smaller or equal to τ. For $k = 2$ and $\tau = EY$ the partial moments are called upper or lower semivariance of Y. The lower semivariance is defined as

$$E\left[(Y - EY)^2 I_{(-\infty, EY]}(Y)\right]. \tag{1.18}$$

The square root of the lower semivariance can be used to replace the standard deviation in the definition of the Sharpe ratio or in the Markowitz criterion. We can define conditional versions of partial moments by changing the expectations to conditional expectations.

1.1.5 Covariance and Correlation Regression

The covariance of random variables Y and Z is defined by

$$\text{Cov}(Y, Z) = E[(Y - EY)(Z - EZ)] = E(YZ) - EY\,EZ.$$

The sample covariance is defined by

$$\widehat{\text{Cov}}(Y, Z) = \frac{1}{n}\sum_{i=1}^{n}(Y_i - \bar{Y})(Z_i - \bar{Z}) = \frac{1}{n}\sum_{i=1}^{n} Y_i Z_i - \bar{Y}\bar{Z},$$

where $Y_1, \ldots, Y_n$ and $Z_1, \ldots, Z_n$ are samples of random variables having identical distributions with Y and Z, $\bar{Y} = n^{-1}\sum_{i=1}^{n} Y_i$, and $\bar{Z} = n^{-1}\sum_{i=1}^{n} Z_i$. The conditional covariance is obtained by changing the expectations to conditional expectations.

We have two methods of estimation of conditional covariance, analogously to two methods of conditional variance estimation based on formulas (1.13) or (1.14). The first method uses $\text{Cov}(Y, Z) = E[(Y - EY)(Z - EZ)]$ and the second method uses $\text{Cov}(Y, Z) = E(YZ) - EY\,EZ$.

The correlation is defined by

$$\text{Cor}(Y, Z) = \frac{\text{Cov}(Y, Z)}{\text{sd}(Y)\,\text{sd}(Z)},$$

where $\text{sd}(Y)$ and $\text{sd}(Z)$ are the standard deviations of Y and Z. The conditional correlation is defined by

$$\text{Cor}(Y, Z \mid X = x) = \frac{\text{Cov}(Y, Z \mid X = x)}{\text{sd}(Y \mid X = x)\,\text{sd}(Z \mid X = x)}, \tag{1.19}$$

where

$$\text{sd}(Y \mid X = x) = \sqrt{\text{Var}(Y \mid X = x)}, \qquad \text{sd}(Z \mid X = x) = \sqrt{\text{Var}(Z \mid X = x)}.$$

We can write

$$\text{Cor}(Y, Z \mid X = x) = \text{Cov}(\tilde{Y}, \tilde{Z} \mid X = x), \tag{1.20}$$

where

$$\tilde{Y} = \frac{Y}{\text{sd}(Y \mid X = x)}, \qquad \tilde{Z} = \frac{Z}{\text{sd}(Z \mid X = x)}.$$

Thus we have two approaches to the estimation of conditional correlation.

1. We can use (1.19). First we estimate the conditional covariance and the conditional standard deviations. Second we use (1.19) to define the estimator of the conditional correlation.

2. We can use (1.20). First we estimate the conditional standard deviations by $\widehat{\text{sd}}_Y(x)$ and $\widehat{\text{sd}}_Z(x)$, and calculate the standardized observations $\tilde{Y}_i = Y_i / \widehat{\text{sd}}_Y(X_i)$ and $\tilde{Z}_i = Z_i / \widehat{\text{sd}}_Z(X_i)$. Second we estimate the conditional correlation using $(X_i, \tilde{Y}_i, \tilde{Z}_i)$, $i = 1, \ldots, n$.

A time series $(Y_t)_{t \in \mathbf{Z}}$ is weakly stationary if $EY_t = EY_{t+h}$ and $EY_t Y_{t+h}$ depends only on h, for all $t, h \in \mathbf{Z}$. For a weakly stationary time series $(Y_t)_{t \in \mathbf{Z}}$, the autocovariance function is defined by

$$\gamma(h) = \text{Cov}(Y_t, Y_{t+h}),$$

and the autocorrelation is defined by

$$\rho(h) = \gamma(h)/\gamma(0),$$

where $h = 0, \pm 1, \ldots$.

A vector time series $(X_t)_{t \in \mathbf{Z}}$, $X_t \in \mathbf{R}^d$, is weakly stationary if $EX_t = EX_{t+h}$ and $EX_t X'_{t+h}$ depends only on h, for all $t, h \in \mathbf{Z}$. For a weakly stationary vector time series $(X_t)_{t \in \mathbf{Z}}$, the autocovariance function is defined by

$$\Gamma(h) = \text{Cov}(X_t, X_{t+h}) = E[(X_t - \mu)(X_{t+h} - \mu)'], \tag{1.21}$$

for $h = 0, \pm 1, \ldots$, where $\mu = EX_t = EX_{t+h}$. Matrix $\Gamma(h)$ is a $d \times d$ matrix which is not symmetric. It holds that

$$\Gamma(h) = \Gamma(-h)'. \tag{1.22}$$

1.1.6 Quantile Regression

A quantile generalizes the median. In quantile regression a conditional quantile is estimated. Quantiles can be used to measure the value at risk (VaR). The expected shortfall is a related measure of dispersion and risk.

Quantile and Conditional Quantile The pth quantile is defined as

$$Q_p(Y) = \inf\{y : P(Y \le y) \ge p\}, \qquad x \in \mathbf{R}^d, \tag{1.23}$$

where $0 < p < 1$. For $p = 1/2$, $Q_p(Y)$ is equal to median $\text{med}(Y)$, defined in (1.8). In the case of a continuous distribution function we have

$$P\left(Y \le Q_p(Y)\right) = p$$

and thus it holds that

$$Q_p(Y) = F_Y^{-1}(p),$$

where $F_Y(y) = P(Y \le y)$ is the distribution function of Y and F_Y^{-1} is the inverse of F_Y. The pth conditional quantile is defined replacing the distribution of Y with the conditional distribution of Y given X:

$$Q_p(Y \mid X = x) = \inf\{y : P(Y \le y \mid X = x) \ge p\}, \qquad x \in \mathbf{R}^d, \tag{1.24}$$

where $0 < p < 1$. Conditional quantile estimation has been considered in Koenker (2005) and Koenker & Bassett (1978).

Estimation of a Quantile and a Conditional Quantile Estimation of quantiles is closely related to the estimation of the distribution function. It is usually possible to derive a method for the estimation of a quantile or a conditional quantile if we have a method for the estimation of a distribution function or a conditional distribution function.

Empirical Quantile Let us define the empirical distribution function, based on the dta $Y_1, \ldots, Y_n$, as

$$\hat{F}(y) = \frac{1}{n} \sum_{i=1}^{n} I_{(-\infty, y]}(Y_i), \qquad y \in \mathbf{R}. \tag{1.25}$$

Now we can define an estimate of the quantile by

$$\hat{Q}_p = \inf\{x : \hat{F}(x) \ge p\}, \tag{1.26}$$

where $0 < p < 1$. Now it holds that

$$\hat{Q}_p = \begin{cases} Y_{(1)}, & 0 < p \le 1/n, \\ Y_{(2)}, & 1/n < p \le 2/n, \\ \vdots & \\ Y_{(n-1)}, & 1 - 2/n < p \le 1 - 1/n, \\ Y_{(n)}, & 1 - 1/n < p < 1, \end{cases} \tag{1.27}$$

where the ordered sample is denoted by $Y_{(1)} \le Y_{(2)} \le \cdots \le Y_{(n)}$. A third description of the empirical estimator of the quantile is given by the following steps:

1. Order the sample from the smallest observation to the largest observation: $Y_{(1)} \le \cdots \le Y_{(n)}$.

2. Let $m = \lceil pn \rceil$, where $\lceil y \rceil$ is the the smallest integer $\ge y$.

3. Set $\hat{Q}_p = Y_{(m)}$.

Standard Deviation-Based Quantile Estimators We can also use an estimate of the standard deviation to derive an estimate for a quantile. Namely, consider the location-scale model

$$Y = \mu + \sigma\,\epsilon,$$

where $\mu \in \mathbf{R}$, $\sigma > 0$, and ϵ is a random variable with a continuous distribution. Now

$$P(Y \le y) = P\left(\epsilon \le \frac{y-\mu}{\sigma}\right) = F_\epsilon\left(\frac{y-\mu}{\sigma}\right),$$

where F_ϵ is the distribution function of ϵ. If ϵ has a continuous distribution, then F_ϵ is monotone increasing and the inverse function F_ϵ^{-1} exists. The pth quantile $Q_p(Y)$ of Y satisfies $P\left(Y \le Q_p(Y)\right) = p$, and we can solve this equation to get

$$Q_p(Y) = \mu + \sigma\, F_\epsilon^{-1}(p).$$

Thus, for a known F_ϵ, we get from the estimates $\hat{\mu}$ of μ and $\hat{\sigma}$ of σ the estimate

$$\hat{Q}_p(Y) = \hat{\mu} + \hat{\sigma}\, F_\epsilon^{-1}(p). \tag{1.28}$$

Standard Deviation-Based Conditional Quantile Estimators To get an estimate for a conditional quantile in the heteroskedastic fixed design model (1.15), we can use

$$\hat{Q}_p(Y \,|\, X = x) = \hat{f}(x) + \hat{\sigma}(x)\, F_\epsilon^{-1}(p). \tag{1.29}$$

Similarly, in the conditional heteroskedasticity model (1.16) we can use

$$\hat{Q}_p(Y_t \,|\, \mathcal{F}_{t-1}) = \hat{\sigma}_t\, F_{\epsilon_t}^{-1}(p). \tag{1.30}$$

We apply in Section 2.5.1 and in Section 3.11.3 three quantile estimators which are based on the standard deviation estimates.

1. First estimator uses the standard normal distribution, which gives the quantile estimator

$$\hat{Q}_p(Y_t \,|\, \mathcal{F}_{t-1}) = \hat{\sigma}_t\, \Phi^{-1}(p), \tag{1.31}$$

   where Φ is the distribution function of the standard normal distribution.

2. Second estimator uses the t-distribution, which gives the quantile estimator

$$\hat{Q}_p(Y_t \,|\, \mathcal{F}_{t-1}) = \sqrt{\frac{\nu-2}{\nu}}\; \hat{\sigma}_t\, t_\nu^{-1}(p), \tag{1.32}$$

   where t_ν is the distribution function of the t-distribution with ν degrees of freedom. If $X \sim t_\nu$, then $\mathrm{Var}(X) = \nu/(\nu-2)$, so that $\sqrt{(\nu-2)/\nu}\; t_\nu^{-1}(p)$ is the p-quantile of the standardized t-distribution, which has unit variance.

3. Third estimator uses the empirical quantiles of the residuals. Now

$$\hat{Q}_p(Y_t \,|\, \mathcal{F}_{t-1}) = \hat{\sigma}_t\, \hat{Q}^{res}(p), \tag{1.33}$$

   where $\hat{Q}^{res}(p)$ is the empirical quantile of the residuals $Y_t/\hat{\sigma}_t$. Empirical quantiles were defined in (1.26). This estimator was suggested in Fan & Gu (2003).

Expected Shortfall The expected shortfall is a measure of risk which aggregates all quantiles in the right tail (or in the left tail). The expected shortfall for the right tail is defined as

$$\mathrm{ES}_p(Y) = \frac{1}{1-p}\int_p^1 Q_u(Y)\,du, \qquad 0 < p < 1.$$

When Y has a continuous distribution function, then

$$\mathrm{ES}_p(Y) = E\left(Y \mid Y \geq Q_p(Y)\right) = \frac{1}{1-p}\,E\left(Y I_{[Q_p(Y),\infty)}(Y)\right); \tag{1.34}$$

see McNeil, Frey & Embrechts (2005, lemma 2.16). We have defined the loss in (1.86) as the negative of the change in the value of the portfolio, and thus the risk management wants to control the right tails of the loss distribution However, we can define the expected shortfall for the left tail as

$$\mathrm{ES}_p(Y) = \frac{1}{p}\int_0^p Q_u(Y)\,du, \qquad 0 < p < 1. \tag{1.35}$$

When Y has a continuous distribution function, then

$$\mathrm{ES}_p(Y) = E(Y \mid Y \leq Q_p(Y)) = \frac{1}{p}\,E\big(Y I_{(-\infty,Q_p(Y)]}(Y)\big).$$

This expression shows that in the case of a continuous distribution function, $p\mathrm{ES}_p(Y)$ is equal to the expectation which is taken only over the left tail, when the left tail is defined as the region which is to the left of a quantile of the distribution.[2]

The expected shortfall can be estimated from the data $Y_1, \ldots, Y_n$ in the case where the expected shortfall is given in (1.34) by using

$$\hat{\mathrm{ES}}_p = \frac{1}{m}\sum_{i=m}^{n} Y_{(i)},$$

where $Y_{(1)} \leq \cdots \leq Y_{(n)}$ and $m = \lceil (1-p)n \rceil$. When the expected shortfall is given by (1.35), then we define

$$\hat{\mathrm{ES}}_p = \frac{1}{m}\sum_{i=1}^{m} Y_{(i)},$$

where $m = \lceil pn \rceil$.

Let us consider the location-scale model

$$Y = \mu + \sigma\,\epsilon,$$

where $\mu \in \mathbf{R}$, $\sigma > 0$, and ϵ is a random variable with a continuous distribution. Now

$$\mathrm{ES}_p(Y) = \mu + \sigma \mathrm{ES}_p(\epsilon).$$

[2]Sometimes the expected shortfall for the left tail is defined as $Q_p(Y) - E[Y I_{(-\infty,Q_p(Y)]}(Y)]$ and the absolute shortfall is defined as $-E[Y I_{(-\infty,Q_p(Y)]}(Y)]$.

Thus the estimate for the expected shortfall can be obtained as

$$\hat{\mathrm{ES}}_p(Y) = \hat{\mu} + \hat{\sigma}\mathrm{ES}_p(\epsilon),$$

where $\hat{\mu}$ is an estimate of μ and σ is an estimate of σ.

If $\epsilon \sim N(0,1)$ and the expected shortfall is defined for the right tail as in (1.34), then

$$\mathrm{ES}_p(\epsilon) = \frac{\phi(\Phi^{-1}(p))}{1-p},$$

where ϕ is the density function of the standard normal distribution and Φ is the distribution function of the standard normal distribution. If $\epsilon \sim t_\nu$, where t_ν is the t-distribution with ν degrees of freedom, and the expected shortfall is defined for the right tail as in (1.34), then

$$\mathrm{ES}_p(\epsilon) = \frac{g_\nu(t_\nu^{-1}(p))}{1-p} \frac{\nu + (t_\nu^{-1}(p))^2}{v-1},$$

where g_ν is the density function of the t-distribution with ν degrees of freedom and t_ν is the distribution function of the t-distribution with ν degrees of freedom.

Expected shortfall is sometimes preferred to the quantiles on the grounds that the expected shortfall satisfies the axiom of subadditivity. Risk measure ϱ is said to be subadditive if $\varrho(X+Y) \le \varrho(X) + \varrho(Y)$, where X and Y are random variables interpreted as portfolio losses. Quantiles do not satisfy subadditivity like the expected shortfall. The other axioms of a coherent risk measure are the monotonicity: if $Y \ge X$, then $\varrho(Y) \ge \varrho(X)$; the positive homogeneity: for $\lambda \ge 0$, $\varrho(\lambda Y) = \lambda\varrho(Y)$; and the translation invariance: for $a \in \mathbf{R}$, $\varrho(Y+a) = \varrho(Y) + a$. For more about coherent risk measures, see McNeil et al. (2005, Section 6.1).

1.1.7 Approximation of the Response Variable

We have defined the regression function in (1.5) as the conditional expectation of the response variable. The conditional expectation can be viewed as an approximation of response variable $Y \in \mathbf{R}$ with the help of explanatory random variables $X_1, \ldots, X_d \in \mathbf{R}$. The approximation is a random variable $f(X_1, \ldots, X_d) \in \mathbf{R}$, where $f : \mathbf{R}^d \to \mathbf{R}$ is a fixed function. This viewpoint leads to generalizations. The best approximation of the response variable can be defined using various loss functions $\rho : \mathbf{R} \to \mathbf{R}$. The best approximation is $f(X_1, \ldots, X_d)$, where f is defined as

$$f = \mathrm{argmin}_{g \in \mathcal{G}} E\rho(Y - g(X)), \qquad X = (X_1, \ldots, X_d), \tag{1.36}$$

where $\mathcal{G}$ is a suitable class of functions $g : \mathbf{R}^d \to \mathbf{R}$. Since f is defined in terms of the unknown distribution of (X, Y), we have to estimate f using statistical data available from the distribution of (X, Y).

Examples of Loss Functions We give examples of different choices of ρ and $\mathcal{G}$.

1. When $\rho(t) = t^2$ and $\mathcal{G}$ is the class of all measurable functions $\mathbf{R}^d \to \mathbf{R}$, then f, defined by (1.36), is equal to the conditional expectation:

$$f(x) = E(Y \mid X = x) = \text{argmin}_{g \in \mathcal{G}} E(Y - g(X))^2.$$

   Indeed,

$$E(g(X) - Y)^2 = E(g(X) - E(Y \mid X))^2 + E(E(Y \mid X) - Y)^2, \quad (1.37)$$

   because $E[(g(X) - E(Y \mid X))(E(Y \mid X) - Y)] = 0$, and thus $E(g(X) - Y)^2$ is minimized with respect to $g : \mathbf{R}^d \to \mathbf{R}$ by choosing $g(x) = E(Y \mid X = x)$.[3] Note also that the expectation EY is the best constant approximation of Y. That is, if we choose $\mathcal{G}$ as the class of constant functions

$$\mathcal{G} = \{g : \mathbf{R}^d \to \mathbf{R} \mid g(x) = \mu \text{ for all } x \in \mathbf{R}, \mu \in \mathbf{R}\},$$

   then

$$EY = \text{argmin}_{g \in \mathcal{G}} E(Y - g(X))^2 = \text{argmin}_{\mu \in \mathbf{R}} E(Y - \mu)^2. \quad (1.38)$$

   Indeed,

$$E(Y - \mu)^2 = E(Y - EY)^2 + (EY - \mu)^2,$$

   and this is minimized with respect to $\mu \in \mathbf{R}$ by choosing $\mu = EY$.

2. When $\rho(t) = |t|$ and $\mathcal{G}$ is the class of all measurable functions $\mathbf{R}^d \to \mathbf{R}$, then f defined by (1.36) is the conditional median:

$$\text{med}(Y \mid X = x) = \text{argmin}_{g \in \mathcal{G}} E|Y - g(X)|, \quad (1.39)$$

   where the conditional median is defined in (1.9). Equation (1.39) is proved in the next item.

3. When ρ is defined as

$$\rho_p(t) = t\,[p - I_{(-\infty,0)}(t))] = \begin{cases} t(p-1), & \text{if } t < 0 \\ tp, & \text{if } t \geq 0, \end{cases} \quad (1.40)$$

   for $0 < p < 1$ and $\mathcal{G}$ is the class of all measurable functions, then the best approximation is the conditional quantile. Figure 1.2 shows the loss function in (1.40) with $p = 0.5$ (black line) and with $p = 0.1$ (red line). We show that if the distribution function F_Y is strictly monotonic, then

$$Q_p(Y) = \text{argmin}_{\theta \in \mathbf{R}} E\rho_p(Y - \theta). \quad (1.41)$$

[3]Note that the conditional expectation defined as $f(x) = E(Y \mid X = x)$ is a real-valued function of x, but $E(X \mid Y)$ is a real-valued random variable which can be defined as $E(X \mid Y) = f(X)$.

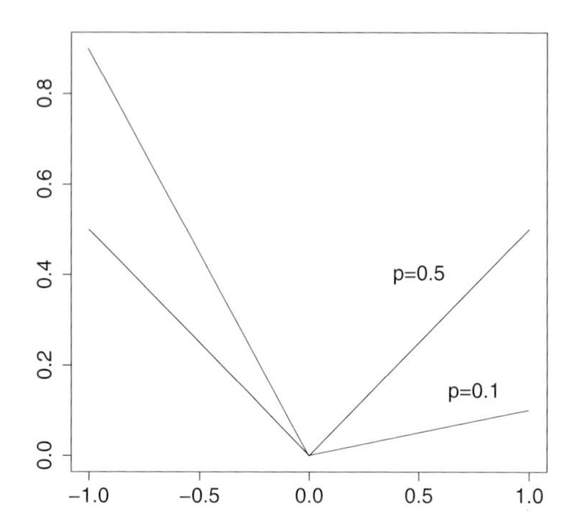

Figure 1.2 *Loss functions for quantile estimation.* Loss function in (1.40) with $p = 0.5$ (black line) and with $p = 0.1$ (red line).

To show (1.41), note that

$$E\rho_p(Y - \theta) = (p - 1) \int_{-\infty}^{\theta} (y - \theta)\, dF_Y(y) + p \int_{\theta}^{\infty} (y - \theta)\, dF_Y(y)$$

and thus

$$\frac{\partial}{\partial \theta} E\rho_p(Y - \theta) = (1 - p) \int_{-\infty}^{\theta} dF_Y(y) - p \int_{\theta}^{\infty} dF_Y(y) = F_Y(\theta) - p.$$

Setting $\partial E\rho_p(Y - \theta)/\partial\theta = 0$, we get (1.41), when F_Y is strictly monotonic. We can prove similarly the case of conditional quantiles:

$$Q_p(Y \mid X = x) = \text{argmin}_{g \in \mathcal{G}} E\rho_p(Y - g(X)),$$

where $\mathcal{G}$ is the class of measurable functions $\mathbf{R}^d \to \mathbf{R}$. When $p = 1/2$, then

$$\rho_p(t) = \frac{1}{2}\, |t|,$$

and we have proved the result (1.39).

Estimation Using Loss Function If a regression function can be characterized as a minimizer of a loss function, then we can use empirical risk minimization with this loss function to define an estimator for the regression function. Empirical risk minimization is discussed in Chapter 5.

For example, conditional expectation $f(x) = E(Y \mid X = x)$ can be estimated minimizing the sum of squared errors:

$$\hat{f} = \mathrm{argmin}_{f \in \mathcal{F}} \sum_{i=1}^{n} (Y_i - f(X_i))^2 ,$$

where $\mathcal{F}$ is a class of functions $f : \mathbf{R}^d \to \mathbf{R}$. For example, $\mathcal{F}$ could be the class of linear functions.

Estimation of quantiles and conditional quantiles can also be done using empirical risk minimization. The estimator of the pth quantile is

$$\hat{Q}_p(Y) = \mathrm{argmin}_{\theta \in \mathbf{R}} \sum_{i=1}^{n} \rho_p(Y_i - \theta)$$

and the estimator of the pth conditional quantile $f(x) = Q_p(Y \mid X = x)$ is

$$\hat{f} = \mathrm{argmin}_{f \in \mathcal{F}} \sum_{i=1}^{n} \rho_p(Y_i - f(X_i)),$$

where $\mathcal{F}$ is a class of functions $f : \mathbf{R}^d \to \mathbf{R}$. A further idea which we will discuss in Section 5.2 is to define an estimator for the conditional quantile using local empirical risk:

$$\hat{f}(x) = \mathrm{argmin}_{\theta \in \mathbf{R}} \sum_{i=1}^{n} p_i(x)\, \rho_p(Y_i - \theta),$$

where $p_i(x) \geq 0$ and $\sum_{i=1}^{n} p_i(x) = 1$. These weights should have the property that $p_i(x)$ is large when X_i is close to x and $p_i(x)$ is small when X_i is far away from x.

1.1.8 Conditional Distribution and Density

Instead of estimating only conditional expectation, conditional variance, or conditional quantile, we can try to estimate the complete conditional distribution by estimating the conditional distribution function or the conditional density function.

Conditional Distribution Function The distribution function of random variable $Y \in \mathbf{R}$ is defined as[4]

$$F_Y(y) = P(Y \leq y), \qquad y \in \mathbf{R}.$$

The conditional distribution function is defined as

$$F_{Y \mid X=x}(y) = P(Y \leq y \mid X = x), \qquad y \in \mathbf{R},\ \ x \in \mathbf{R}^d,$$

[4]This definition can be extended to the multivariate case $Y = (Y_1, \ldots, Y_d)$ by

$$F_Y(y) = P(Y_1 \leq y_1, \ldots, Y_d \leq y_d), \qquad y = (y_1, \ldots, y_d) \in \mathbf{R}^d.$$

where $Y \in \mathbf{R}$ is a scalar random variable and $X \in \mathbf{R}^d$ is a random vector. We have

$$F_{Y\,|\,X=x}(y) = E\left[I_{(-\infty,y]}(Y)\,|\,X = x\right] \tag{1.42}$$

and thus the estimation of the conditional distribution function can be considered as a regression problem, where the conditional expectation of the random variable $I_{(-\infty,y]}(Y)$ is estimated. The random variable $I_{(-\infty,y]}(Y)$ takes only values 0 or 1. The unconditional distribution function can be estimated with the empirical distribution function, which is defined for the data $Y_1, \ldots, Y_n$ as

$$\hat{F}_Y(y) = \frac{1}{n}\sum_{i=1}^{n} I_{(-\infty,y]}(Y_i) = n^{-1}\#\{i : Y_i \le y, i = 1, \ldots, n\}, \tag{1.43}$$

where $\#A$ means the cardinality of set A. The conditional distribution function estimation is considered in Section 3.7, where local averaging estimators are defined.

Conditional Density Conditional density function is defined as

$$f_{Y\,|\,X=x}(y) = \begin{cases} \frac{f_{X,Y}(x,y)}{f_X(x)}, & \text{when } f_X(x) > 0, \\ 0, & \text{otherwise,} \end{cases}$$

for $y \in \mathbf{R}$, where $f_{X,Y} : \mathbf{R}^{d+1} \to \mathbf{R}$ is the joint density of (X, Y) and $f_X : \mathbf{R}^d \to \mathbf{R}$ is the density of X. We mention three ways to estimate the conditional density.

First, we can replace the density of (X, Y) and the density of X with their estimators $\hat{f}_{X,Y}$ and $\hat{f}_X$ and define

$$\hat{f}_{Y\,|\,X=x}(y) = \frac{\hat{f}_{X,Y}(x,y)}{\hat{f}_X(x)},$$

for $\hat{f}_X(x) > 0$. This approach is close to the approach used in Section 3.6, where local averaging estimators of the conditional density are defined.

Second, empirical risk minimization can be used in the estimation of the conditional density, as explained in Section 5.1.3.

Third, sometimes it is reasonable to assume that the conditional density has the form

$$f_{Y\,|\,X=x}(y) = f_{g(x)}(y), \tag{1.44}$$

where f_θ, $\theta \in A \subset \mathbf{R}^k$, is a family of density functions and $g : \mathbf{R}^d \to A$, where $k \ge 1$. Then the estimation of the conditional density reduces to the estimation of the "regression function" g. The mean regression is a special case of this approach when the distribution of errors is known: Assume that

$$Y = f(X) + \epsilon,$$

where ϵ is independent of X, $E\epsilon = 0$, and the density of ϵ is denoted by f_ϵ. Then

$$f_{Y\,|\,X=x}(y) = f_\epsilon(y - f(x)),$$

which is a special case of (1.44), when we take $f_\theta(y) = f_\epsilon(y - \theta)$ and $g(x) = f(x)$. The case of heteroskedastic variance is an other example: Now we assume that

$$Y = f(X) + \sigma(X)\,\epsilon,$$

where ϵ is independent of X, $E\epsilon = 0$, and the density of ϵ is denoted by f_ϵ. Then

$$f_{Y\,|\,X=x}(y) = \sigma(x)^{-1} f_\epsilon((y - f(x))/\sigma(x)),$$

which is a special case of (1.44), when we take $\theta = (\theta_1, \theta_2)$, $f_\theta(y) = \theta_2^{-1} f_\epsilon((y - \theta_1)/\theta_2)$, and $g(x) = (f(x), \sigma(x))$. This approach is used in parametric family regression, explained in Section 1.3.1.

1.1.9 Time Series Data

Regression data are a sequence $(X_1, Y_1), \ldots, (X_n, Y_n)$ of identically distributed copies of (X, Y), where $X \in \mathbf{R}^d$ is the explanatory variable and $Y \in \mathbf{R}$ is the response variable, as we wrote in (1.4). However, we can use regression methods with time series data

$$Z_1, \ldots, Z_T \in \mathbf{R},$$

where the observation Z_t is made at time t, $t = 1, \ldots, T$. In order to apply regression methods we identify the response variable and the explanatory variables. We consider two ways for the choice of the explanatory variables. In the first case the state space of the time series is used as the space of the explanatory variables, and in the second case the time space is used as the space of the explanatory variables.

State–Space Prediction In the state–space prediction an autoregression parameter $k \geq 1$ is chosen and we denote

$$Y_i = Z_{i+1}, \qquad X_i = (Z_i, \ldots, Z_{i-k+1}), \tag{1.45}$$

$i = k, \ldots, T-1$. When the time series $Z_1, \ldots, Z_T$ is stationary, then the sequence (X_i, Y_i), $i = k, \ldots, T-1$, consists of identically distributed random variables and we can denote by (X, Y) a random vector which is identically distributed as (X_i, Y_i).

We define the regression function, as previously, by

$$f(x) = E(Y \,|\, X = x), \qquad x \in \mathbf{R}^k. \tag{1.46}$$

We can estimate this regression function using data (X_i, Y_i), $i = k, \ldots, T-1$. Estimator of the regression function $f : \mathbf{R}^k \to \mathbf{R}$ can be used to predict or explain the next outcome of the time series using k previous observations. For example, let $\hat{f}_T$ be an estimator of the regression function at time T, constructed using data (X_i, Y_i), $i = k, \ldots, T-1$. The prediction of the next outcome is $\hat{f}_T(X_T)$, where $X_T = (Z_T, \ldots, Z_{T-k+1})$.

Let

$$Z_1, \ldots, Z_T \in \mathbf{R}^d$$

be a d-dimensional vector time series. Definition (1.45) generalizes to the setting of vector time series. Define

$$Y_i = g(Z_{i+1}), \qquad X_i = (Z_i, \ldots, Z_{i-k+1}), \tag{1.47}$$

$i = k, \ldots, T-1$, where $g : \mathbf{R}^d \to \mathbf{R}$ is a function with real values. We define the regression function, as previously, by

$$f(x) = E(Y_i \,|\, X_i = x), \qquad x \in \mathbf{R}^{dk}.$$

The regression function is now defined on the higher-dimensional space of dimension kd.

We can predict and explain without autoregression parameter k and take into account all the previous observations and not just the k last observations. However, this approach does not fit into the standard regression approach. Let $Z_1, \ldots, Z_T \in \mathbf{R}$ be a scalar time series and define

$$Y_i = Z_{i+1}, \qquad X_i = (Z_i, \ldots, Z_1),$$

$i = 1, \ldots, T-1$. The sequence of observations $(Y_1, X_1), \ldots, (Y_{T-1}, X_{T-1})$ is not a sequence of identically distributed random vectors. For example, the regression function $f_i(x) = E(Y_i \,|\, X_i = x)$, $x \in \mathbf{R}^{id}$, is defined in a different space for each i.

Time–Space Prediction In time–space prediction the time parameter is taken as the explanatory variable, in contrast to (1.45), where the previous observations in the time series are taken as the explanatory variables. We denote

$$Y_i = Z_i, \qquad X_i = i, \qquad i = 1, \ldots, T. \tag{1.48}$$

The obtained regression model is a fixed design regression model, as described in Section 1.1.1.

Time–space prediction can be used when the time series can be modeled as a nonstationary time series of signal with additive noise:

$$Y_i = \mu_i + \sigma_i \epsilon_i, \qquad i = 1, \ldots, T, \tag{1.49}$$

where $\mu_i \in \mathbf{R}$ is the deterministic signal, $\sigma_i > 0$ are nonrandom values, and the noise ϵ_i is stationary with mean zero and unit variance. For statistical estimation and asymptotic analysis we can use a slightly different model

$$Y_{i,T} = \mu(t_{i,T}) + \sigma(t_{i,T})\, \epsilon_{i,T}, \qquad i = 1, \ldots, T, \tag{1.50}$$

where $t_{i,T} = i/T$, $\mu : [0,1] \to \mathbf{R}$, $\sigma : [0,1] \to (0,\infty)$, and $\epsilon_{i,T}$ is stationary with mean zero and unit variance. Now it can be thought that the observations are coming from a continuous time process $Y(t)$, $t \in [0,1]$, and the sampled discrete time process is obtained as $Y_{i,T} = Y(i/T)$, $i = 1, \ldots, T$. The asymptotics as $T \to \infty$ is called in-fill asymptotics, because points $t_{i,T}$ are filling the interval $[0,1]$ as $T \to \infty$.

1.1.10 Stochastic Control

We consider two types of stochastic control problems. The first type of stochastic control problem appears in option pricing and hedging and the second type of stochastic control problem appears in portfolio selection. The connection of these stochastic control problems to portfolio selection and to option pricing and hedging are explained in Section 1.5.3 and in Section 1.5.4, respectively.

Option-Pricing-Type Stochastic Control Consider the time series

$$X_{t_0}, X_{t_0+1}, \ldots, X_{T-1} \in \mathbf{R}$$

and a random variable $Y_T \in \mathbf{R}$. We are able to choose coefficient $\beta_t \in \mathbf{R}$ at time t, for $t = t_0, \ldots, T-1$ and a constant term $\alpha_{t_0} \in \mathbf{R}$ at time t_0. We want to choose these coefficients in such a way that the mean squared error

$$\mathrm{MSE}\left(\alpha_{t_0}, \beta_{t_0}, \ldots, \beta_{T-1}\right) = E\left(\alpha_{t_0} + \beta_{t_0} X_{t_0} + \cdots + \beta_{T-1} X_{T-1} - Y_T\right)^2$$

is minimized. The optimal coefficients at time t_0 are defined by

$$\left(\alpha^o_{t_0}, \beta^o_{t_0}\right) = \mathrm{argmin}_{\alpha_{t_0}, \beta_{t_0}} \min_{\beta_{t_0+1}, \ldots, \beta_{T-1}} \mathrm{MSE}\left(\alpha_{t_0}, \beta_{t_0}, \ldots, \beta_{T-1}\right), \qquad (1.51)$$

where the minimization is done over coefficients β_t at time t and over coefficient α_{t_0} at time t_0.

Note that at time t_0 the coefficients $\beta_{t_0+1}, \ldots, \beta_{T-1}$ are nuisance coefficients since they are chosen at later times, and at time t_0 we use them only to calculate the optimal values $\alpha^o_{t_0}$ and $\beta^o_{t_0}$. Then, at time $t_0 + 1$ we choose parameters α_{t_0+1} and β_{t_0+1} and parameters $\beta_{t_0+2}, \ldots, \beta_{T-1}$ are nuisance parameters at time $t_0 + 1$.

Note the difference to the usual least squares problem. Namely, in the usual least squares problem we solve the problem

$$\min_{\alpha_{t_0}, \beta_{t_0}, \ldots, \beta_{T-1}} \mathrm{MSE}\left(\alpha_{t_0}, \beta_{t_0}, \ldots, \beta_{T-1}\right)$$

at time $T-1$. That is, all coefficients are chosen at the same time $T-1$ and at that time all values $X_{t_0}, \ldots X_{T-1}$ are known. This problem appears for example in the linear autoregression, where we minimize the expected squared error

$$E\left(\alpha_{t_0} + \beta_{t_0} X_{t_0} + \cdots + \beta_{T-1} X_{T-1} - X_T\right)^2$$

at time $T-1$. In the one-step case the stochastic control and the usual least squares problem are identical, because in the one-step problem we minimize

$$E\left(\alpha_{t_0} + \beta_{t_0} X_{t_0} - Y_{t_0+1}\right)^2 = E\left(\alpha_{T-1} + \beta_{T-1} X_{T-1} - Y_T\right)^2$$

at time $t_0 = T-1$.

If we have n realizations $(X^i_1, \ldots, X^i_{T-t_0}, Y^i_{T-t_0+1})$, $i = 1, \ldots, n$, which have the same distribution as $(X_{t_0}, \ldots, X_{T-1}, Y_T)$, then we can find data-based coefficients as

$$\left(\alpha^o_{t_0}, \beta^o_{t_0}\right) = \mathrm{argmin}_{\alpha_{t_0}, \beta_{t_0}} \min_{b_{t_0+1}, \ldots, b_{T-1}} \mathrm{MSE}_n\left(\alpha_{t_0}, \beta_{t_0}, \ldots, \beta_{T-1}\right),$$

where

$$\begin{aligned}
&\mathrm{MSE}_n\left(\alpha_{t_0}, \beta_{t_0}, \ldots, \beta_{T-1}\right) \\
&= \sum_{i=1}^{n} \left(\alpha_{t_0} + \beta_{t_0} X_1^i + \cdots + \beta_{T-1} X_{T-t_0}^i - Y_{T-t_0+1}^i\right)^2 .
\end{aligned}$$

The connection of this type of stochastic control problem to option pricing is explained in Section 1.5.4.

Portfolio-Selection-Type Stochastic Control Consider the time series

$$X_{t_0+1}, X_{t_0+1}, \ldots, X_T \in \mathbf{R}^d .$$

We are able to choose coefficient $\beta_t \in \mathbf{R}^d$ at time t, for $t = t_0, \ldots, T-1$. We want to choose these coefficients in such a way that

$$\mathrm{W}\left(\beta_{t_0}, \ldots, \beta_{T-1}\right) = Eu\left(\prod_{t=t_0}^{T-1} \beta_t' X_{t+1}\right)$$

is maximized, where $u : \mathbf{R} \to \mathbf{R}$. The optimal coefficients at time t_0 are defined by

$$\beta_{t_0}^o = \mathrm{argmax}_{\beta_{t_0}} \max_{\beta_{t_0+1}, \ldots, \beta_{T-1}} \mathrm{W}\left(\beta_{t_0}, \ldots, \beta_{T-1}\right), \tag{1.52}$$

where the maximization is done over vector β_t at time t. The connection of this type of stochastic control problem to portfolio selection is explained in Section 1.5.3, see (1.97).

1.1.11 Instrumental Variables

The method of instrumental variables is used to estimate causal relationships when it is not possible to make controlled experiments. There are three classical examples of the cases where a need for instrumental variables arises: when there are relevant explanatory variables which are not observed (omitted variables), when the explanatory variables are subject to measurement errors, or when the response variable has a causal influence on one of the explanatory variables (reverse causation).

The method of instrumental variables can be used when we want to estimate structural function $g : \mathbf{R}^d \to \mathbf{R}$ in the model

$$Y = g(X) + U, \tag{1.53}$$

where $Y \in \mathbf{R}$, $X \in \mathbf{R}^d$, and

$$E(U \mid X) \neq 0.$$

Now $g(x)$ is not the conditional expectation $E[Y \mid X = x]$. Estimation of g is possible when we have observations (X_i, Y_i, Z_i), $i = 1, \ldots, n$, where (X_i, Y_i) are distributed as (X, Y) and Z_i are observations from the distribution of an instrumental variable $Z \in \mathbf{R}^d$ that satisfies

$$E(U \mid Z) = 0. \tag{1.54}$$

We give two examples of model (1.53). The first example explains how an omitted variable can lead to (1.53). The second example explains how an error in the explanatory variable can lead to (1.53).

Omitted Variable As an example of a case where model (1.53) can arise, consider the case where X is a variable indicating the type of the treatment a patient receives:

$$X = \begin{cases} 0, & \text{when patient receives treatment A,} \\ 1, & \text{when patient receives treatment B,} \end{cases}$$

and Y is a variable measuring the health of the patient after receiving the treatment. This example is modeled after McClellan, McNeil & Newhouse (1994). We want to estimate the causal influence of X on Y. Let us denote with W the random variable measuring the health of a patient at the time the patient receives the treatment. Also the variable W is influencing Y. In this example W is also affecting X, because the decision about the treatment a patient receives is partially based on the health condition of the patient (if the patient is weak, he will not receive a treatment that is physiologically demanding). Using usual regression methods and observations of X and Y would give a biased estimate of the causal influence of X on Y. (If patients with a weak condition receive treatment A more often, then the estimate would give a pessimistic estimate of the effect of treatment A.)

We have three approaches to estimate the casual influence of X on Y: (1) We can use randomization, so that the value of X is determined by coin tossing, and the influence of W on X is removed. However, in this example this is not possible for ethical reasons. (2) We can estimate the conditional expectation $E(Y \mid X = x,\ W = z)$. However, in this example we have not observed W, so the estimation of this conditional expectation is not possible. (3) We can use the method of instrumental variables. In this example the instrumental variable Z can be chosen as the difference between the shortest distance from a patients home to a hospital giving treatment A and the shortest distance from a patients home to a hospital giving treatment B. Variable Z has an influence on X, because patients had an influence on choice of the treatment they received, and they tended to choose a treatment that was given in the nearest hospital. Variable Z does not have any influence on the health of patients, so it is otherwise external variable, influencing only X. Thus we can use Z to make a pseudo randomization even when a proper randomization was not possible.

We assume an additive model

$$Y = \alpha + f_1(X) + f_2(W) + \epsilon,$$

where $E(\epsilon \mid X = x,\ W = w) = 0$, $Ef_1(X) = 0$, and $Ef_2(W) = 0$. We have observations (Y_i, X_i, Z_i), $i = 1, \ldots, n$, but no observations of W. Using these observations, we can estimate f_1, but not f_2. Estimation of f_1 is enough to give information on the causal influence of X to Y.

Denoting $g(X) = \alpha + f_1(X)$ and $U = f_2(W) + \epsilon$, we have that

$$Y = g(X) + U,$$

where $E(U \mid X) \neq 0$, because $\mathrm{Cov}(X, W) \neq 0$, and $E(U \mid Z) = 0$, because Z is external to the system, having influence only on X. Thus we are in the setting of model (1.53).

Errors in Variables in a Linear Model As an example of model (1.53), consider the case where the linear model

$$Y = \alpha + \beta X^* + U^*$$

holds. However, the explanatory variable X^* is not observed directly but we observe only pairs (Y_i, X_i), $i = 1, \ldots, n$, where

$$X_i = X_i^* + \epsilon_i, \qquad i = 1, \ldots, n.$$

Thus the observed values X_i are contaminated with additive errors. We assume that

$$\mathrm{Cov}(X^*, U^*) = 0, \qquad \mathrm{Cov}(U^*, \epsilon) = 0 \tag{1.55}$$

and

$$\mathrm{Cov}(X^*, \epsilon) = 0. \tag{1.56}$$

We can write the observed response variables as

$$Y = \alpha + \beta X + U^* - \beta\epsilon$$

and the new error term is denoted by

$$U = U^* - \beta\epsilon$$

to get the new linear model

$$Y = \alpha + \beta X + U. \tag{1.57}$$

In this new linear model $E(U \mid X) \neq 0$. Thus we have the same situation as in (1.53), with $g(X) = \alpha + \beta X$.

The fact $E(U \mid X) \neq 0$, follows from $\mathrm{Cov}(X, U) \neq 0$. We have that

$$\begin{aligned} \mathrm{Cov}(X, U) &= \mathrm{Cov}(X, U^*) - \beta \mathrm{Cov}(X, \epsilon) \\ &= -\beta \left[\mathrm{Cov}(X^*, \epsilon) + \mathrm{Cov}(\epsilon, \epsilon)\right] \\ &= -\beta \mathrm{Var}(\epsilon) \\ &\neq 0, \end{aligned}$$

because

$$\mathrm{Cov}(X, U^*) = \mathrm{Cov}(X^*, U^*) + \mathrm{Cov}(\epsilon, U^*) = 0$$

by assumption (1.55) and $\mathrm{Cov}(X^*, \epsilon) = 0$ by assumption (1.56).

Estimation of the Structural Function We give a linear instrumental variables estimator in (2.24). This estimator can be used to estimate parameters α and β in (1.57). The linear instrumental variable estimator is

$$\hat{\beta} = \frac{\sum_{i=1}^n (X_i - \bar{X})(Z_i - \bar{Z})}{\sum_{i=1}^n (X_i - \bar{X})^2}, \qquad \hat{\alpha} = \bar{Y} - \hat{\beta}\bar{X},$$

where

$$\bar{X} = \frac{1}{n}\sum_{i=1}^n X_i, \qquad \bar{Y} = \frac{1}{n}\sum_{i=1}^n Y_i, \qquad \bar{Z} = \frac{1}{n}\sum_{i=1}^n Z_i.$$

Hall & Horowitz (2005) approach the estimation of $g(x)$ in the model (1.53) by deriving an operator equation for g. From (1.54) we obtain

$$E(Y \mid Z) = E(g(X) \mid Z) + E(U \mid Z) = (Kg)(Z),$$

where the operator K is defined as

$$(Kg)(z) = E(g(X) \mid Z = z) = \int f_{X \mid Z=z}(x) g(x)\, dx.$$

The operator K is an integral operator mapping $L_X^2 = \{g : \mathbf{R}^d \to \mathbf{R} \mid E(g^2(X)) < \infty\}$ to $L_Z^2 = \{h : \mathbf{R}^d \to \mathbf{R} \mid E(h^2(Z)) < \infty\}$. By estimating K and estimating $E(Y \mid Z)$ we can find an estimator for g.

1.2 DISCRETE RESPONSE VARIABLE

We introduce first binary response models, where the response variable is a Bernoulli random variable, second we introduce discrete choice models, where the response variable is a categorical random variable, and third we introduce count data models, where the response variable is a Poisson random variable. In Section 1.3 we introduce more general exponential family models which contain as special cases the binary response models, discrete choice models, and Poisson count models.

1.2.1 Binary Response Models

In a binary response model the response variable Y is a Bernoulli distributed random variable, so that it takes only values 0 and 1. When $Y \sim$ Bernoulli(p), where $0 \le p \le 1$, then the probability mass function of Y is

$$f_Y(y) = p^y (1-p)^{1-y}, \qquad y \in \{0, 1\}. \tag{1.58}$$

Now we can construct a model for the conditional distribution of Y given X as

$$f_{Y \mid X=x}(y) = p(x)^y (1 - p(x))^{1-y}, \qquad y \in \{0, 1\}, \ x \in \mathbf{R}^d, \tag{1.59}$$

where $p : \mathbf{R}^d \to [0,1]$ is a function. Note that in the Bernoulli model $EY = p$ and in the conditional Bernoulli model (1.59) the conditional expectation of Y given X is

$$E[Y \mid X = x] = P(Y = 1 \mid X = x) = p(x).$$

Since function p is a conditional expectation, we can use any regression method to estimate p. However, it can happen that a regression function estimate is such that it takes values outside the interval $[0,1]$. For example, a linear regression function estimate takes values outside the range $[0,1]$ for large or small enough values of the explanatory variables. There are several natural estimators for function p:

1. In a generalized linear model it is assumed that

$$p(x) = G(\alpha + \beta' x),$$

where $G : \mathbf{R}^d \to [0,1]$ is a known link function. Generalized linear models in the case of a binary response model are considered in Section 2.3.2.

2. In the single index model it is assumed that

$$p(x) = g(\alpha + \beta' x),$$

where $g : \mathbf{R}^d \to [0,1]$ is an unknown link function. Single link estimators are considered in Section 4.1.

3. We can estimate p with the help of a density function estimator, if vector X has a continuous distribution. If vector X has a continuous distribution, we can write

$$P(Y = 1 \mid X = x) = \frac{P(Y = 1) f_{X \mid Y=1}(x)}{f_X(x)},$$

where $f_{X \mid Y=1}$ is the density of $X \mid Y = 1$ and f_X is the density of X. The prior probability $P(Y = 1)$ can be estimated by

$$\hat{p}_1 = \frac{1}{n} \#\{i = 1, \ldots, n : Y_i = 1\}.$$

The densities $f_{X \mid Y=1}$ and f_X can be estimated by any density estimation method. For example, in kernel density estimation we take

$$\hat{f}_X(x) = \frac{1}{n} \sum_{i=1}^{n} K_h(x - X_i), \qquad \hat{f}_{X \mid Y=1}(x) = \frac{1}{n} \sum_{i=1}^{n} K_h(x - X_i) I_{\{1\}}(Y_i),$$

where $K_h(x) = K(x/h)/h^d$, $K : \mathbf{R}^d \to \mathbf{R}$ is the kernel function, and $h > 0$ is the smoothing parameter. See (3.39) for the definition of the kernel density estimator. Finally, we define the estimator of function $p : \mathbf{R}^d \to [0,1]$ as

$$\hat{p}(x) = \frac{\hat{p}_1 \hat{f}_{X \mid Y=1}(x)}{\hat{f}_X(x)}. \tag{1.60}$$

4. We can estimate function $p : \mathbf{R}^d \to [0,1]$ with a local averaging

$$\hat{p}(x) = \sum_{i=1}^{n} p_i(x) \, Y_i, \tag{1.61}$$

where the weights $p_i(x)$ satisfy $p_i(x) \geq 0$ and $\sum_{i=1}^{n} p_i(x) = 1$. Examples of the local averaging are given in Chapter 3, where regressogram weights, kernel weights, and nearest neighborhood weights are defined. In the case of kernel regression and kernel density estimation formulas (1.60) and (1.61) are equivalent, see (3.37).

1.2.2 Discrete Choice Models

In discrete choice models the response variable is a discrete random variable taking only a finite number of values. We can distinguish the cases where the values of the response variable are unordered and the cases where they are ordered. The random variables whose values are unordered are called nominal or categorical random variables and the random variables whose values are ordered are called ordinal random variables.

Let us consider a discrete choice model with a categorical response variable. A categorical response variable Y has a categorical distribution, taking K distinct values $0, 1, \ldots, K-1$, say. The categorical distribution family generalizes the Bernoulli distribution family, where the variable takes only values 0 and 1. When $Y \sim \text{Categorical}(p_0, \ldots, p_{K-1})$, where $0 \leq p_k \leq 1$, $\sum_{k=0}^{K-1} p_k = 1$, then the probability mass function of Y is

$$f_Y(y) = \sum_{k=0}^{K-1} p_i I_{\{k\}}(y), \qquad y \in \{0, \ldots, K-1\}. \tag{1.62}$$

Now we can construct a model for the conditional distribution of Y given X as

$$f_{Y \mid X=x}(y) = \sum_{k=0}^{K-1} p_k(x) I_{\{k\}}(y), \qquad y \in \{0, \ldots, K-1\}, \;\; x \in \mathbf{R}^d, \tag{1.63}$$

where $p_k : \mathbf{R}^d \to [0,1]$ are functions satisfying $\sum_{k=0}^{K-1} p_k(x) = 1$ for each $x \in \mathbf{R}^d$. Note that now the conditional probability of Y given X is

$$P(Y = k \mid X = x) = p_k(x), \qquad k = 0, \ldots, K-1.$$

There are several reasonable estimators of $p_k(x)$.

1. We can use the parametric form

$$p_k(x) = \frac{e^{\beta'_k x}}{1 + \sum_{i=1}^{K-1} e^{\beta'_i x}},$$

for $k = 1, \ldots, K-1$ and $p_0(x) = 1 - \sum_{i=1}^{K-1} p_i(x)$. A more restrictive form is

$$p_k(x) = \frac{e^{\beta' x}}{\sum_{i=0}^{K-1} e^{\beta' x}}, \tag{1.64}$$

where the conditional probability is the same for all classes. This form is obtained by defining

$$U_i = \beta' X + \epsilon_i$$

and

$$Y = \text{argmax}_{i=0,\ldots,K-1} U_i.$$

Assume that ϵ_i are independent and identically distributed with the Weibull distribution. The distribution function of the Weibull distribution is $F_{\epsilon_i}(x) = \exp\{-e^{-x}\}$. Now $p_k(x) = P(Y = k \,|\, X = x)$ is given by (1.64). The estimation can be done with the maximum likelihood or with the least squares method.

2. We can estimate p with the help of any density function estimator. If vector X has a continuous distribution, then we can write,

$$P(Y = k \,|\, X = x) = \frac{P(Y = k) f_{X \,|\, Y=k}(x)}{f_X(x)},$$

where $k = 0, \ldots, K-1$, $f_{X \,|\, Y=1}$ is the density of $X \,|\, Y = 1$ and f_X is the density of X. The prior probability $P(Y = k)$ can be estimated by

$$\hat{p}_k = \frac{1}{n} \#\{i = 1, \ldots, n : Y_i = k\},$$

where $\#A$ denotes the cardinality of set A. The densities $f_{X \,|\, Y=k}$ and f_X can be estimated by any density estimation method. See (3.39) for the definition of the kernel density estimator. Finally we define the estimator of p as

$$\hat{p}(x) = \frac{\hat{p}_k \hat{f}_{X \,|\, Y=k}(x)}{\hat{f}_X(x)}. \tag{1.65}$$

3. Define K Bernoulli random variables $Y^{(0)}, \ldots, Y^{(K-1)}$ with the definition that $Y^{(k)} = 1$ if and only if $Y = k$. Then

$$p_k(x) = E(Y^{(k)} \,|\, X = x).$$

We can use, for example, kernel regression to estimate $p_k(x)$ using regression data $(X_1, Y_1^{(k)}), \ldots, (X_n, Y_n^{(k)})$, for $k = 0, \ldots, K-1$.

1.2.3 Count Data

Count data occurs when the response variable Y gives the number of occurrences of an event. For instance, Y could give the annual number of bank failures. The count data is such that Y takes values $\{0, 1, 2, \ldots\}$. Count data can be modeled with the Poisson distribution. If $Y \sim \text{Poisson}(\nu)$, then

$$P(Y = y) = e^{-\nu} \frac{\nu^y}{y!}, \qquad y = 0, 1, 2, \ldots,$$

where $\nu > 0$ is the unknown intensity parameter. Now $EY = \nu$ and $\text{Var}(Y) = \nu$. In the Poisson regression the regression function is

$$\nu(x) = E(Y \mid X = x),$$

where $X \in \mathbf{R}^d$ is the vector of explanatory variables. The Poisson regression is a heteroskedastic regression model. A parametric Poisson regression model is obtained if

$$\nu(x) = \exp\{x'\beta\},$$

where $\beta \in \mathbf{R}^d$ is the unknown parameter. This choice guarantees that $\nu(x) > 0$. Besbeas, de Feis & Sapatinas (2004) make a comparative simulation study of wavelet shrinkage estimators for Poisson counts.

1.3 PARAMETRIC FAMILY REGRESSION

We obtain the binary response models, discrete choice models, and Poisson count models, introduced in Section 1.2, as special cases of parametric family regression, introduced in Section 1.3.1. In fact, these are a special cases of exponential family regression, introduced in Section 1.3.2. A different type of parametric family regression is obtained by copula modeling, introduced in Section 1.3.3.

1.3.1 General Parametric Family

Let us consider a family $(P_\theta, \theta \in \Theta)$ of probability measures, where $\Theta \subset \mathbf{R}^p$. Let $Y \in \mathbf{R}$ be a response variable and let $X \in \mathbf{R}^d$ be a vector of explanatory variables that satisfy

$$Y \sim P_{f(X)},$$

where $f : \mathbf{R}^d \to \Theta$ is an unknown function to be estimated. The function f is estimated using identically distributed observations $(X_1, Y_1), \ldots, (X_n, Y_n)$ from the distribution of (X, Y). After estimating function f, we have an estimator of the conditional distribution $Y \mid X = x$, because

$$Y \mid X = x \sim P_{f(x)}. \tag{1.66}$$

The following examples illustrate the model.

1. We obtain a Gaussian mean regression model when $P_\theta = N(\theta, \sigma^2)$, where $\theta \in \Theta = \mathbf{R}$. Now
$$Y \mid X = x \sim N\left(f(x), \sigma^2\right),$$
which follows from
$$Y = f(X) + \epsilon,$$
where $\epsilon \sim N(0, \sigma^2)$.

2. We obtain a Gaussian volatility model, when $P_\theta = N(0, \theta)$, where $\theta \in \Theta = (0, \infty)$. Now
$$Y \mid X = x \sim N(0, f(x)),$$
which follows from
$$Y = f(X)^{1/2}\,\epsilon,$$
where $\epsilon \sim N(0, 1)$.

3. We obtain a Gaussian heteroskedastic mean regression model, when $P_\theta = N(\theta_1, \theta_2)$, where $\theta = (\theta_1, \theta_2)$, and $\Theta = \mathbf{R} \times (0, \infty)$. Now
$$Y \mid X = x \sim N(f_1(x), f_2(x)),$$
which follows from
$$Y = f_1(X) + f_2(X)^{1/2}\epsilon,$$
where $\epsilon \sim N(0, 1)$, and we denote $f = (f_1, f_2)$.

4. We obtain the binary choice model, when $P_\theta = \text{Bernoulli}(\theta)$, where $\theta \in \Theta = [0, 1]$. Then $P(Y = 1) = f(X)$ and $P(Y = 0) = 1 - f(X)$.

Let us assume that the probability measures P_θ are dominated by a σ-finite measure, and denote the density functions of P_θ by $p(y, \theta)$. We use the term density function, although $p(y, \theta)$ can be also a probability mass function, if Y has a discrete distribution. In Section 1.3.2 we make the assumption that $(P_\theta, \theta \in \Theta)$ is an exponential family.

Under the assumption that $(X_1, Y_1), \ldots, (X_n Y_n)$ are i.i.d., the log-likelihood of the sample is
$$\sum_{i=1}^{n} \log p(Y_i, f(X_i)).$$
The log-likelihood can be maximized over collection $\mathcal{F}$ of functions, and we denote $\mathcal{F} = (f_\beta, \beta \in \mathcal{B})$. We have two general approaches.

1. The first possibility is to define
$$\hat{f} = \text{argmax}_{\beta \in \mathcal{B}} \sum_{i=1}^{n} \log p(Y_i, f_\beta(X_i)),$$
where $(f_\beta, \beta \in \mathcal{B})$ is a large collection of functions, like the collection of linear functions: $f_\beta(x) = \beta_0 + \beta_1 x_1 + \cdots + \beta_d x_d$.

2. A second possibility is to maximize a local log-likelihood and define

$$\hat{f}(x) = \text{argmax}_{f\in\mathcal{F}} \sum_{i=1}^{n} \log p(Y_i, f(X_i))\, p_i(x), \tag{1.67}$$

where $p_i(x)$ are weights, for example $p_i(x) = K_h(x - X_i)$, where $K_h(x) = K(x/h)/h^d$, $K : \mathbf{R}^d \to \mathbf{R}$, and $h > 0$. Now we can take f_β to be a constant function: $f_\beta(x) = \beta$, where $\beta \in \mathbf{R}$. The local likelihood approach has been covered in Spokoiny (2010).

1.3.2 Exponential Family Regression

An exponential family is a collection $\mathcal{P} = (P_\theta, \theta \in \Theta)$ of probability measures. The probability measures in $\mathcal{P}$ are dominated by a σ-finite measure. In a one-parameter exponential family the density functions have the form

$$p(y, \theta) = p(y) \exp\{yc(\theta) - b(\theta)\},$$

where $\theta \in \Theta \subset \mathbf{R}$, and $y \in \mathcal{Y} \subset \mathbf{R}$. The functions c and b are nondecreasing functions on Θ and function $p : \mathcal{Y} \to \mathbf{R}$ is nonnegative. In the exponential family with the canonical parameterization the density functions are

$$p(y, v) = p(y) \exp\{yv - d(v)\}. \tag{1.68}$$

The canonical parameterization is obtained by putting $v = c(\theta)$ and $d(v) = b(\theta)$. Examples of exponential families include the family of Gaussian, Bernoulli, Poisson, and gamma distributions. An exposition of exponential families is given by Brown (1986).

We use the modeling approach in (1.66), and assume that the conditional distribution of Y given $X = x$ belongs to an exponential family and the parameter of the conditional distribution is $v = f(x)$:

$$Y \mid X = x \sim p(y, f(x)), \tag{1.69}$$

for a function $f : \mathbf{R}^d \to \mathcal{V}$, where we use the natural parameterization in (1.68), and $\mathcal{V}$ is the parameter space of the natural parameter.

If the parameterization is natural, and d is continuously differentiable, then

$$E_v Y = d'(v), \tag{1.70}$$

where $Y \sim f(y, v)$. Indeed,

$$\frac{\partial}{\partial v} \log p(y, v) = y - d'(v).$$

On the other hand,

$$E_v \frac{\partial}{\partial v} \log p(Y, v) = 0,$$

under regularity assumptions.[5] Thus, (1.70) holds. Under the assumption (1.69), we get

$$E(Y \mid X = x) = d'(f(x)).$$

If the parameterization is natural, and d is two times continuously differentiable, then

$$\mathrm{Var}_v(Y) = d''(v). \tag{1.71}$$

Indeed,

$$\begin{aligned} \mathrm{Var}_v(Y) &= E\left(Y - d'(v)\right)^2 = E\left[\frac{\partial}{\partial v} \log p(Y, v)\right]^2 = -E\,\frac{\partial^2}{\partial v^2} \log p(Y, v) \\ &= d''(v). \end{aligned}$$

Under the assumption (1.69), we get

$$\mathrm{Var}(Y \mid X = x) = d''(f(x)).$$

Brown, Cai & Zhou (2010) suggest a reduction method where the exponential family regression can be transformed to the Gaussian regression by binning and variance stabilizing transform.

1.3.3 Copula Modeling

Let (Y_1, Y_2) be a random vector with a continuous distribution function

$$F(y_1, y_2) = P(Y_1 \le y_1, Y_2 \le y_2),$$

where $y_1, y_2 \in \mathbf{R}$. We can write the distribution function uniquely as

$$F(y_1, y_2) = C(F_1(y_1), F_2(y_2)), \tag{1.72}$$

where $F_1(y_1) = P(Y_1 \le y_1)$ and $F_2(y_2) = P(Y_2 \le y_2)$ are the distribution functions of Y_1 and Y_2. The function $C : [0,1]^2 \to \mathbf{R}$ is the copula of the distribution of (Y_1, Y_2). Function C is a distribution function whose marginals are uniform on $[0, 1]$. The copula is defined by

$$C(u_1, u_2) = F\left(F_1^{-1}(u_1), F_2^{-1}(u_2)\right),$$

where $u_1, u_2 \in [0, 1]^2$. These facts were proved in Sklar (1959). See also Nelsen (1999).

For example, a Gaussian two dimensional copula is a normal distribution with unit marginal standard deviations. The family of Gaussian two dimensional copulas C_θ has the parameter $\theta \in (-1, 1)$, where θ is the correlation coefficient between Y_1 and Y_2.

[5] We have that $E_v \frac{\partial}{\partial v} \log p(Y, v) = E_v \frac{\partial p(Y,v)/\partial v}{p(Y,v)} = \int \frac{\partial}{\partial v} p(y, v)\, dy = \frac{\partial}{\partial v} \int p(y, v)\, dy = \frac{\partial}{\partial v} 1 = 0$, if the order of derivation and integration can be changed.

The copula representation of the distribution as in (1.72) gives a useful way to construct models and to estimate the unknown parameters of the model. Let $(c_\theta, \theta \in \Theta)$ be a family of copula densities, where $\Theta \subset \mathbf{R}^p$. This leads to a semiparametric model with densities

$$f(y_1, y_2; \theta, f_1, f_2) = c_\theta\left(F_1(y_1), F_2(y_2)\right) f_1(y_1)\, f_2(y_2),$$

where $\theta \in \Theta$ and $f_1, f_2 \in \mathcal{F}$, where $\mathcal{F}$ is a nonparametric collection of univariate density functions. The estimation of θ, f_1, f_2 can be done with the two stage approach. In the first stage we estimate nonparametrically the marginal distributions f_1 and f_2. In the second stage we estimate the copula parameter θ.

Assume that $X \in \mathbf{R}^d$ is a vector of explanatory variables and we want to estimate the conditional distribution $(Y_1, Y_2) \mid X = x$. We assume that the conditional distribution function is

$$\begin{aligned} F_{Y_1,Y_2|X=x}(y_1, y_2) &= P(Y_1 \le y_1, Y_2 \le y_2 \mid X = x) \\ &= C_{\theta(x)}\left(F_{Y_1|X=x}(y_1), F_{Y_2|X=x}(y_2)\right), \end{aligned}$$

where $\theta : \mathbf{R}^d \to \mathbf{R}$. The conditional density is

$$f_{Y_1,Y_2|X=x}(y_1, y_2) = c_{\theta(x)}\left(F_{Y_1|X=x}(y_1), F_{Y_2|X=x}(y_2)\right) f_{Y_1|X=x}(y_1)\, f_{Y_2|X=x}(y_2).$$

In the first stage we estimate nonparametrically the conditional distribution functions and get the estimates $\hat{F}_{Y_1|X=x}(y_1)$ and $\hat{F}_{Y_2|X=x}(y_2)$. In the second stage we estimate the function $\theta(x)$. This can be done analogously to (1.67), and we get the locally constant likelihood estimator as

$$\hat{\theta}(x) = \mathrm{argmin}_{\theta \in \Theta} \sum_{i=1}^{n} p_i(x) \log c_\theta\left(\hat{F}_{Y_1|X=x}(y_1), \hat{F}_{Y_2|X=x}(y_2)\right).$$

This method has been studied in Abegaz, Gijbels & Veraverbeke (2012).

We have defined in (1.72) the standard copula decomposition. This decomposition can be inconvenient because the copula density c has the support inside $[0,1]^2$, and the estimation is typically complicated with the boundary effects. Alternatively, we can make the copula decomposition

$$F(x_1, x_2) = C\left(\Phi^{-1}(F_1(x_1)), \Phi^{-1}(F_2(x_2))\right),$$

where $\Phi : \mathbf{R} \to \mathbf{R}$ is the distribution function of the standard Gaussian distribution. Now C is a distribution function whose marginals are standard Gaussian, and C is defined by

$$C(u, v) = F\left(F_1^{-1}\left(\Phi(u)\right), F_2^{-1}\left(\Phi(v)\right)\right), \qquad u, v \in \mathbf{R}.$$

1.4 CLASSIFICATION

Let the sequence $(X_1, Y_1), \ldots, (X_n, Y_n)$ consist of identically distributed random vectors. Let (X, Y) be distributed as (X_i, Y_i), for $i = 1, \ldots, n$. Let the possible values of Y be $\{0,\ldots,K-1\}$. We want to find a classification function $g : \mathbf{R}^d \to \{0,\ldots,K-1\}$. The classification function is interpreted as such function that if we observe a new random variable X_{n+1}, distributed as X, then $g(X_{n+1})$ guesses the class label of X_{n+1}, that is, we decide that X_{n+1} comes from the distribution of $X \mid Y = k$, if $g(X_{n+1}) = k$.

In the case of classification Y can take only a finite number of values (as many values as there are classes), since the values of the response variable Y indicate the class label. In the case of regression analysis the response variable Y can in many cases take as values any real number. However, in Section 1.2.1 we have considered binary response models, where the response variable takes only two values and in Section 1.2.2 we have considered discrete choice models, where the response variable takes a finite number of values. In binary response models and in discrete choice models we are, however, interested in estimating the conditional expectation $f(x) = E(Y \mid X = x)$, $f : \mathbf{R}^d \to \mathbf{R}$, whereas in the case of classification, we want to estimate the classification function $g : \mathbf{R}^d \to \{0,\ldots,K-1\}$, which predicts the class label of a future observation. As an example, consider the case where $K = 2$, so that there are two classes, and thus Y is a Bernoulli distributed random variable. Now the regression function is

$$f(x) = E(Y \mid X = x) = P(Y = 1 \mid X = x). \tag{1.73}$$

Thus $f(X_{n+1}) \in [0,1]$, but we would like to find a classification function g such that $g(X_{n+1}) \in \{0,1\}$.

We have explained the classification in the case where the number of observations in each class is a random number. There also exist cases where the observation number in each class can be chosen by the designer of the experiment. Then we have fixed numbers $n_0,\ldots,n_{K-1}$ and observations $X_{k1},\ldots,X_{kn_k} \in \mathbf{R}^d$ are coming from the kth distribution, $k = 0,\ldots,K-1$. We will consider only the case where the class frequencies are random.

1.4.1 Bayes Risk

In the random design regression we can motivate the estimation of conditional expectation $f(x) = E(Y \mid X = x)$ by noting that the conditional expectation minimizes the mean squared error: $f = \text{argmin}_g E(Y - g(X))^2$; see Section 1.1.7. Similarly, in the case of classification, we can find a population quantity which minimizes a natural criterion. This criterion is the probability of misclassification, or Bayes risk,

$$R(g) = P(g(X) \neq Y). \tag{1.74}$$

Let

$$g^* = \text{argmin}_g R(g),$$

where the minimization is done over all classification functions $g : \mathbf{R}^d \to \{0,\ldots,K-1\}$. The classification function g^* which minimizes the probability of misclassification is called the Bayes rule. It can be proved that

$$g^*(x) = \text{argmax}_{k=0,\ldots,K-1} P(Y = k \mid X = x). \tag{1.75}$$

The proof of (1.75) for the case $K = 2$ can be found in Györfi et al. (2002, Lemma 1.1, p. 6). It holds that

$$g^*(x) = \text{argmax}_{k=0,\ldots,K-1} P(Y = k)\, f_{X \mid Y=k}(x), \tag{1.76}$$

where $f_{X \mid Y=k} : \mathbf{R}^d \to \mathbf{R}$ is the density function of $X \mid Y = k$.

1.4.2 Methods of Classification

We shall mention four principles to construct classification functions: classification using regression function estimates, classification using density estimates, classification using empirical risk minimization, and classification using nearest neighbors.

Classification by Regression Function Estimation A classification function can be constructed from a regression function estimate. In the two class case we can take the data $(X_1, Y_1), \ldots, (X_n, Y_n)$ as if it would originate from the binary response model, and in the multiclass case we can take the data as if it would originate from the discrete choice model with a categorical response variable. In Section 1.2.1 we introduced binary response models, and in Section 1.2.2 we introduced discrete choice models.

In a discrete choice model we estimate the class posterior probabilities

$$p_k(x) = P(Y = k \mid X = x), \qquad k = 0, \ldots, K-1.$$

A natural classification function is

$$g^*(x) = \text{argmax}_{k=0,\ldots,K-1} p_k(x). \tag{1.77}$$

In fact, we note in (1.75) that the classification function g^* is in a sense the optimal classification function. Let us denote the estimators of the class posterior probabilities by $\hat{p}_k(x)$, and let us define an estimator of the classification function by

$$\hat{g}(x) = \text{argmax}_{k=0,\ldots,K-1}\, \hat{p}_k(x). \tag{1.78}$$

We can find the estimators $\hat{p}_k(x)$ in the following way. We define K response variables, that are the indicators of the class labels:

$$Y_i^{(k)} = I_{\{k\}}(Y_i), \qquad i = 1, \ldots, n, \;\; k = 0, \ldots, K-1. \tag{1.79}$$

Let $\hat{p}_k(x)$ be a regression function estimator of the posterior probability

$$p_k(x) = E(Y^{(k)} \mid X = x) = P(Y = k \mid X = x). \tag{1.80}$$

Estimator $\hat{p}_k(x)$ is constructed using regression data $(X_1, Y_1^{(k)}), \ldots, (X_n, Y_n^{(k)})$, for $k = 0, \ldots, K-1$.

In the two class case, when $Y \in \{0, 1\}$, we do not have to use (1.79), because Y is already a class indicator. In the two class case we can write the empirical decision rule in a simplified form. Let $\hat{f} : \mathbf{R}^d \to \mathbf{R}$, be any regression function estimator constructed using regression data $(X_1, Y_1), \ldots, (X_n, Y_n)$, and define

$$\hat{g}(x) = \begin{cases} 1, & \text{if } \hat{f}(x) \geq 1/2, \\ 0, & \text{otherwise,} \end{cases} \tag{1.81}$$

which estimates the natural classification function

$$g(x) = \begin{cases} 1, & \text{if } P(Y = 1 \,|\, X = x) \geq P(Y = 0 \,|\, X = x), \\ 0, & \text{otherwise.} \end{cases} \tag{1.82}$$

Classification by Density Estimation A classification function can be constructed from density estimates of the class densities. We assume now that X is a random vector with a continuous distribution. Let us consider the classification rule $g^*(x) = \text{argmax}_{k=0,\ldots,K-1} p_k(x)$, defined in (1.77). We can write

$$p_k(x) = P(Y = k \,|\, X = x) = \frac{P(Y = k) f_{X|Y=k}(x)}{f_X(x)},$$

where $k = 0, \ldots, K-1$, and $x \in \mathbf{R}^d$. Thus,

$$\text{argmax}_{k=0,\ldots,K-1} p_k(x) = \text{argmax}_{k=0,\ldots,K-1} P(Y = k) f_{X|Y=k}(x).$$

An estimator for the classification function, based on data $(X_1, Y_1), \ldots, (X_n, Y_n)$, is obtained as

$$\hat{g}(x) = \text{argmax}_{k=0,\ldots,K-1} \hat{p}_k \hat{f}_{X|Y=k}(x), \tag{1.83}$$

where $\hat{f}_{X|Y=k}$ is a density estimator of the class density function $f_{X|Y=k}$ and $\hat{p}_k$ is an estimator of the class prior probability $P(Y = k)$. We can define

$$\hat{p}_k = \frac{1}{n} \#\{i = 1, \ldots, n : Y_i = k\}.$$

Classification by Empirical Risk Minimization A classification function can be constructed using empirical risk minimization. In (1.78), classification is reduced to regression function estimation (in binary response models or in discrete choice models). In (1.83), classification is reduced to density estimation. However, according to Vapnik's principle, we should not try to estimate more than is needed, and thus we should also consider the direct construction of a classification function, without reducing the problem to regression function estimation or to density function estimation.

We define a classifier by

$$\hat{g} = \text{argmin}_{g \in \mathcal{G}} \gamma_n(g), \tag{1.84}$$

where $\mathcal{G}$ is a class of functions $g : \mathbf{R}^d \to \{0, \ldots, K-1\}$ and $\gamma_n(g)$ is the empirical error of classifier g. We get different classifiers depending on the choice of the empirical error $\gamma_n(g)$ and depending on the choice of class $\mathcal{G}$.

We can define the empirical error of a classifier g by

$$\gamma_n(g) = \#\{i = 1, \ldots, n : g(X_i) \neq Y_i\}. \tag{1.85}$$

Quantity $\gamma_n(g)$ is equal to the number of misclassifications in the learning sample. We can also decompose the number of misclassifications according to the class labels. Let

$$\gamma_n^{(k)}(g) = \#\{i = 1, \ldots, n : g(X_i) \neq k,\ Y_i = k\},$$

where $k = 0, \ldots, K-1$. Then we can define the empirical error of the classifier as a weighted sum of the single class misclassification errors:

$$\gamma_n(g) = \sum_{k=0}^{K-1} w_k \gamma_n^{(k)}(g).$$

For $w_k \equiv 1$ we get the overall classification error (1.85).

In the two class case it has also been suggested to use class labels $Y \in \{-1, 1\}$ and not the labels $\{0, 1\}$, consider classifiers $h : \mathbf{R}^d \to \mathbf{R}$, and define the classification function $g(x) = \text{sign}(h(x))$. The empirical risk is defined as

$$\gamma_n(g) = \sum_{i=1}^{n} \phi(Y_i h(X_i)),$$

where $\phi : \mathbf{R} \to (0, \infty)$ is a convex nonincreasing function with $\phi(u) \geq I_{(-\infty,0]}(u)$ for $u \in \mathbf{R}$. We can take the hinge loss $\phi(u) = \max\{0, 1-u\}$, the exponential loss $\phi(u) = \exp\{-u\}$, or the logit loss $\phi(u) = \log_2(1 + e^{-u})$. Support vector machines, mentioned in Section 5.3, use the hinge loss and a penalized empirical risk.

An example for the choice of class $\mathcal{G}$ is given in (2.84). In this example the class $\mathcal{G}$ is chosen so that the classification functions are linear.

Classification by Nearest Neighbors The nearest-neighbor rule defines the class estimate to be that class label that occurs most often among the k nearest neighbors. That is, for an integer $k \in \{1, 2, \ldots\}$, define the k nearest neighbors, based on observations $(X_1, Y_1), \ldots, (X_n, Y_n)$, as the set

$$\mathcal{Y}(x) = \{Y_i : \|X_i - x\| \leq r_k(x)\},$$

where

$$r_k(x) = \min\{r > 0 : \#\{X_i \in B_r(x)\} = k\},$$

where $B_r(x) = \{z \in \mathbf{R}^d : \|z - x\| \leq r\}$. Now we can define the classifier[6]

$$\hat{g}(x) = \text{argmax}_{y=0,\ldots,K-1} \#\{Y_i \in \mathcal{Y}(x) : Y_i = y\}.$$

Hastie et al. (2001, Section 13) use the term "prototype methods" to denote classifiers which classify the new observation to the class whose observed values are most similar to the new observation.

[6]We denote now the class labels by $y = 0, \ldots, K-1$, because the symbol k is used to denote the number of nearest neighbors.

1.5 APPLICATIONS IN QUANTITATIVE FINANCE

Portfolio selection, risk management, and option pricing belong to the main branches of quantitative finance. Estimation of conditional variances and conditional quantiles can be applied in risk management. Estimation of conditional expectations can be applied in portfolio selection. Option pricing is related to optimal control.

Other applications are described in later sections. Section 2.1.7 explains how linear regression can be applied to estimate the beta of an asset, the beta of a portfolio, the alpha of a portfolio, and the alpha of a hedge fund. Section 2.2.2 explains how varying coefficient regression can be applied in hedge fund replication and in performance measurement. Data sets are described in Section 1.6.

1.5.1 Risk Management

The process of portfolio selection tries to address the problem of balancing the risk and return, but it is useful to have an independent risk management to make an evaluation of the risk of the portfolios at a daily basis.

The economic capital can roughly be defined to mean the amount of money which is needed to secure survival of a company in a worst case scenario. The definition of the economic capital can be made precise with the concept of a value at risk. The economic capital can be used in portfolio selection to calculate return distributions. The regulatory capital is the capital required by the regulators that financial institutions should maintain. The regulatory capital is often defined in terms of value at risk.

Variance trading can be used in speculation, but variance swaps can also be used in risk management to adjust the overall exposure of a portfolio to the volatility.

Value-at-Risk Quantiles can be used to measure the risk of a portfolio. The distribution of the change in the value of the portfolio is called the profit-and-loss distribution: If we denote by V_t the value of the portfolio at time t and by V_u the value of the portfolio at a later time, then the distribution of $V_u - V_t$ is called the profit-and-loss distribution for the time period from t to u. We define the loss as the negative of the change in the value of the portfolio

$$L_u = -(V_u - V_t). \tag{1.86}$$

The upper quantiles of the loss distribution are called the value at risk or VaR:

$$\text{VaR}_p = Q_p(L_u), \tag{1.87}$$

where p is equal to 0.99 or 0.999, for example. A larger value of VaR_p indicates that the portfolio is more risky, because VaR_p is such threshold that the probability that the loss is larger than VaR_p is smaller or equal to $1 - p$. We can write

$$L_u = -V_t R_u,$$

where R_u is the return of the portfolio,

$$R_u = \frac{V_u - V_t}{V_t}.$$

Thus, if we have the quantile $Q_p(R_u)$ of the return distribution, the VaR_p is obtained by the formula

$$\mathrm{VaR}_p = -V_t\, Q_p(R_u).$$

Quantile as risk measure takes into account the number of exceedances of the VaR threshold, but it does not take into account the largeness of the exceedances. Expected shortfall takes also the largeness of the exceedances into account.

Economic Capital in Portfolio Selection Let us consider a bank which wants to choose among a collection of investment proposals. The investment with the best return distribution will be chosen. The problem is to calculate the return distribution since many investments do not require any initial capital, and we cannot calculate the return by dividing by the initial investment.

First we have to construct a profit–loss distribution for each investment proposal. These profit–loss distributions may be very difficult to estimate, because one has to take into account each possible future state of affairs and its probability. In order to estimate the probabilities of the states, one has to take into account all current investments of the bank and consider the interaction of the new investment with the current investments. For example, when we write a call option, the maximum loss is in general infinite; but if we already own the underlying stock, then the loss is bounded.

We want to set aside enough capital to cover adverse events with a given probability of occurrence. The frequency of default for AA-rated companies over a one-year period has been roughly one in three thousand. Thus one could choose the 0.0003th quantile of the profit–loss distribution ($1/3000 \approx 0.0003 = 0.03\%$), which could be for example a loss of 1 million Euros, and set aside enough capital to cover this loss. This capital is called the economical capital. The return on investment is calculated by dividing by the economical capital. That is, we get the return distribution from the profit–loss distribution by dividing with the economical capital. See Rebonato (2007, Chapter 9).

Finally, we choose the best return distribution by the maximization of the expected utility or by the maximization of the variance penalized expected return.

Variance as a Risk Measure The Sharpe ratio of a portfolio is defined as

$$\frac{E(R-r)}{\mathrm{sd}(R-r)}, \tag{1.88}$$

where R is the return of the portfolio for a given time period, r is the return of a risk-free rate for the same time period, and $\mathrm{sd}(R-r)$ is the standard deviation. The Sharpe ratio belongs to the class of performance measures having the form

$$\frac{\text{expected return}}{\text{risk}}.$$

The basic idea is that in measuring the quality of a portfolio we have to take the risk into account and not only the return. In the definition of the Sharpe ratio the expected return and the standard deviation is defined using the *excess return*, which is the return of the portfolio minus the return of a risk-free return.

In portfolio selection the risk aversion can be taken into account by using the Markowitz criterion

$$E(R-r) - \frac{\lambda}{2} \cdot \mathrm{sd}(R-r), \tag{1.89}$$

where $\lambda \geq 0$ is the risk aversion parameter. The Markowitz criterion has the general structure of a risk-penalized expected return:

$$\text{expected return} - \frac{\lambda}{2} \cdot \text{risk}.$$

The Sharpe ratio and the Markowitz criterion use the standard deviation of the excess returns as the risk measure. The standard deviation does not take into account the possibility of a nonsymmetric distribution. It penalizes from a positive skewness of the return distribution. Thus, we can consider replacing the standard deviation by the square root of the partial variance in the definition of the Sharpe ratio and the Markowitz criterion. The partial variance is defined in (1.18).

1.5.2 Variance Trading

Variance estimation can be applied in quantile estimation, because standard deviation estimates can be used to construct quantile estimates; see (1.28)–(1.30). Variance estimation can be applied in portfolio performance measurement and in portfolio selection, see (1.88) and (1.89). A third application for variance estimation comes from the volatility trading.

Volatility can be traded with variance and volatility swaps. A variance swap is a forward contract that pays

$$V_T - K$$

at the expiration date T, where K is the delivery price, and V_T is the realized variance, defined by

$$V_T = \sum_{t=t_0+1}^{T} [\log(S_t/S_{t-1})]^2,$$

where t_0 is the starting day of the contract, and S_t are the prices of a financial asset. The volatility swap pays at the expiration

$$\sqrt{V_T} - L,$$

where L is the delivery price.

Variance and volatility swaps are traded over the counter (OTC), but the Chicago Board Options Exchange (CBOE) offers variance futures for the variance of the S&P 500 index, calculated with the daily returns of the index.

Variance swaps open an opportunity to covariance trading if we have an access to a variance swap of an index and to variance swaps of its constituents. Let us consider an index whose returns are

$$R_t = pR_t^{(1)} + qR_t^{(2)},$$

where $R_t^{(i)}$ are the log returns of the index constituents and p and q are the weights of of the constituents. Let us define the realized covariance as

$$C_T = \sum_{t=t_0+1}^{T} R_t^{(1)} R_t^{(2)}.$$

Thus,

$$C_T = \frac{1}{2pq}\left(V_T - p^2 V_T^{(1)} - q^2 V_T^{(2)}\right),$$

where V_T is the realized variance of the index and $V_T^{(i)} = \sum_{t=t_0}^{T}(R_t^{(i)})^2$ are the realized variances of the index constituents.

1.5.3 Portfolio Selection

Basic Concepts of Portfolio Selection Let

$$S_t = (S_t^1, \ldots, S_t^N), \qquad t = 0, 1, 2, \ldots, T,$$

be a vector time series of N asset prices. Asset prices satisfy $0 < S_t^i < \infty$, $i = 1, \ldots, N$. A portfolio vector $b_t = (b_t^1, \ldots, b_t^N) \in \mathbf{R}^N$ determines how the wealth is allocated among the assets at time t. A portfolio vector b_t satisfies

$$\sum_{i=1}^{N} b_t^i = 1. \tag{1.90}$$

When $0 \le b_t^i \le 1$ for all $i = 1, \ldots, N$, then the portfolio is called a long only portfolio and the value b_t^i is equal to the proportion of wealth is invested in asset S_t^i at time t. Negative values of b_t^i are interpreted as short selling.[7] One of the assets can be a bank account, and selling a bank account short is interpreted as borrowing. For example, when $N = 2$, $b_t^1 = -1$, and $b_t^2 = 2$, this means that at time t we sell short asset 1 with an amount which equals all our wealth and simultaneously buy asset 2 with all our wealth and with the proceedings obtained from selling short the asset 1.

We define a new vector time series of gross returns (price relatives) by

$$R_t = \frac{S_t}{S_{t-1}} = \left(\frac{S_t^1}{S_{t-1}^1}, \ldots, \frac{S_t^N}{S_{t-1}^N}\right), \qquad t = 1, 2, \ldots, T. \tag{1.91}$$

It is reasonable to assume that the time series $R_1, \ldots, R_T$ is approximately stationary. In statistical portfolio selection we have available, besides the historical returns R_t of the assets, also other information Z_t. The variables in vector Z_t can be macroeconomic variables, like the term premium, default premium, and dividend yield.[8] The problem of portfolio selection can now be described as a problem of choosing a portfolio vector b_T at time T using data (R_t, Z_t), $t = 1, \ldots, T$.

[7] Selling short an asset means that we borrow the asset and then sell it, that is, we sell an asset that we do not own. Naked short selling means that an asset is sold before it is borrowed, or before making sure that it can be borrowed.

[8] The term premium is the difference between the long-term and short-term interest rates. For example, the term premium can be the difference between the annualized yields of a portfolio of 10-year U.S.

Single-Period Portfolio Selection Let $W_T > 0$ be the wealth available at time T. When the portfolio vector is b_T, then the gross return of the portfolio for the time period from T to $T+1$ is

$$\frac{W_{T+1}}{W_T} = \sum_{i=1}^{N} b_T^i \frac{S_{T+1}^i}{S_T^i} = b_T' R_{T+1}. \tag{1.92}$$

In the single-period portfolio selection the optimal portfolio vector can be defined as

$$b_T^o = \text{argmax}_{b_T \in B_N} E_T u\left(b_T' R_{T+1}\right), \tag{1.93}$$

where $u : (0, \infty) \to \mathbf{R}$ is a utility function and

$$B_N = \left\{ (b^1, \ldots, b^N) : \sum_{i=1}^{N} b^i = 1 \right\}. \tag{1.94}$$

Note that $0 < b_T' R_{T+1} < \infty$. The notation E_T means that the expectation is taken at time T, using information available at time T. If the available information is contained in the historical returns R_t and in the historical values of the variables Z_t, then the expectation E_T can be taken as the conditional expectation, conditional on the previous returns and previous values of variables Z_t:

$$E_T u\left(b_T' R_{T+1}\right) = E\left[u\left(b_T' R_{T+1}\right) \mid R_1, Z_1, \ldots, R_T, Z_T\right].$$

In the maximization problem (1.93) we apply utility function u to the one-period gross return, given in (1.92).

A utility function $u : (0, \infty) \to \mathbf{R}$ is an increasing function (the derivative is positive) that is concave (the second derivative is negative). The power utility functions are defined by

$$u_\gamma(t) = \begin{cases} \frac{t^{1-\gamma}}{1-\gamma}, & \text{if } \gamma > 1, \\ \log_e t, & \text{if } \gamma = 1, \end{cases} \tag{1.95}$$

for $t > 0$. The power utility functions are called constant relative risk aversion utility functions (CRRA). A utility function is used instead of the pure return, because through it we take also risk into account and do not optimize the pure return.[9]

government bonds and a 90-day Treasury bill. The default premium is the difference between the interest rate of a lower grade bond and a higher grade bond. For example, the default premium can be the difference between the annualized yields of Moody's Baa and Aaa rated bonds. The dividend yield is the dividend payment of a company divided by its market capitalization, when the market capitalization is the value of the stock multiplied by the number of stocks.

[9]It does not matter whether we take the utility from the wealth or from the gross return. Indeed, for $\gamma > 0$,

$$u\left(W_T b_T' U_{T+1}\right) = W_T^{1-\gamma} \cdot u(b_T' U_{T+1})$$

and for $\gamma = 0$,

$$u\left(W_T b_T' U_{T+1}\right) = u(W_T) + u(b_T' U_{T+1}).$$

Thus the optimal portfolio vector is the same regardless of the initial wealth W_T.

Parameter $\gamma \geq 1$ is the risk aversion parameter, and larger γ means larger risk aversion. A gross return equal to zero would mean that we have made a bankruptcy, and thus the utility of zero gross return should be equal to minus infinity. Thus the utility function makes a severe penalization of returns near zero. Also, the utility of a positive return does not grow linearly but is a concave function of the return.

Multiperiod Portfolio Selection When we start with wealth W_T at time T and use portfolio weights $b_T, \ldots, b_{T_1-1}$, then the wealth at time T_1 is

$$W_T \prod_{t=T}^{T_1-1} b_t' R_{t+1}.$$

The gross return of the portfolio for the time period from T to T_1 is

$$\prod_{t=T}^{T_1-1} b_t' R_{t+1}. \tag{1.96}$$

In the multiperiod portfolio selection, assuming that our investment horizon extends from T to a future time T_1, and we are able to change the portfolio weights at all times $T, \ldots, T_1 - 1$, the optimal portfolio weights at time T are defined by

$$b_T^o = \operatorname{argmax}_{b_T} \max_{b_{T+1},\ldots,b_{T_1-1}} E_T u\left(\prod_{t=T}^{T_1-1} b_t' R_{t+1}\right). \tag{1.97}$$

In the maximization problem (1.97) we apply utility function u to the multiperiod gross return, given in (1.96). The single period case is obtained as a special case when $T_1 = T + 1$. The optimization problem (1.97) is of the same type as the optimization problem of the stochastic control in (1.52).

Portfolio Selection and Regression Function Estimation We describe how regression function estimation can be used in portfolio selection. We consider the single period portfolio selection and want to choose a portfolio vector $b_T = (b_T^1, \ldots, b_T^N) \in \mathbf{R}^N$ at time T so that the expected utility of the wealth is maximized at time $T + 1$, as in the optimization problem (1.93). We can define, for a fixed portfolio vector $b \in \mathbf{R}^N$, with $\sum_{i=1}^N b^i = 1$, the response and the explanatory variables

$$Y_{b,t} = u\left(b' R_{t+1}\right), \qquad X_t \in \mathbf{R}^d,$$

$t = 1, \ldots, T - 1$. We assume that $(Y_{b,t}, X_t)$, $t = 1, \ldots, T - 1$, are identically distributed, and denote by (Y_b, X) a random vector which has the same distribution as $(Y_{b,t}, X_t)$. The data can be used to estimate the regression function

$$f_b(x) = E(Y_b \,|\, X = x), \qquad x \in \mathbf{R}^d,$$

where b is a fixed portfolio vector. This regression function gives a prediction for the utility of the gross return of the portfolio. The prediction can be inaccurate; but the collection of all predictions, for all values of the portfolio vector b, gives a way to choose the optimal portfolio vector. Namely, at time T we use the data

$$(Y_{b,t}, X_t), \qquad t = 1, \ldots, T-1,$$

to estimate the regression function. Let us denote this estimate by

$$\hat{f}_{b,T} : \mathbf{R}^d \to \mathbf{R}.$$

We choose the optimal portfolio vector $\hat{b}_T$ at time T by

$$\hat{b}_T = \mathrm{argmax}_{b \in B} \hat{f}_{b,T}(X_T), \tag{1.98}$$

where $B \subset B_N$, where B_N is the sphere in $\mathbf{R}^N$, defined in (1.94). Thus we choose the portfolio vector for which the prediction of the utility of the return of the portfolio is the highest. Since T is the current time, we use $\hat{b}_T$ to allocate the current wealth, and the portfolio vectors $\hat{b}_t$, $t = 1, \ldots, T-1$, can be used to analyze the statistical properties of the portfolio selection method.

We can also describe the procedure by defining function $b : \mathbf{R}^d \to B$ by

$$b(x) = \mathrm{argmax}_{b \in B} f_b(x).$$

This function is estimated at time T by

$$\hat{b}_T(x) = \mathrm{argmax}_{b \in B} \hat{f}_{b,T}(x).$$

At time T we choose the portfolio vector $\hat{b}_T(X_T)$.

We can use the idea of (1.47) to transform the time series (1.91) to regression data and we can define the explanatory variables

$$X_t = (R_t, \ldots, R_{t-k+1}) \in \mathbf{R}^{Nk}, \tag{1.99}$$

$t = k, \ldots, T-1$. The explanatory variable X_t is defined as a vector of length k of past gross returns. This choice can be justified if the past returns contain all relevant information available to predict the future returns. Clearly it is possible that the quality of predictions can be improved if we make some transformation of the past returns. Possible transformations are discussed in Section 1.7. If the time series $R_1, \ldots, R_T$ is stationary, then $(Y_{b,t}, X_t)$, $t = k, \ldots, T-1$, are identically distributed.

An application of regression function estimation in portfolio selection has been made by Brandt (1999), Aït-Sahalia & Brandt (2001), and Györfi, Lugosi & Udina (2006). See also Györfi & Schäfer (2003), Györfi, Urbán & Vajda (2007), Györfi, Udina & Walk (2008), and Györfi, Ottucsác & Walk (2012).

Portfolio Selection and Classification We assume to have data (R_t, X_t), $t = 1, \ldots, T$, where $R_t \in \mathbf{R}^N$ is the gross return vector defined in (1.91) and $X_t \in \mathbf{R}^d$ is the the vector of explanatory variables observed at time t.

Let $B = \{b_0, \ldots, b_{K-1}\} \subset \mathbf{R}^N$ be a finite class of portfolio vectors. Define the class labels Y_t by

$$Y_t = k \Leftrightarrow b_k = \text{argmax}_{b \in B} b' R_{t+1}, \tag{1.100}$$

where $k = 0, \ldots, K-1$. Now $b_k \in B$ is the portfolio vector chosen at time t that gave the best return at time $t+1$, among all the portfolio vectors in B.

We have now defined classification data (X_t, Y_t), $t = 1, \ldots, T-1$, which is used at time T to estimate the classification function. The estimated classification function $\hat{g}$ chooses one of the portfolio vectors in B. Thus we define the portfolio vector which is chosen at time T by

$$\hat{b}_T = \hat{g}(X_T).$$

With the classification approach we are not able to introduce a risk aversion parameter, as in the case of regression approach, where a utility transformed return was predicted. The portfolios obtained by classification correspond to using the risk aversion parameter $\gamma = 1$.

Andriyashin, Härdle & Timofeev (2008) use a classification based approach to portfolio selection. They make for each stock in DAX 30 a decision to either buy, sell, or stay neutral, and the final portfolio is an equally weighted portfolio of the individual decisions for each stock.

Mean-Variance Preferences Portfolio choice with mean-variance preferences was proposed by Markowitz (1952) and Markowitz (1959). This method provides an alternative to the use of the maximization of the expected utility. The optimal portfolio vector in the mean-variance sense maximizes the penalized expected return

$$E\left(b' R_{T+1}\right) - \frac{\gamma}{2} \text{Var}\left(b' R_{T+1}\right), \tag{1.101}$$

where $\gamma \geq 0$ is the coefficient of risk aversion and

$$R_{T+1} = \left(S_{T+1}^1 / S_T^1, \ldots, S_{T+1}^N / S_T^N\right)$$

is the vector of the gross returns of the N portfolio components, see (1.91) and (1.92). The minimization is done over a space of portfolio vectors $B \subset B_N$, where B_N is the sphere in $\mathbf{R}^N$, defined in (1.94). We have

$$E\left(b' R_{T+1}\right) = b' E R_{T+1}, \qquad \text{Var}\left(b' R_{T+1}\right) = b' \text{Var}(R_{T+1}) b,$$

where $\text{Var}(R_{T+1})$ is the $N \times N$ covariance matrix of R_{T+1}. We have to estimate the vector of expected returns ER_{T+1} and the covariance matrix $\text{Var}(R_{T+1})$.

We shall consider in Section 3.12.3 an example of portfolio selection with two risky assets. Let us derive the optimal portfolio vector for that case. Let us denote the portfolio vector $b = (b^1, b^2) = (1-w, w)$, where $w \in \mathbf{R}$. That is, we put proportion $1-w$ to the first asset and the proportion w to the second asset. Now

$$b' R_{T+1} = (1-w) R_{T+1}^1 + w R_{T+1}^2.$$

Let the expected returns of the stocks be $ER^1_{T+1} = \mu_1$, $ER^2_{T+1} = \mu_2$ and the variances of the returns $\mathrm{Var}(R^1_{T+1}) = \sigma_1^2$, $\mathrm{Var}(R^2_{T+1}) = \sigma_2^2$. Denote the covariance of the returns by $\mathrm{Cov}(R^1_{T+1}, R^2_{T+1}) = \sigma_{12}$. We have

$$
\begin{aligned}
& E(b'R_{T+1}) - \frac{\gamma}{2}\,\mathrm{Var}(b'R_{T+1}) \\
&= \mu_1 + w(\mu_2 - \mu_1) - \frac{\gamma}{2}\left[(1-w)^2\sigma_1^2 + w^2\sigma_2^2 + 2(1-w)w\sigma_{12}\right] \\
&= \mu_1 - \frac{\gamma}{2}\sigma_1^2 + w\left[\mu_2 - \mu_1 - \gamma(\sigma_{12} - \sigma_1^2)\right] - w^2\frac{\gamma}{2}(\sigma_1^2 + \sigma_2^2 - 2\sigma_{12}).
\end{aligned}
$$

Setting the derivative with respect to w to zero and solving for w gives

$$
w = \frac{1}{\gamma}\,\frac{\mu_2 - \mu_1 - \gamma(\sigma_{12} - \sigma_1^2)}{\sigma_1^2 + \sigma_2^2 - 2\sigma_{12}}, \tag{1.102}
$$

when $\gamma > 0$. For $\gamma = 0$, as much as possible is invested to the asset for which the expected return μ_i is larger.

1.5.4 Option Pricing and Hedging

We consider an European option written at time t_0 (today), whose expiration is at a future time T. The option has value H_T at the expiration time and this value is a function of the stock price S_T. For example, in the case of a call option $H_T = \max\{0, S_T - K\}$, where K is the strike price. We need to determine a fair price H_{t_0} for the option at the current time t_0.

The price can be determined as the initial wealth needed to finance a hedging of the option. Hedging is done through a self financing trading using the stock S_t and the bond B_t. We take the interest rate equal to zero so that we can take $B_t = 1$ for all t. We consider the discrete time model, where trading is done at the time points $t_0, t_0+1, \ldots, T-1$. Let W_t be the wealth at time t used to buy stocks and bonds. Let ξ_t be the number of stocks bought at time $t-1$, and kept until time t, where a rebalancing is made. Let a_t be the number of bonds bought at time $t-1$ and kept until time t. Since the portfolio is self financing, the quantities ξ_t and a_t have to satisfy

$$
W_{t-1} = a_t + \xi_t S_{t-1}.
$$

The wealth at time t is then

$$
W_t = a_t + \xi_t S_t,
$$

which is again distributed among the stock and the bond by choosing a_{t+1} and ξ_{t+1}. Thus,

$$
\begin{aligned}
W_t &= a_t + \xi_t S_t \\
&= (a_t + \xi_t S_{t-1}) + \xi_t(S_t - S_{t-1}) \\
&= W_{t-1} + \xi_t(S_t - S_{t-1}).
\end{aligned} \tag{1.103}
$$

We get inductively[10]

$$W_T = W_{t_0} + \sum_{t=t_0}^{T-1} \xi_{t+1}(S_{t+1} - S_t).$$

We can use two slightly different heuristics to define the fair price.

1. We consider the fair price to be the initial wealth W_{t_0} that minimizes the minimal difference between the final wealth and the payout of the option. That is, we want to minimize
$$E(W_T - H_T)^2$$
over all initial wealths W_0 and over all hedging strategies.

2. The writer of the option receives the premium H_{t_0} at time t_0, hedges his position at time points $t = t_0, \ldots, T-1$ with initial wealth $W_{t_0} = 0$, and pays H_T at the expiration to the holder of the option. Thus the wealth of the option writer at the expiration time T is equal to
$$\tilde{W}_T = H_{t_0} + \sum_{t=t_0}^{T-1} \xi_{t+1}(S_{t+1} - S_t) - H_T.$$
We want to find H_{t_0} and $\xi_{t_0+1}, \ldots, \xi_T$ so that $\tilde{W}_T$ is as close to zero as possible and the corresponding value for H_{t_0} can be considered as a fair value of the option. That is, we want to minimize
$$E\tilde{W}_T^2,$$
where the mean squared error measures closeness to zero.

Both heuristics lead to the following definition of the fair price and the optimal hedging coefficient. Denote

$$Y = H_T, \qquad (X_1, \ldots, X_d) = (S_{t_0+1} - S_{t_0}, \ldots, S_T - S_{T-1}),$$

where $d = T - t_0$. We define the fair price and the optimal hedging coefficient at time t_0 as

$$(H_{t_0}, \xi_{t_0+1}) = \text{argmin}_{a \geq 0, b_1 \in \mathbf{R}} \min_{b_2, \ldots, b_d \in \mathbf{R}} E\rho(a + b_1 X_1 + \cdots + b_d X_d - Y), \quad (1.104)$$

where $\rho(t) = t^2$ or ρ is some other loss function as in (1.36). We have obtained a problem of stochastic control as described in (1.51).

[10] When interest rate for one period is $r > 0$, so that $B_{t+1} = (1+r)B_t$, we get the expression

$$W_T = (1+r)^{T-t_0} \left(W_{t_0} + \sum_{t=t_0}^{T-1} \xi_{t+1}(Z_{t+1} - Z_t) \right),$$

where $Z_t = (1+r)^{t_0 - t} S_t$.

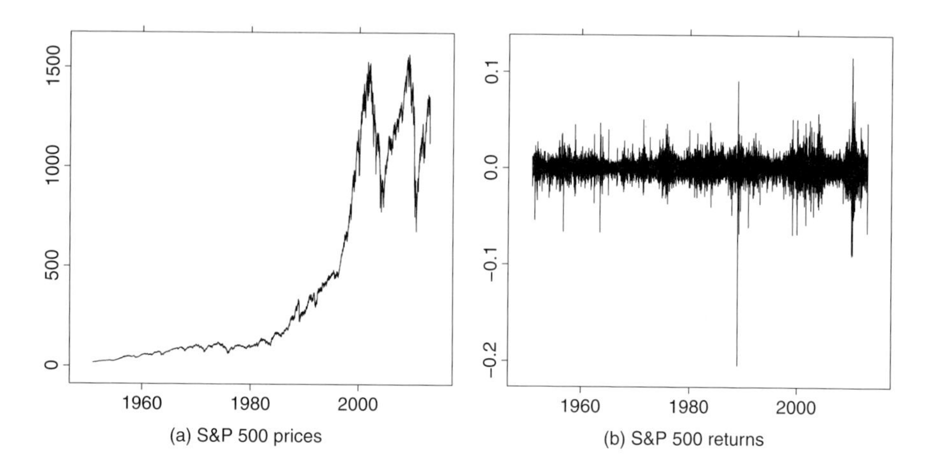

Figure 1.3 *S&P 500 index.* (a) The prices of S&P 500. (b) The net returns of S&P 500.

1.6 DATA EXAMPLES

We use two data sets as the main examples to illustrate the methods of regression and classification. The first data set is a time series of S&P 500 returns, described in Section 1.6.1. The second data set is a vector time series of S&P 500 and Nasdaq-100 returns, described in Section 1.6.2.

We use also other data sets as examples. In Section 2.1.7 a vector time series of DAX 30 and Daimler returns is used to illustrate an application of linear regression to the calculation of the beta of an asset. In Section 2.2.2 a time series of a hedge fund index returns is used to illustrate an application of varying coefficient regression in hedge fund replication. In Section 6.2 density estimation is illustrated with monthly S&P 500 data and U.S. Treasury 10-year bond data. In Section 6.3.2 a time series of DAX 30 returns is used to illustrate multidimensional scaling.

1.6.1 Time Series of S&P 500 Returns

The S&P 500 index data consist of the daily closing prices of the S&P 500 index during the period from 1950-01-03 until 2013-04-25, which makes 15 930 observations. The data are provided by Yahoo, where the index symbol is ^GSPC.

Figure 1.3 shows the prices and the net returns of the S&P 500 index. The net return is defined as

$$Y_t = \frac{P_t - P_{t-1}}{P_{t-1}},$$

where P_t is the price of the index at the end of day t.

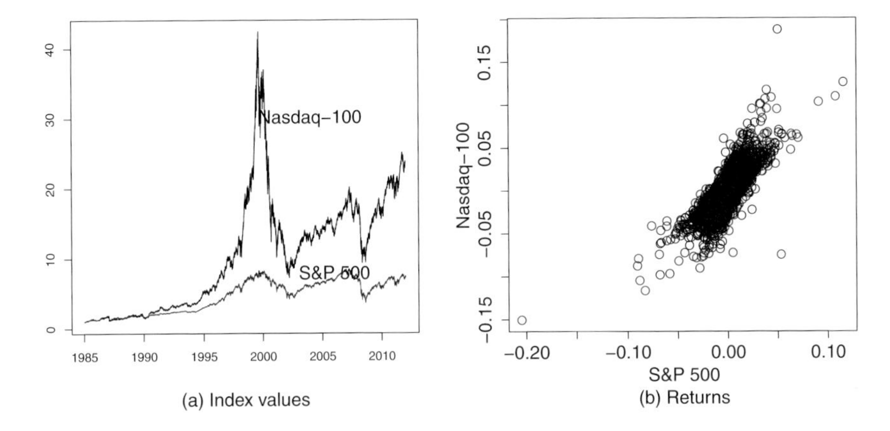

Figure 1.4 *S&P 500 and Nasdaq-100 indexes.* (a) The normalized values of the S&P 500 and Nasdaq-100 indexes. (b) The scatter plot of the net returns.

1.6.2 Vector Time Series of S&P 500 and Nasdaq-100 Returns

The S&P 500 and Nasdaq-100 index data consist of the daily closing prices of the S&P 500 index and the Nasdaq-100 index starting at 1985-10-01 and ending at 2013-03-19, which makes 6925 days of observations. The data are provided by Yahoo, where the index symbols are ^GSPC and ^NDX.

Figure 1.4 shows the S&P 500 and Nasdaq-100 indexes over the observation period. Panel (a) shows the time series of normalized index values. The index values are normalized so that they both have the value one at 1985-10-01. Panel (b) shows the scatter plot of the net returns of the indexes.

1.7 DATA TRANSFORMATIONS

In regression function estimation it is often useful to transform the variables before estimating the regression function. A transformation of the explanatory variables is important when the regression function is estimated with a method of local averaging, defined in Chapter 3. If the local neighborhood of a local averaging estimator is spherically symmetric, as is the case when we use kernel estimation with a spherically symmetric kernel function and with a single smoothing parameter for each variable, then the scales of the explanatory variables should be compatible. For example, if one variable takes values in $[0, 1]$ and an other variable takes values in $[0, 100]$, then the variable with the shorter range would effectively be canceled out when using spherically symmetric neighborhoods.

First, we define data sphering, which is a transformation of the explanatory variables that makes the variances of the explanatory variables equal and the covariance matrix of the explanatory variables diagonal. Second, we define a copula transformation that makes the marginal distributions of the explanatory variables approximately standard Gaussian, or uniform on $[0,1]$, but keeps the copula of the explanatory variables unchanged. Third, we define transformations of the response variable.

1.7.1 Data Sphering

We can make the scales of variables compatible by normalizing observations so that the sample variances of the variables are equal to one. Let $X_i = (X_{i1}, \ldots, X_{id})$, $i = 1, \ldots, n$, be the original observations. The transformed observations are

$$Z_i = \left(\frac{X_{i1}}{s_1}, \ldots, \frac{X_{id}}{s_d} \right), \qquad i = 1, \ldots, n,$$

where the sample variances are

$$s_k^2 = \frac{1}{n} \sum_{i=1}^{n} \left(X_{ik} - \bar{X}_k \right)^2, \qquad k = 1, \ldots, d,$$

with the arithmetic mean $\bar{X}_k = n^{-1} \sum_{i=1}^n X_{ik}$. We can also make the ranges of the variables equal by defining the transformed observations as $Z_i = (Z_{i1}, \ldots, Z_{id})$, $i = 1, \ldots, n$, where

$$Z_{ik} = \frac{X_{ik} - \min_{i=1,\ldots,n} X_{ik}}{\max_{i=1,\ldots,n} X_{ik} - \min_{i=1,\ldots,n} X_{ik}}, \qquad k = 1, \ldots, d.$$

Data sphering is a more extensive transformation than just standardizing the sample variances equal to one; we make such linear transformation of data that the covariance matrix becomes the identity matrix. The sphering is almost the same as the principal component transformation. In the principal component transformation the covariance matrix is diagonalized but it is not made the identity matrix.

1. Sphering of a random vector $X \in \mathbf{R}^d$ means that we make a linear transform of X so that the new random variable has expectation zero and the identity covariance matrix. Let

$$\Sigma = E\left[(X - EX)(X - EX)' \right]$$

be the covariance matrix and make the spectral representation of Σ:

$$\Sigma = A \Lambda A',$$

where A is orthogonal and Λ is diagonal. Then

$$Z = \Lambda^{-1/2} A' (X - EX)$$

is the sphered random vector, having the property[11]

$$\text{Cov}(Z) = I_d.$$

[11]The orthogonality of A means that $A'A = AA' = I_d$. Thus $A'\Sigma A = \Lambda$ and $\text{Cov}(Z) = \Lambda^{-1/2} A' \text{Cov}(X) A \Lambda^{-1/2} = \Lambda^{-1/2} A' \Sigma A \Lambda^{-1/2} = I_d$.

2. Data sphering means that the data are transformed so that the arithmetic mean of the observations is zero and the empirical covariance matrix is the unit matrix. Let Σ_n be the empirical covariance matrix,

$$\Sigma_n = \frac{1}{n}\sum_{i=1}^{n}(X_i - \bar{X})(X_i - \bar{X})',$$

where $\bar{X} = n^{-1}\sum_{i=1}^{n} X_i$ is the $d \times 1$ column vector of arithmetic means. We find the spectral representation of Σ_n,

$$\Sigma_n = A_n \Lambda_n A_n',$$

where A_n is orthogonal and Λ_n is diagonal. Define the transformed observations

$$Z_i = \Lambda_n^{-1/2} A_n' (X_i - \bar{X}), \qquad i = 1, \ldots, n.$$

The sphered data matrix is the $n \times d$ matrix $\mathbb{Z}_n$ defined by

$$\mathbb{Z}_n' = \Lambda_n^{-1/2} A_n' \left(\mathbb{X}_n' - \bar{X}_n 1_{1\times n}\right),$$

where $\mathbb{X}_n = (X_1, \ldots, X_n)'$ is the original $n \times d$ data matrix, and $1_{1\times n}$ is the $1 \times n$ row vector of ones.

1.7.2 Copula Transformation

Copula modeling was explained in Section 1.3.3. Copula modeling leads also to useful data transformations. A copula transformation changes the marginal distributions but keeps the copula (the joint distribution) the same.

1. The copula transformation of random vector $X = (X_1, \ldots, X_d)$, when X has a continuous distribution, gives random variable $Z = (Z_1, \ldots, Z_d)$ whose marginals have the uniform distribution on $[0, 1]$, or some other suitable distribution. Let $F_{X_k}(t) = P(X_k \le t)$, $k = 1, \ldots, d$, be the distribution functions of the components of X. Now

$$Z = (F_{X_1}(X_1), \ldots, F_{X_d}(X_d))$$

is a random vector whose marginal distributions are uniform on $[0, 1]$.[12] The distribution function of this random vector is called the copula of the distribution of $X = (X_1, \ldots X_d)$. Often the copula with uniform marginals is inconvenient due to boundary effects. We may get statistically more tractable distribution by defining

$$Z = \left(\Phi^{-1}(F_{X_1}(X^1)), \ldots, \Phi^{-1}(F_{X_d}(X_d))\right),$$

[12]Random variable $F_{X_k}(X_k)$ has the uniform distribution on $[0, 1]$, because $P(F_{X_k}(X_k) \le t) = P(X_k \le F_{X_k}^{-1}(t)) = F_{X_k}(F_{X_k}^{-1}(t)) = t$.

where Φ is the distribution function of the standard Gaussian distribution. The components of Z have the standard Gaussian distribution.[13]

2. The copula transformation of data $X_1, \ldots, X_n$ means that the data are transformed so that the marginal distributions are approximately uniform, or have approximately some other suitable distribution. Let the rank of observation X_{ik}, $i = 1, \ldots, n$, $k = 1, \ldots, d$, be

$$\text{rank}(X_{ik}) = \#\{X_{jk} : X_{jk} \leq X_{ik}, j = 1, \ldots, n\}.$$

We normalize the ranks to get observations with approximately uniform distribution on $[0, 1]$:

$$Z_i = \left(\frac{\text{rank}(X_{i1})}{n+1}, \ldots, \frac{\text{rank}(X_{id})}{n+1} \right),$$

for $i = 1, \ldots, n$. Often the standard Gaussian distribution is more convenient and we define

$$Z_i = \left(\Phi^{-1}\left(\frac{\text{rank}(X_{i1})}{n+1} \right), \ldots, \Phi^{-1}\left(\frac{\text{rank}(X_{id})}{n+1} \right) \right), \qquad (1.105)$$

for $i = 1, \ldots, n$.

Figure 1.5 shows scatter plots of S&P 500 and Nasdaq-100 copula transformed net returns. The data is described in Section 1.6.2. Panel (a) shows the case where the marginals are transformed to be approximately standard Gaussian. Panel (b) shows the case where the marginals are transformed to be approximately uniformly distributed in $[0, 1]$. We have used in scatter plots histogram smoothing with 70^2 bins, as explained in Section 6.1.1. Uniform marginals make the data concentrate on the lower left and on the upper right corners, which can make the estimation difficult due to the boundary effects. The Gaussian marginals make the distribution of the data have tails which decrease smoothly to zero.

1.7.3 Transformations of the Response Variable

The transformation of the response variable can be used to obtain a more normal distribution or to remove heteroskedasticity by stabilizing variance. See Efron (1982).

The power transformations are called the Box–Cox transformations and defined for $\lambda \in \mathbf{R}$ by

$$Z_i^{(\lambda)} = \begin{cases} \frac{Y_i^\lambda - 1}{\lambda}, & \lambda \neq 0, \\ \log Y_i, & \lambda = 0, \end{cases}$$

where we assume that $Y_i \geq 0$. Box–Cox transformations were defined in Box & Cox (1962). Tukey (1957) considered the power transformation Y_i^λ for $\lambda \neq 0$.

[13] Random variable $\Phi^{-1}(U)$, where U has the uniform distribution on $[0, 1]$, has the standard Gaussian distribution because $P(\Phi^{-1}(U) \leq t) = P(U \leq \Phi(t)) = \Phi(t)$.

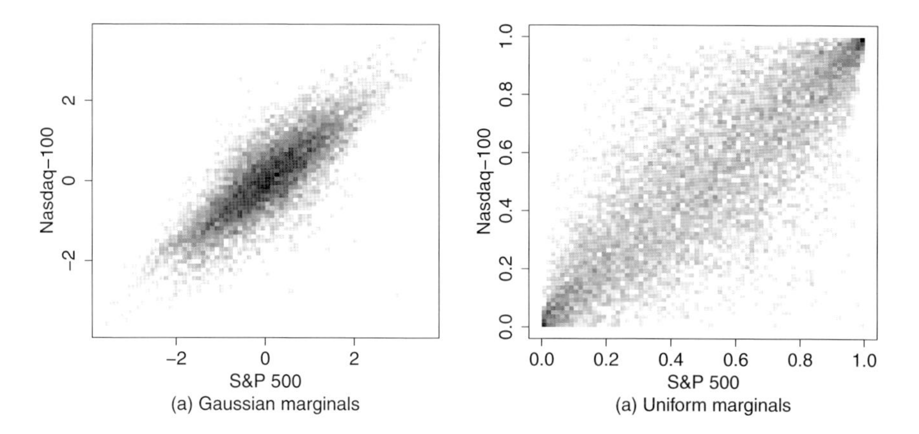

Figure 1.5 *Copula transform.* Scatter plots of S&P 500 and Nasdaq-100 returns are shown. (a) Gaussian marginals. (b) Uniform marginals.

The natural exponential family was defined in (1.68). In the natural exponential family

$$E_v(Y) = \mu(v) = d'(v), \qquad \mathrm{Var}_v(Y) = V(v) = d''(v).$$

A subclass of natural exponential families consists of the families with a quadratic variation function. Now we have

$$\mathrm{Var}_v(Y) = V(v) = a_0 + a_1\mu(v) + a_2\mu(v)^2,$$

where $\mu(v) = d'(v)$. The examples are normal, gamma, NEF-GHS (the natural exponential family generated by the generalized hyperbolic secant distribution), binomial, negative binomial, and Poisson. Denote $\mathrm{Var}_v(Y) = V(\mu(v))$. Define a function $G : \mathbf{R} \to \mathbf{R}$ to be such that

$$G'(\mu) = V^{-1/2}(\mu).$$

By the central limit theorem, we obtain

$$n^{1/2}\left(\bar{Y} - \mu(v)\right) \xrightarrow{d} N(0, V(\mu(v))),$$

as $n \to \infty$, where $\bar{Y} = n^{-1}\sum_{i=1}^n Y_i$, and $Y_1, \ldots, Y_n$ are assumed i.i.d. By the delta method, we have

$$n^{1/2}\left(G(\bar{Y}) - G(\mu(v))\right) \xrightarrow{d} N(0, 1),$$

as $n \to \infty$. Thus we call the transformation G a variance stabilizing transform.

1.8 CENTRAL LIMIT THEOREMS

A central limit theorem is needed to test the difference between two prediction methods; see Section 1.9.1. A central limit theorem is also needed to derive asymptotic distributions for estimators; see Section 2.1.4.

1.8.1 Independent Observations

Let $Y_1, Y_2, \ldots$ be a sequence of real-valued i.i.d. random variables with Var$(Y_i) = \sigma^2$, where $0 < \sigma^2 < \infty$. According to the central limit theorem, we have

$$n^{-1/2} \sum_{i=1}^{n} (Y_i - EY_i) \xrightarrow{d} N\left(0, \sigma^2\right),$$

as $n \to \infty$. Let $X_1, X_2, \ldots$ be an i.i.d. sequence of random vectors with Cov$(X_i) = \Sigma$, where the diagonal elements of Σ are finite and positive. According to the central limit theorem, we have

$$n^{-1/2} \sum_{i=1}^{n} (X_i - EX_i) \xrightarrow{d} N(0, \Sigma),$$

as $n \to \infty$.

1.8.2 Dependent Observations

We need a central limit theorem for dependent observations. Let $(Y_t)_{t \in \mathbf{Z}}$ be a strictly stationary time series. We define the weak dependence in terms of a condition on the α-mixing coefficients. Let $\mathcal{F}_i^j$ denote the sigma algebra generated by random variables $Y_i, \ldots, Y_j$. The α-mixing coefficient is defined as

$$\alpha_n = \sup_{A \in \mathcal{F}_{-\infty}^0, B \in \mathcal{F}_n^\infty} |P(A \cap B) - P(A)P(B)|,$$

where $n = 1, 2, \ldots$. Now we can state the central limit theorem. Let $E|Y_t|^\delta < \infty$ and $\sum_{j=1}^\infty \alpha_j^{1-2/\delta} < \infty$ for some constant $\delta > 2$. Then,

$$n^{-1/2} \sum_{i=1}^{n} (Y_i - EY_i) \xrightarrow{d} N(0, \sigma^2), \tag{1.106}$$

where

$$\sigma^2 = \sum_{j=-\infty}^{\infty} \gamma(j) = \gamma(0) + 2 \sum_{j=1}^{\infty} \gamma(j),$$

$\gamma(j) = \mathrm{Cov}(X_t, X_{t+j})$, and we assume that $\sigma^2 > 0$.

Ibragimov & Linnik (1971, Theorem 18.4.1) gave necessary and sufficient conditions for a central limit theorem under α-mixing conditions A proof for our statement of the central limit theorem in (1.106) can be found in Peligrad (1986); see also Fan & Yao (2005, Theorem 2.21) and Billingsley (2005, Theorem 27.4)

Let us state the central limit theorem for the vector time series $(X_t)_{t\in\mathbf{Z}}$, where $X_t \in \mathbf{R}^d$. If the time series $(a'X_t)_{t\in\mathbf{Z}}$ satisfies the conditions for the univariate central limit theorem for all $a \in \mathbf{R}^d$, then[14]

$$n^{-1/2}\sum_{i=1}^{n}(X_i - EX_i) \xrightarrow{d} N(0,\Sigma), \tag{1.107}$$

where

$$\Sigma = \sum_{j=-\infty}^{\infty}\Gamma(j) = \Gamma(0) + \sum_{j=1}^{\infty}(\Gamma(j)+\Gamma(j)'),$$

and the autocovariance matrix $\Gamma(j)$ was defined in (1.21) as

$$\Gamma(j) = \mathrm{Cov}(X_t, X_{t+j}).$$

Note that we used the property (1.22) $\Gamma(j) = \Gamma(-j)'$.

Let us explain the expression for the asymptotic variance σ^2 in the univariate central limit theorem (1.106). Let us assume that $EY_i = 0$. The variance of the normalized sum is

$$\mathrm{Var}\left(n^{-1/2}\sum_{i=1}^{n}Y_i\right) = n^{-1}\sum_{i=1}^{n}\mathrm{Var}(Y_i) + n^{-1}\sum_{i\neq j}\mathrm{Cov}(Y_i,Y_j).$$

Thus, for an i.i.d. time series we have that

$$\mathrm{Var}\left(n^{-1/2}\sum_{i=1}^{n}Y_i\right) = \mathrm{Var}(Y_1) = \gamma(0).$$

For a weakly stationary time series we have

$$\begin{aligned}
\mathrm{Var}\left(n^{-1/2}\sum_{i=1}^{n}Y_i\right) &= n^{-1}\sum_{i=1}^{n}\mathrm{Var}(Y_i) + 2n^{-1}\sum_{i=1}^{n-1}\sum_{j=i+1}^{n}\mathrm{Cov}(Y_i,Y_j)\\
&= \gamma(0) + 2n^{-1}\sum_{i=1}^{n-1}(n-i)\gamma(i)\\
&= \sum_{i=-(n-1)}^{n-1}\left(1-\frac{|i|}{n}\right)\gamma(i).
\end{aligned}$$

Thus, in order that $\mathrm{Var}\big(n^{-1/2}\sum_{i=1}^{n}Y_i\big) \to c$, for a finite positive constant c, we need that $\gamma(n) \to 0$ sufficiently fast, as $n\to\infty$. A sufficient condition is that $\sum_{j=1}^{\infty}|\gamma(j)| < \infty$.

[14] Cramér–Wold theorem states that $Y_n \xrightarrow{d} Y$ if and only if $a'Y_n \xrightarrow{d} a'Y$ for all $a \in \mathbf{R}^d$, as $n\to\infty$, where Y_n and Y are random vectors.

1.8.3 Estimation of the Asymptotic Variance

In the applications we have to estimate the asymptotic variance and the asymptotic covariance matrix. For i.i.d. data we can use the sample variance and the sample covariance matrix. For dependent data the estimation is more complicated. Let us discuss the estimation of the variance σ^2 in (1.106) using the observations $Y_1, \ldots, Y_n$, and the estimation of the covariance matrix Σ in (1.107) using the observations $X_1, \ldots, X_n$.

Let us start with the estimation of σ^2 in (1.106). An application of the sample covariances would lead to the estimator

$$\tilde{\sigma}^2 = \hat{\gamma}(0) + 2 \sum_{j=1}^{n-1} \hat{\gamma}(j),$$

where

$$\hat{\gamma}(j) = \frac{1}{n} \sum_{i=1}^{n-j} (Y_i - \bar{Y})(Y_{i+j} - \bar{Y}),$$

for $j = 0, \ldots, n-1$. Note that for large j only few observations are used in the estimator $\hat{\gamma}(j)$. For example, when $j = n-1$ the estimator uses only one observation: $\hat{\gamma}(n-1) = Y_1 Y_n / n$, which is a very imprecise estimator. We can use weighting to remove the imprecise estimators and define

$$\hat{\sigma}^2 = \hat{\gamma}(0) + 2 \sum_{j=1}^{n-1} w(j) \hat{\gamma}(j), \tag{1.108}$$

where

$$w(j) = \left(1 - \frac{j}{h}\right)_+,$$

where $1 \le h \le n-1$ is a chosen smoothing parameter. We can generalize the estimator to other weights and define

$$w(j) = K(j/h), \tag{1.109}$$

where $K : \mathbf{R} \to \mathbf{R}$ is a kernel function satisfying $K(x) = K(-x)$, $K(0) = 1$, $|K(x)| \le 1$ for all x, and $K(x) = 0$ for $|x| > 1$.

To estimate Σ in (1.107) we use

$$\hat{\Sigma} = \hat{\Gamma}(0) + \sum_{j=1}^{n-1} w(j) \left(\hat{\Gamma}(j) + \hat{\Gamma}(j)'\right), \tag{1.110}$$

where

$$\hat{\Gamma}(j) = \frac{1}{n} \sum_{i=1}^{n-j} \left(X_i - \bar{X}\right)\left(X_{i+j} - \bar{X}\right)',$$

for $j = 0, \ldots, n-1$. We will apply weights in an estimator of an asymptotic covariance matrix in (2.44).

The weighting we have used is related to the smoothing in the estimation of the spectral density. The unnormalized spectral density function of a weakly stationary time series, having autocorrelation coefficients $\gamma(k)$ with $\sum_{j=-\infty}^{\infty} |\gamma(j)| < \infty$, is defined by

$$g(\omega) = \frac{1}{2\pi} \sum_{j=-\infty}^{\infty} \gamma(j)\, e^{-ij\omega},$$

where $\omega \in [-\pi, \pi]$; see Brockwell & Davis (1991, Section 4.3). The lag window spectral density estimator, based on data $Y_1, \ldots, Y_n$, is defined by

$$\hat{g}(\omega) = \frac{1}{2\pi} \sum_{|j| \le h} K(j/h)\, \hat{\gamma}(j)\, e^{-ij\omega},$$

where $\hat{\gamma}(j)$ are the sample autocorrelation coefficients, $h = 1, 2, \ldots, n-1$, and K is similar as in (1.109); see Brockwell & Davis (1991, Section 10.4). Now we have

$$\hat{g}(0) = \frac{1}{2\pi} \sum_{|j| \le h} K(j/h)\, \hat{\gamma}(j) = \frac{1}{2\pi}\, \hat{\sigma}^2,$$

where $\hat{\sigma}^2$ is defined in (1.108) with the weights defined in (1.109).

1.9 MEASURING THE PERFORMANCE OF ESTIMATORS

We discuss measuring the performance of regression function estimators, conditional variance, covariance, and quantile estimators, estimators of the expected shortfall, and classifiers.

1.9.1 Performance of Regression Function Estimators

We denote by $\hat{f}(x)$ an estimator of the conditional expectation $f(x) = E(Y \mid X = x)$. We define theoretical performance measures, which are used to compare estimators of f under given theoretical assumptions. After that we define empirical performance measures, which try to estimate the performance of estimate $\hat{f}$ using the available data.

Theoretical Performance Measures Theoretical performance measures can be divided into global risk functionals, like the mean integrated squared error, and into pointwise risk functionals, like the mean squared error.

Global Error We can use the mean integrated squared error (MISE) or the mean averaged squared error to measure the goodness of regression function estimators $\hat{f}$ globally, when we want to recover the complete curve and not its value at a single point $x \in \mathbf{R}^d$.

The prediction error of regression function f can be measured by

$$E(f(X) - Y)^2.$$

This measure of prediction is natural since $f(x) = E(Y \mid X = x)$ and the conditional expectation minimizes the mean squared error, as shown in (1.37). When we have an estimator $\hat{f}$ of f, then we can measure the prediction error of the estimator by

$$E\left(\hat{f}(X) - Y\right)^2.$$

Now the expectation is with respect to the distribution of

$$(X, Y), (X_1, Y_1), \ldots, (X_n, Y_n),$$

because $\hat{f}$ is a random function depending on the sample $(X_1, Y_1), \ldots, (X_n, Y_n)$. We have that

$$\begin{aligned} & E\left[\left(\hat{f}(X) - Y\right)^2 \,\middle|\, (Y_1, X_1), \ldots, (Y_n, X_n)\right] \\ = & \int_{\mathbf{R}^d} \left(\hat{f}(x) - f(x)\right)^2 f_X(x)\, dx + E(f(X) - Y)^2, \qquad (1.111) \end{aligned}$$

where f_X is the density function of X. The minimization of expression (1.111) with respect to estimator $\hat{f}$ is equivalent to the minimization of the expression

$$\int_{\mathbf{R}^d} \left(\hat{f}(x) - f(x)\right)^2 f_X(x)\, dx.$$

This calculation can be used to justify the mean integrated error, defined in (1.112).

The Mean Integrated Squared Error The mean integrated squared error is defined as

$$\begin{aligned} \text{MISE}(\hat{f}, f) &= E\left(\hat{f}(X) - f(X)\right)^2 \\ &= EE\left[\left(\hat{f}(X) - f(X)\right)^2 \,\middle|\, (Y_1, X_1), \ldots, (Y_n, X_n)\right] \\ &= E\int_{\mathbf{R}^d} \left(\hat{f}(x) - f(x)\right)^2 f_X(x)\, dx, \qquad (1.112) \end{aligned}$$

where X is independent of $(Y_1, Y_1), \ldots, (Y_n, X_n)$ and f_X is the density function of X. Using the short hand notation we write the mean integrated error as

$$\text{MISE}(\hat{f}, f) = E\left\|\hat{f} - f\right\|_{2,X}^2, \qquad (1.113)$$

where $\|f\|_{2,X}^2 = \int_{\mathbf{R}^d} f(x)^2\, dP_X(x)$, and P_X is the probability distribution of random vector X. We can generalize (1.113) to

$$E\int_{\mathbf{R}^d} \left(\hat{f}(x) - f(x)\right)^2 w(x)\, dP_X(x),$$

where $w : \mathbf{R}^d \to \mathbf{R}$ is a weight function The weight function could be $w \equiv 1$, to get (1.113). We can choose $w(x) = 1/f_X(x)$, to get the L_2 error with respect to the Lebesgue measure. The weight function $w(x)$ could also be used to trim away boundary effects.

The Mean Averaged Squared Error The mean averaged squared error is defined as

$$\text{MASE}(\hat{f}, f) = E\left[\frac{1}{n} \sum_{i=1}^{n} \left(\hat{f}(X_i) - f(X_i) \right)^2 \middle| X_1, \ldots, X_n \right]. \tag{1.114}$$

Using the short hand notation we write the mean averaged squared error as

$$\text{MASE}(\hat{f}, f) = E_{X^{(n)}} \left\| \hat{f} - f \right\|_{2,X^{(n)}}^2,$$

where

$$\|f\|_{2,X^{(n)}}^2 = \int_{\mathbf{R}^d} f(x)^2 \, dP_X^{(n)}(x) = \frac{1}{n} \sum_{i=1}^{n} f(X_i)^2,$$

$P_{X^{(n)}}$ is the empirical probability distribution of the sample $(X_1, \ldots, X_n)$, and $E_{X^{(n)}}$ is the conditional expectation under the condition $(X_1, \ldots, X_n)$. We can generalize the mean averaged squared error by defining $\|f\|_{2,X^{(n)}}^2 = n^{-1} \sum_{i=1}^{n} f(X_i)^2 w(X_i)$, where $w : \mathbf{R}^d \to \mathbf{R}$ is a weight function.

Pointwise Error Pointwise performance measures quantify how well the value of f is recovered at a single point $x \in \mathbf{R}^d$. We can use mean squared error (MSE) either unconditionally or conditionally.

- The unconditional mean squared error at point $x \in \mathbf{R}^d$ is defined as

$$\text{MSE}(\hat{f}(x), f(x)) = E\left(\hat{f}(x) - f(x) \right)^2,$$

  where f is the true regression function.

- The conditional mean squared error at point $x \in \mathbf{R}^d$ is defined as

$$\text{MSE}(\hat{f}(x), f(x)) = E\left[\left(\hat{f}(x) - f(x) \right)^2 \middle| X_1, \ldots, X_n \right],$$

  where f is the true regression function.

The Use of Theoretical Performance Measures Theoretical performance measures can be used to compare estimators in a given model. A model is a collection of probability distributions for the distribution of (X, Y) and on the distribution of the sample $(X_1, Y_1), \ldots, (X_n, Y_n)$. We can describe a model also as a collection of regression functions $\mathcal{F}$ together with the additional assumptions on the distribution of (X, Y) and on the distribution of the sample $(X_1, Y_1), \ldots, (X_n, Y_n)$. To compare estimators, we use the supremum risk

$$\sup_{f \in \mathcal{F}} \text{MISE}(\hat{f}, f).$$

We use the supremum risk, because it is necessary to require that an estimator performs uniformly well over a model, because for a single regression function f it is trivial to define the best estimator; this is the regression function f itself: $\hat{f} = f$.

Empirical Performance Measures Empirical performance measures can be used to estimate the performance of an estimator and to compare estimators. Empirical performance measures are calculated using the available regression data $(X_1, Y_1), \ldots, (X_n, Y_n)$.

Empirical Performance Measures for Cross-Sectional Data The mean integrated squared error

$$\text{MISE}(\hat{f}, f) = E\left(\hat{f}(X) - f(X)\right)^2,$$

defined in (1.113), cannot be approximated by $n^{-1}\sum_{i=1}^{n}(\hat{f}(X_i) - Y_i)^2$. This approximation fails, because we are using the same data to construct the estimator and to estimate the prediction error. Using the same learning data and the test data leads to overly optimistic evaluation of the performance. However, we can avoid the problem using sample splitting or cross-validation.

1. *Sample Splitting* Let $\hat{f}^*$ be the regression function estimator constructed from the data $(X_1, Y_1), \ldots, (X_{n^*}, Y_{n^*})$, where $1 \le n^* < n$, and typically $n^* = [n/2]$. Then we use

$$\text{MISE}_n(\hat{f}) = \frac{1}{n - n^*} \sum_{i=n^*+1}^{n} \left(\hat{f}^*(X_i) - Y_i\right)^2 \qquad (1.115)$$

   to estimate the mean integrated squared error.

2. *Cross Validation* Let $\hat{f}_{-i}$ be a regression function estimator constructed from the other data points but not (X_i, Y_i). Then we use

$$\text{MISE}_n(\hat{f}) = \frac{1}{n} \sum_{i=1}^{n} \left(\hat{f}_{-i}(X_i) - Y_i\right)^2 \qquad (1.116)$$

   to estimate the mean integrated squared error.

Cross validation is discussed in Section 3.2.7 in the case of kernel estimation.

Empirical Performance Measures in the Time Series Setting In the time series setting we have observations $(X_1, Y_1), \ldots, (X_T, Y_T)$ that are observed at consecutive time instants. We can construct regression function estimator $\hat{f}_t$ using data $(X_1, Y_1), \ldots, (X_t, Y_t)$ that is observed until time t, and define the mean of squared prediction errors by

$$\mathrm{MSPE}_T(\hat{f}) = \frac{1}{T-1} \sum_{t=1}^{T-1} \left(\hat{f}_t(X_t) - Y_{t+1} \right)^2, \tag{1.117}$$

which is analogous to the estimate of the mean integrated squared defined in (1.116). We will use later in Section 3.12.1 the mean of absolute prediction errors

$$\mathrm{MAPE}_T(\hat{f}) = \frac{1}{T-1} \sum_{t=1}^{T-1} \left| \hat{f}_t(X_t) - Y_{t+1} \right|. \tag{1.118}$$

Diebold & Mariano (1995) proposed a test for testing the equality of forecast accuracy. Let us have two predictors $\hat{f}_t(X_{t+1})$ and $\hat{g}_t(X_{t+1})$ and the corresponding losses

$$F_t = \left(\hat{f}_t(X_{t+1}) - Y_{t+1} \right)^2, \qquad G_t = \left(\hat{g}_t(X_{t+1}) - Y_{t+1} \right)^2.$$

The losses do not have to be squared prediction errors, but we can also use absolute prediction errors, for example. We get the time series of loss differentials

$$d_t = F_t - G_t.$$

The null hypothesis and the alternative hypothesis are

$$H_0 : Ed_t = 0, \qquad H_1 : Ed_t \neq 0.$$

We apply the central limit theorem as stated in (1.106). Under the null hypothesis and under the assumptions of the central limit theorem, we have

$$(T - t_0 + 1)^{-1/2} \sum_{t=t_0}^{T} d_t \xrightarrow{d} N(0, \sigma^2),$$

as $T \to \infty$, where

$$\sigma^2 = \sum_{k=-\infty}^{\infty} \gamma(k), \qquad \gamma(k) = Ed_0 d_k.$$

We can use the estimate

$$\hat{\sigma}^2 = \sum_{k=-(T-1)}^{T-1} w(k) \hat{\gamma}(k),$$

where $w(k)$ is defined in (1.109). Let us choose the test statistics

$$D = \hat{\sigma}^{-1}(T - t_0 + 1)^{-1/2} \sum_{t=t_0}^{T} d_t.$$

When we observe $|D| = d_{obs}$, then the p-value is calculated by $P(|D| > d_{obs}) \approx 2(1 - \Phi(d_{obs}))$, where Φ is the distribution function of the standard normal distribution.

1.9.2 Performance of Conditional Variance Estimators

Theoretical Performance Measures Theoretical performance measures can be generalized from the case of regression function estimation to the case of conditional variance estimators. For example, when $f(x) = \mathrm{Var}(Y \mid X = x)$ and $\hat{f}(x)$ is an estimator of $f(x)$, then we can measure the performance of $\hat{f}$ by

$$E \int_{\mathbf{R}^d} \left(\hat{f}(x) - f(x) \right)^2 w(x)\, dP_X(x), \tag{1.119}$$

where $w : \mathbf{R}^d \to \mathbf{R}$ is a weight function.

Empirical Performance Measures We define the empirical performance measures first for cross-sectional data and then for time series data.

Cross-Sectional Data Empirical performance measures of conditional variance estimators can be found naturally in the case where

$$E(Y \mid X = x) = 0,$$

so that

$$f(x) = \mathrm{Var}(Y \mid X = x) = E(Y^2 \mid X = x).$$

For example, we can use sample splitting. Let $\hat{f}^*$ be an estimator of f, constructed from the data $(X_1, Y_1), \ldots, (X_{n^*}, Y_{n^*})$, where $1 \le n^* < n$. Then we can use

$$\frac{1}{n - n^*} \sum_{i=n^*+1}^{n} \left| \hat{f}^*(X_i) - Y_i^2 \right| \tag{1.120}$$

to measure the performance of the estimator.

Time Series Data We use slightly different notation in the case of state space smoothing and in the case of time space smoothing.

State–Space Smoothing When we have identically distributed time series observations $(X_1, Y_1), \ldots, (X_T, Y_T)$, then we can construct an estimator $\hat{f}_t$ of the conditional variance using data $(X_1, Y_1), \ldots, (X_t, Y_t)$ and calculate the mean of absolute prediction errors

$$\mathrm{MAPE}_T(\hat{f}) = \frac{1}{T - t_0} \sum_{t=t_0}^{T-1} \left| \hat{f}_t(X_{t+1}) - Y_{t+1}^2 \right|, \tag{1.121}$$

where t_0 is the initial estimation period, $1 \leq t_0 \leq T - 1$. We start to evaluate the performance of the estimator after t_0 observations are available, because any estimator can behave erratically when only few observations are available. Mean absolute prediction error is sometimes called the mean absolute deviation error (MADE).

Time–Space Smoothing In autoregressive time–space smoothing methods, like in the GARCH models studied in Section 3.9.2, the explanatory variables are the previous observations, and the estimate $\hat{\sigma}_t^2$ of $E(Y_t^2 \mid \mathcal{F}_{t-1})$ is calculated using observations $Y_1, \ldots, Y_{t-1}$. Now we have

$$\mathrm{MAPE}_T\left(\hat{\sigma}^2\right) = \frac{1}{T - t_0 + 1} \sum_{t=t_0}^{T} \left| \hat{\sigma}_t^2 - Y_t^2 \right|. \tag{1.122}$$

Spokoiny (2000) proposes to take the square roots and use the mean square root prediction error criterion as the performance measure:

$$\mathrm{MSqPE}_T\left(\hat{\sigma}^2\right) = \frac{1}{T - t_0 + 1} \sum_{t=t_0}^{T} \left| \hat{\sigma}_t^2 - Y_t^2 \right|^{1/2}. \tag{1.123}$$

The mean square root prediction error is such that outliers do not have a strong influence on the results. Fan & Gu (2003) propose to measure the performance with the mean absolute deviation error:

$$\mathrm{MADE}_T\left(\hat{\sigma}^2\right) = \frac{1}{T - t_0 + 1} \sum_{t=t_0}^{T} \left| \sqrt{\frac{2}{\pi}}\, \hat{\sigma}_t - |Y_t| \right|, \tag{1.124}$$

where the factor $\sqrt{2/\pi}$ comes from the fact that for a standard normal random variable $Z \sim N(0, 1)$, we have $E|Z| = \sqrt{2/\pi}$.

We can generalize the performance measures (1.122)–(1.124) and define a class of performance measures by

$$\mathrm{MDE}_T^{(p,q)}\left(\hat{\sigma}^2\right) = \frac{1}{T - t_0 + 1} \sum_{t=t_0}^{T} \left| E|Z|^p \hat{\sigma}_t^p - |Y_t|^p \right|^{1/q}, \tag{1.125}$$

where $Z \sim N(0, 1)$. For $p > -1$, we have

$$E|Z|^p = \frac{2^{p/2} \Gamma((p + 1)/2)}{\sqrt{\pi}}. \tag{1.126}$$

The combinations $(p = 2, q = 1)$, $(p = 2, q = 2)$, $(p = 1, q = 1)$, and $(p = 1, q = 2)$ are of special interest. In Section 3.11.1 we illustrate the differences between the various combinations of p and q; see Figures 3.22 and 3.23. We use $\mathrm{MDE}_T^{(p,q)}$ with $p = 1$ and $q = 2$ in Section 3.11.1 to compare GARCH(1,1) and the exponentially weighted moving average.

Another useful performance measure is the mean of absolute ratio errors

$$\mathrm{MARE}_T^{(p)}(\hat{\sigma}^2) = \frac{1}{T - t_0 + 1} \sum_{t=t_0}^{T} \left| \frac{|Y_t|^p}{E|Z|^p \hat{\sigma}_t^p} - 1 \right|, \tag{1.127}$$

where $p > 0$ and $Z \sim N(0,1)$. We use $\mathrm{MARE}_T^{(p)}$ with $p = 2$ in Section 3.11.1 to compare GARCH(1,1) and the exponentially weighted moving average.

Prediction of Realized Volatility Above we have measured the performance of one step ahead predictions. We can also measure the performance of h-step ahead predictions, for $h = 1, 2, \ldots$. However, sometimes we are interested in estimating the realized volatility. Define the h-step realized volatility by

$$V_{t,h} = Y_{t+1}^2 + \cdots + Y_{t+h}^2.$$

Let $\hat{f}_{t,h}(X_{t+1})$ be a prediction of $V_{t,h}$. We can use the mean square root prediction error as in (1.123). We modify (1.121) to obtain

$$\mathrm{MSqE}_{T,h}(\hat{f}, f) = \frac{1}{T - h - t_0 + 1} \sum_{t=t_0}^{T-h} \left| \hat{f}_{t,h}(X_{t+1}) - V_{t+h} \right|^{1/2}.$$

We can consider $\hat{f}_{t,h}(X_{t+1})$ as an estimate of $E\left(Y_{t+1}^2 + \cdots + Y_{t+h}^2 \,|\, \mathcal{F}_t\right)$.

1.9.3 Performance of Conditional Covariance Estimators

Let us discuss measuring the performance of estimators of conditional covariance $f(x) = \mathrm{Cov}(Y, Z \,|\, X = x)$. Empirical performance measures of conditional covariance estimators can be found naturally in the case where

$$E(Y \,|\, X = x) = 0, \qquad E(Z \,|\, X = x) = 0,$$

so that

$$f(x) = \mathrm{Cov}(Y, Z \,|\, X = x) = E(YZ \,|\, X = x).$$

For example, we can use sample splitting, similarly as in (1.120), where a performance measure for the case of measuring the performance of a conditional variance estimator was given. Let $\hat{f}^*$ be an estimator of f, constructed from the data $(X_1, Y_1, Z_1), \ldots, (X_{n^*}, Y_{n^*}, Z_{n^*})$, where $1 \le n^* < n$. Then we can use

$$\frac{1}{n - n^*} \sum_{i=n^*+1}^{n} \left| \hat{f}^*(X_i) - Y_i Z_i \right|$$

to measure the performance of the estimator.

In autoregressive time–space smoothing methods, like in the MGARCH models and exponential moving average methods studied in Section 3.10.2, the explanatory variables are the previous observations, and the estimate $\hat{\gamma}_t$ of $E(Y_t Z_t \mid \mathcal{F}_{t-1})$ is calculated using observations $(Y_1, Z_1), \ldots, (Y_{t-1}, Z_{t-1})$, and now we define the mean deviation error by

$$\mathrm{MDE}_T^{(q)}(\hat{\gamma}) = \frac{1}{T - t_0 + 1} \sum_{t=t_0}^{T} |\hat{\gamma}_t - Y_t Z_t|^{1/q}, \tag{1.128}$$

where $q > 0$.

1.9.4 Performance of Quantile Function Estimators

Theoretical performance measures for the estimators of the conditional quantile

$$f(x) = \mathrm{Q}_p(Y \mid X = x)$$

can be defined similarly as in the case of conditional variance estimators. For example, using (1.119).

Empirical performance measures can be found in the case of continuous distribution of Y by using the fact

$$\begin{aligned} p &= P\left(Y \le \mathrm{Q}_p(Y \mid X = x) \,\middle|\, X = x\right) \\ &= E\left[I_{(-\infty, \mathrm{Q}_p(Y \mid X=x)]}(Y) \,\middle|\, X = x\right], \end{aligned}$$

where $x \in \mathbf{R}^d$. Let $(X_1, Y_1), \ldots, (X_n, Y_n)$ be regression data and let

$$\hat{q}_i(x) = \hat{\mathrm{Q}}_{p,-i}(Y \mid X = x)$$

be a conditional quantile estimate constructed using the other data but not the ith observation. Let the cross validation quantity be

$$\hat{p} = \frac{1}{n-1} \sum_{i=1}^{n} I_{(-\infty, \hat{q}_i(X_i)]}(Y_i).$$

Finally, the performance is measured by the difference

$$p - \hat{p}.$$

Let us consider the time series setting, where we have observations $Y_1, \ldots, Y_T$. Then we can construct a conditional quantile estimator

$$\hat{q}_t = \hat{\mathrm{Q}}_p(Y_t \mid Y_{t-1}, \ldots)$$

using data $Y_1, \ldots, Y_{t-1}$, and calculate

$$\hat{p} = \frac{1}{T - t_0} \sum_{t=t_0+1}^{T} I_{(-\infty, \hat{q}_t]}(Y_t), \tag{1.129}$$

where $1 \le t_0 \le T-1$. We start to evaluate the performance of the estimator after t_0 observations are available, because any estimator can behave erratically when only a couple of observations are available.

Even when we would know the true quantiles, there is random fluctuation in the numbers $\hat{p}$. The random variables

$$Z_t = I_{(-\infty, q_p]}(Y_{t+1}), \qquad t = t_0, \ldots, T-1,$$

are Bernoulli random variables with $P(Z_t = 1) = p$, where q_p is the true quantile. If random variables Y_t are independent, then random variables Z_t are independent, and

$$M = \sum_{t=t_0}^{T-1} Z_t$$

is a binomial random variable with the distribution $\mathrm{Bin}(n, p)$, where $n = T - t_0$. The probability mass function of M is

$$P(M = i) = \binom{n}{i} p^i (1-p)^{n-i},$$

for $i = 0, \ldots, n$. We can now calculate the numbers c_0 and c_1 such that

$$P(c_0 \le p - \tilde{p} \le c_1) \ge 1 - \alpha, \tag{1.130}$$

where $0 < \alpha < 1$ and $\tilde{p} = M/n$. We have

$$c_0 = p - n^{-1} z_{\alpha/2}, \qquad c_1 = p - n^{-1} z_{1-\alpha/2}, \tag{1.131}$$

where $z_{\alpha/2}$ and $z_{1-\alpha/2}$ are such that $P(z_{\alpha/2} \le M \le z_{1-\alpha/2}) \ge 1 - \alpha$.

If $\hat{p} > p$, this means that the quantile estimates were in average larger than the true quantiles. When we are estimating the left tail, so that p is close to zero, then the relation $\hat{p} > p$ means that the true distribution has a heavier left tail than the quantile estimates would indicate. When we are estimating the right tail, so that p is close to one, then this relation reverses, and the relation $\hat{p} > p$ means that the true distribution has a lighter left tail than the quantile estimates would indicate.

We will show the performance of quantile estimators by plotting the difference

$$R(p, \hat{p}) = \begin{cases} p - \hat{p}, & \text{when } p \le 0.5, \\ \hat{p} - p, & \text{when } p > 0.5. \end{cases} \tag{1.132}$$

Thus, the difference $R(p, \hat{p})$ being negative means that the true distribution has a heavier tail than the quantile estimates would indicate. The difference $R(p, \hat{p})$ being positive means that the true distribution has a lighter tail than the quantile estimates would indicate.

Figure 1.6 illustrates the performance measurement of quantile estimators. We estimate the quantiles of the S&P 500 returns Y_t using the S&P 500 index data described in Section 1.6.1. Let $\hat{q}_t^e$ be the empirical quantile, defined in (1.26),

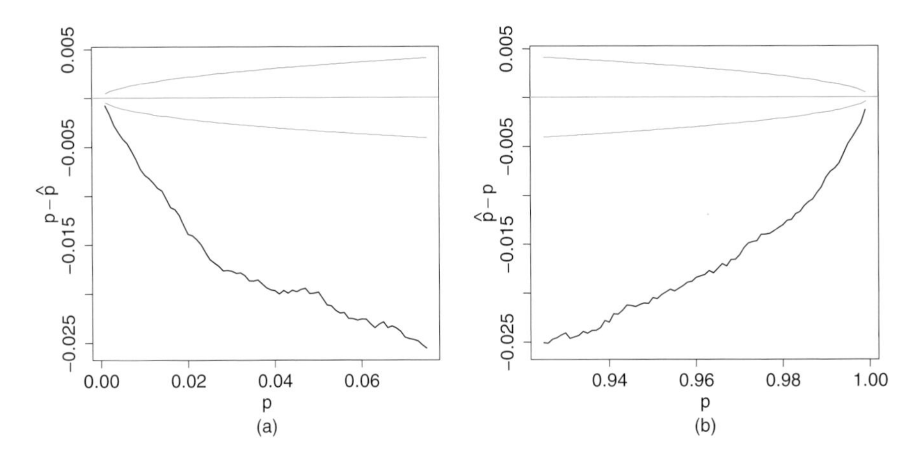

Figure 1.6 *Quantile estimator performance.* Function $p \mapsto R(p, \hat{p})$, defined in (1.132), is plotted with the black curves, when the quantile estimator is the empirical quantile. Panel (a) shows the range $p \in [0.001, 0.075]$, and panel (b) shows the range $p \in [0.925, 0.999]$. The green lines show level $\alpha = 0.05$ fluctuation bands.

and calculated using the data $Y_1, \ldots, Y_t$. We plot the function $p \mapsto R(p, \hat{p})$ in black. Panel (a) shows the range $p \in [0.001, 0.075]$, and panel (b) shows the range $p \in [0.925, 0.999]$. A green line is drawn at level 0, and it is accompanied by the level $\alpha = 0.05$ fluctuation bands, defined in (1.130)–(1.131). Figure 1.6 indicates that the true distribution has heavier tails than the empirical quantile estimates would indicate.

1.9.5 Performance of Estimators of Expected Shortfall

To derive a performance measure for estimators of expected shortfall, we can use the fact that for a continuous distribution of Y, we obtain

$$E\big[(Y - \mathrm{ES}_p(Y))\, I_{(-\infty, q_p]}(Y)\big] = 0.$$

Indeed, for a continuous distribution of Y, we have

$$\mathrm{ES}_p(Y) = \frac{1}{p}\, E\left[Y\, I_{(-\infty, q_p]}(Y)\right]$$

and

$$E\big[I_{(-\infty, q_p]}(Y)\big] = p.$$

If we are in the time series setting and have identically distributed observations $(X_1, Y_1), \ldots, (X_T, Y_T)$, then we can construct an estimator of the expected shortfall

$$\widehat{\mathrm{ES}}_{p,t}$$

using data $(X_1, Y_1), \ldots, (X_t, Y_t)$ and calculate the performance measure

$$\frac{1}{T - t_0} \sum_{t=t_0}^{T-1} \left(Y_{t+1} - \widehat{\mathrm{ES}}_{p,t} \right)^2 I_{(-\infty, \hat{q}_t]}(Y_{t+1}),$$

where $\hat{q}_t = \hat{Q}_{p,t}$ is a quantile estimator and $1 \leq t_0 \leq T - 1$.

1.9.6 Performance of Classifiers

Theoretical Performance Measures Let $g : \mathbf{R}^d \to \{0, \ldots, K-1\}$ be a classification function. The probability of the classification error is

$$R(g) = P(g(X) \neq Y),$$

and this can be used to measure the goodness of g. The goodness of an empirical classification rule $\hat{g}$, calculated from data $(X_1, Y_1), \ldots, (X_n, Y_n)$, is measured by

$$R(\hat{g}) = P\left(\hat{g}(X) \neq Y\right),$$

where P is the probability measure of $(X, Y), (X_1, Y_1), \ldots, (X_n, Y_n)$. We can write the probability of the misclassification more transparently. We have that

$$R(\hat{g}) = \sum_{k=0}^{K-1} P(Y = k) \int_{\hat{G}_k^c} f_{X|Y=k},$$

where $f_{X|Y=k} : \mathbf{R}^d \to \mathbf{R}$ is the density function of $X \mid Y = k$, and

$$\hat{G}_k = \{x \in \mathbf{R}^d : \hat{g}(x) = k\}, \qquad k = 0, \ldots, K-1,$$

is the subset of the sample space, where the classification function $\hat{g}$ chooses class k.

When we analyze the asymptotic performance of the classification functions, we should note that $R(\hat{g})$ does not converge to zero, but at best we can hope that it converges to the minimal classification error $R(g^*)$, which is the classification error of the Bayes rule g^*, defined in (1.75). Thus we should study the rate of convergence to zero of $R(\hat{g}) - R(g^*)$. Let us consider the two-class case $K = 2$ with the equal class priors $P(Y = 0) = P(Y = 1) = 1/2$. Then,

$$R(g^*) = \frac{1}{2} \int_{\mathbf{R}^d} \min\{f_{X|Y=0}(x), f_{X|Y=1}(x)\}\, dx$$

and

$$R(\hat{g}) - R(g^*) = \frac{1}{2} d_{f_{X|Y=0}, f_{X|Y=1}}(\{x : \hat{g}(x) = 1\}, \{x : g^*(x) = 1\}),$$

where

$$d_{g_1, g_2}(G_1, G_2) = \int_{G_1 \Delta G_2} |g_1 - g_2|,$$

with

$$G_1 \Delta G_2 = (G_1^c \cap G_2) \cup (G_1 \cap G_2^c)$$

the symmetric difference of G_1 and G_2. The rate of convergence has been studied in Mammen & Tsybakov (1999).

Empirical Performance Measures The frequency of misclassification can be used as an empirical performance measure of a classification method. We can use sample splitting as in the case of regression function estimation, see (1.115). Let us have classification data $(X_1, Y_1), \ldots, (X_n, Y_n)$ and let us construct classifier $\hat{g}^*$ using the first part $(X_1, Y_1), \ldots, (X_{n_1}, Y_{n_1})$ of data, where $1 \le n_1 < n$, and typically $n_1 = [n/2]$. We can use

$$\frac{1}{n - n_1} \sum_{i=n_1+1}^{n} I_{\{\hat{g}^*(X_i)\}^c}(Y_i)$$

as an estimator of $P(\hat{g}(X) \neq Y)$, where $\hat{g}$ is constructed from the whole sample.

We can also use cross validation, as in the case of regression function estimation in (1.116). In the time series setting, when we have regression data $(X_1, Y_1), \ldots, (X_T, Y_T)$, it is natural to measure the performance of classification method by

$$\frac{1}{T - t_0} \sum_{t=t_0}^{T-1} I_{\{\hat{g}_t^*(X_{t+1})\}^c}(Y_{t+1}), \tag{1.133}$$

where $\hat{g}_t^*$ is a classifier constructed using the data $(X_1, Y_1), \ldots, (X_t, Y_t)$, and t_0 is chosen so large that the first classifier $\hat{g}_{t_0}^*$ in the sequence is already a reasonable classifier. We can divide the classification error into K components

$$\frac{1}{T - t_0} \sum_{t=t_0}^{T-1} I_{\{\hat{g}_t^*(X_{t+1})\}^c}(k)\, I_{\{k\}}(Y_{t+1}), \tag{1.134}$$

where $k = 0, \ldots, K - 1$, which estimate $P(\hat{g}(X) \neq Y \mid Y = k)$.

1.10 CONFIDENCE SETS

We give first several definitions of a confidence interval for regression function estimation. Then we define confidence bands.

1.10.1 Pointwise Confidence Intervals

A pointwise confidence interval $[L, U]$ for the estimation of regression function $f : \mathbf{R}^d \to \mathbf{R}$ at point $x \in \mathbf{R}^d$, with the confidence level $1 - \alpha$, is such that for all $P \in \mathcal{P}$, for all x in a suitable subset of $\mathbf{R}^d$, we have

$$P\left(L \le f(x) \le U\right) = 1 - \alpha,$$

where $\mathcal{P}$ is a collection of distributions of (X, Y). Typically we can give asymptotic confidence intervals of type

$$P\left(L_n \le f(x) \le U_n\right) \longrightarrow 1 - \alpha,$$

when $n \to \infty$, where $[L_n, U_n]$, is a sequence of intervals. Asymptotic pointwise confidence intervals can typically be derived from the asymptotic distribution of the estimator. If we have that

$$n^a \left(\hat{f}(x) - f(x)\right) \xrightarrow{d} N(\mu, \sigma^2),$$

where symbol "$\xrightarrow{d}$" denotes the convergence in distribution, then we can choose

$$L_n = \hat{f}(x) - n^{-a}(\mu + z_{1-\alpha/2}\sigma)$$

and

$$U_n = \hat{f}(x) + n^{-a}(\mu + z_{1-\alpha/2}\sigma),$$

where we denote $z_\alpha = \Phi^{-1}(\alpha)$, and Φ is the distribution function of the standard normal distribution. That is, z_p is the p-quantile of the $N(0,1)$ distribution: For $Z \sim N(0,1)$, we have $P(z_{\alpha/2} \le Z \le z_{1-\alpha/2}) = 1 - \alpha$.

More generally, we can use the term "level $1 - \alpha$ confidence interval" if the inequality

$$P(L \le f(x) \le U) \ge 1 - \alpha$$

holds for all $P \in \mathcal{P}$. We can use the term "asymptotic level $1-\alpha$ confidence interval" if

$$\liminf_{n\to\infty} P(L_n \le f(x) \le U_n) \ge 1 - \alpha$$

for all $P \in \mathcal{P}$. Note that in the asymptotic case is is important to distinguish a uniform asymptotic level $1 - \alpha$ confidence interval, which satisfies

$$\liminf_{n\to\infty} \inf_{P\in\mathcal{P}} P(L_n \le f(x) \le U_n) \ge 1 - \alpha.$$

As pointed out by Wasserman (2005, p. 6), it is better to have uniform confidence intervals.

We give an example of a confidence interval in Section 3.2.10, for the case of kernel regression. As mentioned in Ruppert et al. (2003, Section 6.2) we can derive an approximate confidence interval for linear estimators under some assumptions. We noted in (1.2) that many estimators can be written as linear estimators

$$\hat{f}(x) = \sum_{i=1}^{n} l_i(x)\, Y_i = l(x)'\mathbf{y},$$

where $l(x) = (l_1(x), \ldots, l_n(x))'$ and $\mathbf{y} = (Y_1, \ldots, Y_n)'$. Let us assume that $\hat{f}(x) \sim N(f(x), \mathrm{Var}(\hat{f}(x)))$. If

$$\mathrm{Cov}(\mathbf{y}) = \sigma^2 I_n,$$

then

$$\mathrm{Var}\left(\hat{f}(x)\right) = l(x)'\mathrm{Cov}(\mathbf{y})l(x) = \sigma^2 \|l(x)\|^2.$$

Estimating σ^2 with $\hat{\sigma}^2 = n^{-1}\sum_{i=1}^{n}(Y_i - \hat{f}(X_i))^2$ leads to the confidence interval

$$\left[\hat{f}(x) - \hat{\sigma}\|l(x)\|z_{1-\alpha/2},\ \hat{f}(x) + \hat{\sigma}\|l(x)\|z_{1-\alpha/2}\right],$$

where α is the confidence level, $0 < \alpha < 1$, and $z_{1-\alpha/2}$ is the quantile of the standard normal distribution.

1.10.2 Confidence Bands

A confidence band $(L(x), U(x))$, $x \in A$, for the estimation of regression function $f : \mathbf{R}^d \to \mathbf{R}$, for the set $A \subset \mathbf{R}^d$, with the confidence level $1 - \alpha$, is such that

$$P\left(L(x) \le f(x) \le U(x), \text{ for all } x \in A\right) = 1 - \alpha. \tag{1.135}$$

Confidence bands are called also simultaneous confidence bands, confidence envelopes, or variability bands. The confidence statement of the type

$$P\left(\sup_{x \in A} |f(x) - \hat{f}(x)| \le c_n\right) = 1 - \alpha$$

is equivalent to (1.135) if

$$L(x) = \hat{f}(x) - c_n, \qquad U(x) = \hat{f}(x) + c_n.$$

We can replace the supremum norm with some other function space norm to obtain confidence balls. For example, the L_2 confidence ball with the confidence level $1 - \alpha$ satisfies

$$P\left(\|f(x) - \hat{f}(x)\|_2 \le c_n\right) = 1 - \alpha.$$

A confidence band in the linear model is mentioned in Section 2.1.5.

1.11 TESTING

In the linear regression model

$$Y = \alpha + \beta_1 X_1 + \cdots + \beta_d X_d + \epsilon$$

the typical tests are the tests of restrictions

$$H_0 : \beta_k = 0, \tag{1.136}$$

for $k = 1, \ldots, d$, and

$$H_0 : \beta_1 = \cdots = \beta_d = 0. \tag{1.137}$$

Testing of these hypothesis is considered in Section 2.1.5. There are several ways to generalize these tests to a nonparametric setting, where

$$Y = f(X) + \epsilon.$$

The hypothesis in (1.137) can be generalized to the hypothesis

$$H_0 : f(x) \equiv 0,$$

when we assume that $EY = 0$. We can use a test statistics $T = \|\hat{f}\|$, where $\hat{f}$ is a nonparametric estimate of f. The norm $\|\cdot\|$ can be the L_2 norm, a weighted

L_2 norm, or some other function space norm. Large values of the test statistics T lead to the rejection of the null hypothesis. For the linear regression function $f(x) = \alpha + \beta_1 x_1 + \cdots + \beta_d x_d$, it holds that

$$\frac{\partial}{\partial x_k} f(x) = \beta_k.$$

Thus we can generalize the parameter restriction hypothesis (1.136) to the nonlinear case by

$$H_0 : \frac{\partial}{\partial x_k} f(x) \equiv 0, \tag{1.138}$$

for $k = 1, \ldots, d$. We can generalize the parameter restriction hypothesis (1.137) to the nonlinear case by

$$H_0 : \frac{\partial}{\partial x_1} f(x) \equiv 0, \ldots, \frac{\partial}{\partial x_d} f(x) \equiv 0.$$

We can test the null hypothesis (1.138) with the test statistics

$$T = \left\| \frac{\partial}{\partial x_k} f(x) \right\|,$$

where $\hat{f}$ is a nonparametric estimator of f and $\|\cdot\|$ is a function space norm.

The distribution of the test statistics can be approximated by bootstrap. Generate first B bootstrap samples from the original sample $(X_1, Y_1), \ldots, (X_n, Y_n)$. Based on a bootstrap sample $(X_1^*, Y_1^*), \ldots, (X_n^*, Y_n^*)$, the test statistics T^* is calculated. We obtain a sequence $T_1^*, \ldots, T_B^*$ of values of the test statistics. Let $q_{1-\alpha}$ be the empirical quantile of the sequence of the values of the test statistics. Then we reject the null hypothesis at level $0 < \alpha < 1$, if the observed value t of the test statistics satisfies $t > q_{1-\alpha}$.

Härdle & Mammen (1993) have proposed the wild bootstrap. First the regression function f is estimated with $\hat{f}$ (under the null hypothesis). Then the residuals $\hat{\epsilon}_i = Y_i - \hat{f}(X_i)$ are calculated. Finally, the bootstrap residual ϵ_i^* is generated from a distribution which satisfies $E\epsilon_i^* = 0$, $E(\epsilon_i^*)^2 = \hat{\epsilon}_i^2$, and $E(\epsilon_i^*)^3 = \hat{\epsilon}_i^3$. The bootstrap sample is $(X_1, Y_1^*), \ldots, (X_n, Y_n^*)$, where $Y_i^* = \hat{f}(X_i) + \epsilon_i^*$.

CHAPTER 2

LINEAR METHODS AND EXTENSIONS

In linear regression the conditional expectation is approximated by a linear function:

$$E(Y \mid X = x) \approx \alpha + \beta_1 x_1 + \cdots + \beta_d x_d, \tag{2.1}$$

where $x = (x_1, \ldots, x_d)$. Section 2.1 covers several methods to find the regression coefficients of the linear regression function: the least squares method, the generalized method of moments with instrumental variables, and the ridge regression which uses a penalized least squares criterion. We consider various extensions of linear regression. Section 2.2 discusses varying coefficient linear regression. In the varying coefficient linear regression model

$$E(Y \mid X = x) \approx \alpha(z) + \beta_1(z)\, x_1 + \cdots + \beta_d(z)\, x_d,$$

where each coefficient is a function of variable $Z = z$, which could be taken to be equal to the explanatory variable: $Z = X$. Section 2.3 covers generalized linear models, where nonlinearity is introduced with the help of a link function:

$$E(Y \mid X = x) \approx G(\alpha + \beta_1 x_1 + \cdots + \beta_d x_d),$$

where $G : \mathbf{R} \to \mathbf{R}$. We get a generalization of the linear model if we approximate the conditional expectation with a linear function of transformations of the original

Multivariate Nonparametric Regression and Visualization. By Jussi Klemelä

variables. For example, we may approximate

$$E(Y \mid X = x) \approx \alpha + \beta_1 x_1^2 + \cdots + \beta_d x_d^2.$$

This is a special case of the approximation with the help of a series estimator using basis functions. Section 2.4 covers the series estimators, where

$$E(Y \mid X = x) \approx \alpha + \beta_1 g_1(x) + \cdots + \beta_M g_M(x),$$

where $g_k : \mathbf{R}^d \to \mathbf{R}$. Section 2.5 covers conditional variance estimation with time series observations and, in particular, ARCH models are covered. In the case of the conditional variance estimation, it is a natural idea to use a linear function on the squares of lagged observations:

$$\text{Var}\left(Y_t \mid Y_{t-1} = y_{t-1}, \ldots, Y_{t-d} = y_{t-d}\right) \approx \alpha + \beta_1 y_{t-1}^2 + \cdots + \beta_d y_{t-d}^2.$$

Section 2.1 is about linear regression. Besides linear regression, we discuss the generalized method of moments estimator and the ridge regression. Asymptotic distributions, tests, and confidence intervals are given for the linear regression, and variable selection is considered. As applications of linear regression we mention (a) the measurement of the beta of an asset and of a portfolio and (b) the measurement of the alpha of a portfolio and of a hedge fund.

Section 2.2 defines a varying coefficient regression estimator. An application to hedge fund index replication is given. Section 2.3 covers generalized linear models and binary response models. Section 2.4 considers series estimators. Section 2.5 considers linear estimators for the conditional variance and defines the ARCH model.

Section 2.6 contains applications of linear methods to volatility and quantile estimators with the S&P 500 return data. First we set some benchmarks for quantile estimation using sequential estimators. Then the conditional volatility and quantiles are estimated with least squares regression and the ARCH model, and conditional volatility is estimated with ridge regression.

Section 2.7 defines linear regression-based classifiers, density-based classifiers, and empirical risk minimization-based classifiers.

2.1 LINEAR REGRESSION

In linear regression the conditional expectation $f(x) = E(Y \mid X = x)$ is approximated with a linear function:

$$f(x) \approx \alpha + \beta' x, \qquad x \in \mathbf{R}^d,$$

where $\alpha \in \mathbf{R}$ and $\beta \in \mathbf{R}^d$. The regression function is estimated with

$$\hat{f}(x) = \hat{\alpha} + \hat{\beta}' x, \qquad x \in \mathbf{R}^d, \tag{2.2}$$

where the parameter estimates $\hat{\alpha} \in \mathbf{R}$ and $\hat{\beta} \in \mathbf{R}^d$ are calculated using the regression data $(X_1, Y_1), \ldots, (X_n, Y_n)$.

2.1.1 Least Squares Estimator

The least squares estimator $\hat{f}$ for the conditional expectation is defined by (2.2), where $\hat{\alpha} \in \mathbf{R}$ and $\hat{\beta} \in \mathbf{R}^d$ are defined as the minimizers of the least squares criterion

$$\sum_{i=1}^{n} (Y_i - \alpha - \beta' X_i)^2 . \tag{2.3}$$

The solution can be written as

$$\hat{\alpha} = \bar{Y} - \hat{\beta}' \bar{X}, \tag{2.4}$$

and

$$\hat{\beta} = \left[\sum_{i=1}^{n} (X_i - \bar{X})(X_i - \bar{X})' \right]^{-1} \sum_{i=1}^{n} (X_i - \bar{X})(Y_i - \bar{Y})', \tag{2.5}$$

where

$$\bar{X} = \frac{1}{n} \sum_{i=1}^{n} X_i, \qquad \bar{Y} = \frac{1}{n} \sum_{i=1}^{n} Y_i.$$

In the case $d = 1$ we have

$$\hat{\alpha} = \bar{Y} - \hat{\beta} \bar{X}, \qquad \hat{\beta} = \frac{\sum_{i=1}^{n} (X_i - \bar{X})(Y_i - \bar{Y})}{\sum_{i=1}^{n} (X_i - \bar{X})^2} . \tag{2.6}$$

It is often convenient to use notation where the intercept is included in the vector β. This can be done by choosing the first component of the explanatory variables as the constant one. Thus we observe

$$(X_i, Y_i), \qquad X_i = (1, X_{i,2,}, \ldots, X_{i,d+1}) \in \mathbf{R}^{d+1}, \;\; Y_i \in \mathbf{R}, \tag{2.7}$$

where $i = 1, \ldots, n$. The estimator is now

$$\hat{f}(x) = \hat{\beta}' x = \hat{\beta}_1 + \hat{\beta}_2 x_2 + \cdots + \hat{\beta}_{d+1} x_{d+1}, \tag{2.8}$$

where $x = (1, x_2, \ldots, x_{d+1})$. We use below the notation

$$K = d + 1. \tag{2.9}$$

With this notation the least squares estimator of parameter β can be written as

$$\hat{\beta} = (\mathbf{X}'\mathbf{X})^{-1} \mathbf{X}' \mathbf{y}, \tag{2.10}$$

where $\mathbf{X} = (X_1, \ldots, X_n)'$ is the $n \times K$ matrix whose rows are X_i', and $\mathbf{y} = (Y_1, \ldots, Y_n)'$ is the $n \times 1$ vector. The solution (2.10) can be found by writing the least squares criterion (2.3) with the matrix notation as

$$(\mathbf{y} - \mathbf{X}\beta)'(\mathbf{y} - \mathbf{X}\beta) = \mathbf{y}'\mathbf{y} - 2\beta' \mathbf{X}' \mathbf{y} + \beta' \mathbf{X}' \mathbf{X} \beta.$$

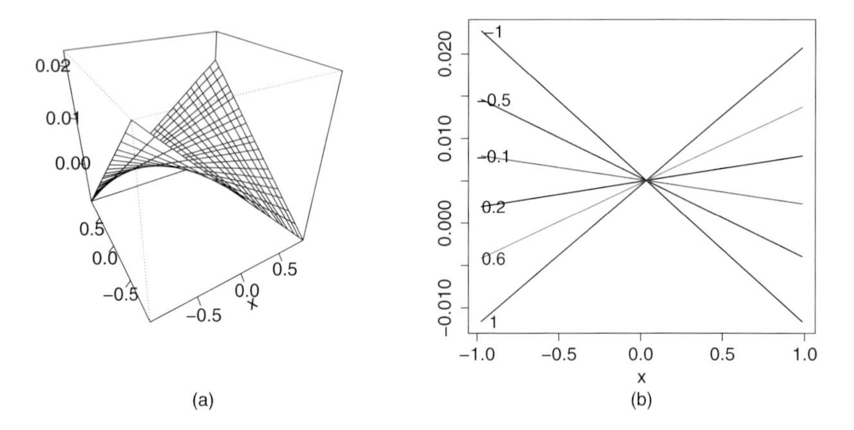

Figure 2.1 *Weights in linear regression.* (a) The function $(x, X_i) \mapsto l_i(x)$. (b) The eleven slices $x \mapsto l_i(x)$ for the choices $X_i = -1, -0.5, \ldots, 1$.

Derivating this with respect to β, and setting the gradient to zero, we get the equations

$$\mathbf{X}'\mathbf{X}\beta = \mathbf{X}'\mathbf{y},$$

which leads to the solution (2.10).

We can write the least squares estimator as

$$\hat{f}(x) = \sum_{i=1}^{n} l_i(x)\, Y_i, \tag{2.11}$$

where

$$l_i(x) = X_i'(\mathbf{X}'\mathbf{X})^{-1}x. \tag{2.12}$$

Note that $\hat{f}(x) = l(x)'\mathbf{y}$, where $l(x) = (l_1(x), \ldots, l_n(x))'$ is the $n \times 1$ vector of weights, defined by

$$l(x) = \mathbf{X}(\mathbf{X}'\mathbf{X})^{-1}x. \tag{2.13}$$

A large class of regression function estimators can be written as a linear function of $Y_1, \ldots, Y_n$ similarly to (2.11), as was noted in (1.2). For example, local averaging can be written as a linear function of $Y_1, \ldots, Y_n$, see (3.1).

Figure 2.1 illustrates the vector $l(x)$ of weights in the case of one-dimensional explanatory variable. Panel (a) shows a perspective plot of the function $(x, X_i) \mapsto l_i(x)$, where $X_1, \ldots, X_n$ is a simulated sample of size $n = 200$ from the uniform distribution on $[-1, 1]$. Panel (b) shows the six functions $x \mapsto l_i(x)$ for the choices $X_i = -1, -0.5, \ldots, 1$. That is, panel (b) shows six slices of the function in panel (a). Panel (b) shows that the functions $x \mapsto l_i(x)$ are linear, and $l_i(x)$ can take negative values.

2.1.2 Generalized Method of Moments Estimator

We define the generalized method of moments (GMM) estimator in the linear regression model

$$Y = \beta' X + \epsilon, \tag{2.14}$$

where $\beta \in \mathbf{R}^K$, $X = (X_1, \ldots, X_K)'$, $Y \in \mathbf{R}$, and $\epsilon \in \mathbf{R}$. Note that we now use notation (2.7)–(2.9), where the intercept is included in the model by choosing $X_1 \equiv 1$. The generalized method of moments estimator was analyzed in Hansen (1982) and White (1982).

Method of Moments Multiplying (2.14) with vector X, we get

$$XY = XX'\beta + X\epsilon.$$

If $E(X\epsilon) = 0$, then

$$E(XY) = E(XX')\beta. \tag{2.15}$$

If $E(XX')$ is invertible, then

$$\beta = [E(XX')]^{-1} E(XY). \tag{2.16}$$

We see that the replacing the expectations with the sample means leads to the least squares estimator. Indeed, let us have identically distributed observations $(X_1, Y_1), \ldots, (X_n, Y_n)$ which have the same distribution as (X, Y). When we replace the expectations in (2.16) with the sample means, we get the estimator

$$\hat{\beta} = \left(\frac{1}{n} \sum_{i=1}^{n} X_i X_i' \right)^{-1} \frac{1}{n} \sum_{i=1}^{n} X_i Y_i. \tag{2.17}$$

This estimator is the same as the least squares estimator in (2.10), as can be seen by noting that

$$\mathbf{X}'\mathbf{X} = \sum_{i=1}^{n} X_i X_i', \qquad \mathbf{X}'\mathbf{y} = \sum_{i=1}^{n} X_i Y_i.$$

Generalized Method of Moments The deduction leading to the estimator (2.17) can be seen as a special case of the generalized method of moments. The matrix equation (2.15) contains K linear equations. Let us write (2.15) with the general notation

$$Eg(X, Y, \beta) = 0, \tag{2.18}$$

where

$$g(X, Y, \beta) = XY - XX'\beta. \tag{2.19}$$

Let us denote $g(X, Y, \beta) = (g_1(X, Y, \beta), \ldots, g_K(X, Y, \beta))'$, where g_k are real-valued functions. Now we replace the expectation in (2.18) with the sample mean to get the equation

$$\sum_{i=1}^{n} g(X_i, Y_i, \beta) = 0. \tag{2.20}$$

The solution of (2.20) is $\hat{\beta}$, given in (2.17). The same solution can be obtained as

$$\hat{\beta} = \text{argmin}_{\beta \in \mathbf{R}^K} \left(\sum_{i=1}^{n} g(X_i, Y_i, \beta) \right)' W \left(\sum_{i=1}^{n} g(X_i, Y_i, \beta) \right), \tag{2.21}$$

where W is a positive definite symmetric $K \times K$ matrix of weights. This does not lead to a new estimator with the choice (2.19), but we consider next the instrumental variables estimator, where the generalized method of moments is useful.

Instrumental Variables Estimator Let us remove the assumption that $E(X\epsilon) = 0$, but assume that there are $L = K$ instrumental variables $Z = (Z_1, \ldots, Z_L)'$ for which $E(Z\epsilon) = 0$. Multiplying the linear model equation (2.14) with the vector Z we get the L linear equations

$$ZY = ZX'\beta + Z\epsilon,$$

and taking the expectation gives

$$E(ZY) = E(ZX')\beta. \tag{2.22}$$

Equation (2.22) contains L linear equations and there are K parameters, but we have assumed $L = K$. If $E(ZX')$ is invertible, we get

$$\beta = [E(ZX')]^{-1} E(ZY). \tag{2.23}$$

Let us have the identically distributed observations $(X_1, Y_1, Z_1), \ldots, (X_n, Y_n, Z_n)$, which have the same distribution as (X, Y, Z). When we replace the expectations in (2.23) with the sample means, we get the estimator

$$\hat{\beta}_{inst} = \left(\frac{1}{n} \sum_{i=1}^{n} X_i Z_i' \right)^{-1} \frac{1}{n} \sum_{i=1}^{n} Z_i Y_i = (\mathbf{X}'\mathbf{Z})^{-1} \mathbf{Z}'\mathbf{y}, \tag{2.24}$$

where

$$\mathbf{X}'\mathbf{Z} = \sum_{i=1}^{n} X_i Z_i', \qquad \mathbf{Z}'\mathbf{y} = \sum_{i=1}^{n} Z_i Y_i,$$

when $\mathbf{X} = (X_1, \ldots, X_n)'$ is the $n \times K$ matrix whose rows are X_i', $\mathbf{Z} = (Z_1, \ldots, Z_n)'$ is the $n \times L$ matrix whose rows are Z_i', and $\mathbf{y} = (Y_1, \ldots, Y_n)'$ is the $n \times 1$ vector.

Instrumental Variables Estimator with GMM Now we assume that there are more instruments than parameters: $L > K$. In this case, the generalized method of moments can be used to define an instrumental variable estimator. We do not assume that $E(X\epsilon) = 0$, but assume that there are $L > K$ instrumental variables $Z = (Z_1, \ldots, Z_L)'$ for which $E(Z\epsilon) = 0$. We will need that the rank of $E(XZ')$ is K. Now equation (2.22) cannot be solved, because there are K parameters and $L > K$ equations. Let us use the GMM estimator and apply formula (2.21). We denote

$$g(X, Y, Z, \beta) = ZY - ZX'\beta,$$

and define the estimator by

$$\hat{\beta}_{gmm} = \text{argmin}_{\beta \in \mathbf{R}^K}\, g_n(\beta)' W g_n(\beta), \tag{2.25}$$

where

$$g_n(\beta) = \sum_{i=1}^{n} g(X_i, Y_i, Z_i, \beta),$$

and W is a symmetric positive definite $L \times L$ matrix. The solution $\hat{\beta}_{gmm}$ of (2.25) can be written in the matrix notation as

$$\hat{\beta}_{gmm} = (\mathbf{X}'\mathbf{Z}W\mathbf{Z}'\mathbf{X})^{-1}\mathbf{X}'\mathbf{Z}W\mathbf{Z}'\mathbf{y}. \tag{2.26}$$

2SLS Estimator If we choose the weighting matrix as

$$W_n = \left(\frac{1}{n}\mathbf{Z}'\mathbf{Z}\right)^{-1} = \left(\frac{1}{n}\sum_{i=1}^{n} Z_i' Z_i\right)^{-1},$$

then we get the estimator

$$\hat{\beta}_{2sls} = \left(\mathbf{X}'\mathbf{Z}(\mathbf{Z}'\mathbf{Z})^{-1}\mathbf{Z}'\mathbf{X}\right)^{-1}\mathbf{X}'\mathbf{Z}(\mathbf{Z}'\mathbf{Z})^{-1}\mathbf{Z}'\mathbf{y}. \tag{2.27}$$

The estimator is called the two-step least squares estimator, because we can construct it in two steps: First the X variables are explained with the Z variables, and the projection $\hat{X}$ is obtained; then the variable Y is explained with the projection $\hat{X}$. The first step gives the fitted values

$$\hat{\mathbf{X}} = \mathbf{Z}\hat{\gamma} = \mathbf{Z}(\mathbf{Z}'\mathbf{Z})^{-1}\mathbf{Z}'\mathbf{X},$$

where $\hat{\gamma} = (\mathbf{Z}'\mathbf{Z})^{-1}\mathbf{Z}'\mathbf{X}$. The second step gives

$$\hat{\beta}_{2sls} = (\hat{\mathbf{X}}'\hat{\mathbf{X}})^{-1}\hat{\mathbf{X}}'\mathbf{y},$$

which is equal to (2.27).

The Optimal Weighting Matrix It can be shown that the optimal weighting matrix is

$$W = \Lambda^{-1}, \qquad \Lambda = E(\epsilon^2 ZZ').$$

The estimator of Λ^{-1} can be taken as $\hat{\Lambda}^{-1}$, where

$$\hat{\Lambda} = \sum_{i=1}^{n}\left(Y_i - \hat{\beta}_{pre}' X_i\right)^2 Z_i Z_i',$$

where $\hat{\beta}_{pre}$ is a preliminary consistent estimator of β. We can take the preliminary estimator to be the two-step least squares estimator: $\hat{\beta}_{pre} = \hat{\beta}_{2sls}$. The optimality means the optimality of the covariance matrix V_W in the limit distribution

$$\sqrt{n}\left(\hat{\beta}_{gmm} - \beta\right) \xrightarrow{d} N(0, V_W),$$

as $n \to \infty$, where V_W is a $K \times K$ covariance matrix depending on the weight matrix W. We want to find the weight matrix W so that $V_W \le V_{W_0}$ for all weight matrices W_0, where the inequality $V_W \le V_{W_0}$ means that $V_{W_0} - V_W$ is a positive semidefinite matrix. For the statement of the convergence result, see (2.46), where the formula for V_W is given. For a proof of the optimality of $W = \Lambda^{-1}$, see Wooldridge (2005, Section 8.3.3).

Population Formulas The population formulas given in (2.16) are more instructive if we write the linear model as

$$Y = \alpha + \beta' X + \epsilon,$$

where $\alpha \in \mathbf{R}$, $\beta \in \mathbf{R}^d$, $X = (X_1, \ldots, X_1)'$, $Y \in \mathbf{R}$, and $\epsilon \in \mathbf{R}$. If $EX\epsilon = 0$, then

$$\alpha = EY - \beta' EX \tag{2.28}$$

and

$$\beta = \mathrm{Cov}(X)^{-1}\, E[(X - EX)(Y - EY)], \tag{2.29}$$

where

$$\mathrm{Cov}(X) = E\left[(X - EX)(X - EX)'\right].$$

and we assume additionally that $\mathrm{Cov}(X)$ is invertible.[15]

In the one-dimensional case $d = 1$ we have

$$\alpha = EY - \beta\, EX, \qquad \bar{\beta} = \frac{\mathrm{Cov}(X, Y)}{\mathrm{Var}(X)}. \tag{2.30}$$

In the two-dimensional case $d = 2$ we have $\alpha = EY - \beta_1' EX_1 - \beta_2 EX_2$,

$$\beta_1 = \frac{1}{\sigma_1^2 \sigma_2^2 - \sigma_{12}^2} \left(\sigma_2^2\, \mathrm{Cov}(X_1, Y) - \sigma_{12}\, \mathrm{Cov}(X_2, Y)\right),$$

and

$$\beta_2 = \frac{1}{\sigma_1^2 \sigma_2^2 - \sigma_{12}^2} \left(\sigma_1^2\, \mathrm{Cov}(X_2, Y) - \sigma_{12}\, \mathrm{Cov}(X_1, Y)\right),$$

where $\sigma_1^2 = \mathrm{Var}(X_1)$, $\sigma_2^2 = \mathrm{Var}(X_2)$, and $\sigma_{12} = \mathrm{Cov}(X_1, X_2)$.

2.1.3 Ridge Regression

Let us consider the linear regression model

$$Y = \beta' X + \epsilon,$$

[15] We get the same solution by minimizing $E(\alpha + \beta' X - Y)^2$. Derivating with respect to α and setting the derivative equal to zero we get $\alpha = EY - \beta' EX$. Then we find $\hat{\beta}$ by minimizing $E(\beta'(X - EX) - (Y - EY))^2$. Derivating with respect to elements of β and setting these derivatives to zero, we get $E(X - EX)(X - EX)'\beta = E(X - EX)(Y - EY)$, which leads to (2.29).

where $\beta \in \mathbf{R}^d$, $X = (X_1, \ldots, X_d)'$, $Y \in \mathbf{R}$, and $\epsilon \in \mathbf{R}$. We assume that $EY = 0$, $EX = 0$, and $\mathrm{Var}(X_k) = 1$, for $k = 1, \ldots, d$. If this does not hold, we normalize the observations $(X_1, Y_1), \ldots, (X_n, Y_n)$. The intercept is not included in the model and that is why we denote in this section by d the number of variables, instead of using the notation K.

In the ridge regression the least squares criterion is replaced by a penalized least squares criterion. The ridge estimator $\hat{\beta}^{ridge}$ is defined as the minimizer of

$$\sum_{i=1}^{n}(Y_i - X_i'\beta)^2 + \lambda \sum_{k=1}^{d} \beta_k^2, \tag{2.31}$$

over $\beta \in \mathbf{R}^d$, where $\lambda \geq 0$ is the penalization parameter. An related way to define a ridge estimator is to define $\hat{\beta}^{ridge}$ as a solution to a constrained minimization problem: Define $\hat{\beta}^{ridge}$ as the minimizer of $\sum_{i=1}^{n} (Y_i - \beta' X_i)^2$ over $\beta \in B_r(0)$, where $B_r(0) = \{\beta \in \mathbf{R}^d : \sum_{k=1}^{d} \beta_k^2 \leq r^2\}$.

Parameter Estimator The parameter estimator $\hat{\beta}^{ridge}$ minimizing (2.31) is

$$\hat{\beta}^{ridge} = (\mathbf{X}'\mathbf{X} + \lambda I)^{-1}\mathbf{X}'\mathbf{y}, \tag{2.32}$$

where $\mathbf{X} = (X_1, \ldots, X_n)'$ is the $n \times d$-matrix, $\mathbf{y} = (Y_1, \ldots, Y_n)'$ is the $n \times 1$ vector, and I is the $d \times d$ identity matrix. The solution (2.32) is found similarly as the ordinary least squares regression estimator (2.10). Indeed, we write the least squares criterion (2.31) with the matrix notation as

$$(\mathbf{y} - \mathbf{X}\beta)'(\mathbf{y} - \mathbf{X}\beta) + \lambda\beta'\beta = \mathbf{y}'\mathbf{y} - 2\beta'\mathbf{X}'\mathbf{y} + \beta'\mathbf{X}'\mathbf{X}\beta + \lambda\beta'\beta.$$

Derivating this with respect to β, and setting the gradient to zero, we get the equations

$$(\mathbf{X}'\mathbf{X} + \lambda I)\,\beta = \mathbf{X}'\mathbf{y},$$

which leads to the solution (2.32). In the case $d = 1$ we get

$$\hat{\beta}^{ridge} = \frac{\sum_{i=1}^{n} X_i Y_i}{\lambda + \sum_{i=1}^{n} X_i^2}.$$

We can write the ridge regression estimator as

$$\hat{f}(x) = \sum_{i=1}^{n} l_i(x)\, Y_i,$$

where

$$l_i(x) = X_i'(\mathbf{X}'\mathbf{X} + \lambda I)^{-1}x.$$

We can write also that $\hat{f}(x) = l(x)'\mathbf{y}$, where $l(x) = (l_1(x), \ldots, l_n(x))'$ is the $n \times 1$ vector of weights, defined by $l(x) = \mathbf{X}(\mathbf{X}'\mathbf{X} + \lambda I)^{-1}x$.

The least squares estimator $\hat{\beta}^{lse}$ is not defined if $\mathbf{X}'\mathbf{X}$ is not an invertible matrix. However, $\mathbf{X}'\mathbf{X} + \lambda I$ is invertible and the ridge regression estimator can be used. This was the motivation to define the ridge regression estimator in Hoerl & Kennard (1970), who invented the name "ridge regression."

It is possible to choose the parameter λ using cross validation, as defined in (1.116). In the cross validation we minimize

$$\text{MISE}(\lambda) = \sum_{i=1}^{n} \left(\hat{f}_{-i}^{ridge}(X_i) - Y_i \right)^2,$$

over $\lambda \geq 0$, where $\hat{f}_{-i}^{ridge}(x) = x'\hat{\beta}_{-i}^{ridge}$, and $\hat{\beta}_{-i}^{ridge}$ is the ridge estimator calculated with the other observations but the ith observation.

Shrinkage Estimators Let us consider fixed design regression with $\mathbf{X}'\mathbf{X} = nI$ and $E\epsilon = 0$. It can be shown that for the least squares estimator

$$\hat{\beta}^{lse} = (\mathbf{X}'\mathbf{X})^{-1}\mathbf{X}'\mathbf{y}$$

we have

$$E\hat{\beta}^{lse} = \beta, \qquad \text{Var}\left(\hat{\beta}^{lse}\right) = \sigma^2(\mathbf{X}\mathbf{X})^{-1},$$

where $\sigma^2 = \text{Var}(\epsilon)$. Then,

$$E\left\|\hat{\beta}^{lse} - \beta\right\|_2^2 = \sum_{k=1}^{d} \text{Var}(\hat{\beta}_k^{lse}) = \frac{d\sigma^2}{n}, \tag{2.33}$$

where we denote $\|\beta\|_2^2 = \sum_{k=1}^{d} \beta_k^2$.

A ridge regression estimator can have a smaller mean squared error than the ordinary least squared estimator. Since $\mathbf{X}'\mathbf{X} = nI$, then

$$\hat{\beta}^{ridge} = \frac{n}{n+\lambda}\hat{\beta}^{lse}. \tag{2.34}$$

We use (2.33) and (2.34) to show that the mean squared error of the ridge regression estimator is

$$E\left\|\hat{\beta}^{ridge} - \beta\right\|_2^2 = d\sigma^2 \frac{n}{(n+\lambda)^2} + \left(\frac{\lambda}{n+\lambda}\right)^2 \|\beta\|_2^2.$$

The minimum of the above expression with respect to λ is achieved by

$$\lambda = \frac{d\sigma^2}{\|\beta\|_2^2},$$

which depends on the unknown β. With this λ,

$$\hat{\beta}^{ridge} = \left(1 - \frac{d\sigma^2}{n\|\beta\|_2^2 + d\sigma^2}\right)\hat{\beta}^{lse}$$

and

$$E\left\|\hat{\beta}^{ridge}-\beta\right\|_2^2=\frac{d\sigma^2}{n+d\sigma^2/\|\beta\|^2}.$$

We can compare the obtained shrinkage factor to the shrinkage factor of James & Stein (1961), who chose

$$\hat{\beta}^{js}=\left(1-\frac{(d-2)\,\sigma^2}{n\|\hat{\beta}^{lse}\|_2^2}\right)\hat{\beta}^{lse},$$

where $\hat{\beta}_k^{lse}$ is the least squares estimator. We have that

$$E\left\|\hat{\beta}^{js}-\beta\right\|_2^2=\frac{2\sigma^2}{n};$$

see Wasserman (2005). The mean squared error of the least squares estimator is larger than the mean squared error of the James–Stein estimator for $d\geq 3$.

LASSO Least absolute shrinkage and selection operator (LASSO) defines the parameter estimator $\hat{\beta}^{lasso}$ as the minimizer of

$$\sum_{i=1}^{n}(Y_i-X_i'\beta)^2+\lambda\sum_{k=1}^{d}|\beta_k|,$$

where $\lambda\geq 0$. The estimator was defined in Tibshirani (1996). The motivation for the definition comes from the fact that with an l_1-penalty we obtain variable selection, in addition to shrinkage: in many cases the most of the coefficients will be set equal to zero. Indeed, if $\mathbf{X}'\mathbf{X}=nI$, then

$$\hat{\beta}_k^{lasso}=\text{sign}(\hat{\beta}_k^{lse})\,(|\hat{\beta}_k^{lse}|-\gamma)_+,$$

for some $\gamma\geq 0$ and $k=1,\ldots,d$, where $(x)_+=\max\{x,0\}$; see Wasserman (2005, Theorem 7.42).

2.1.4 Asymptotic Distributions for Linear Regression

Let us study the regression model

$$Y=\beta'X+\epsilon, \tag{2.35}$$

where $\beta,X\in\mathbf{R}^K$ and $Y,\epsilon\in\mathbf{R}$. We use the notation (2.7)–(2.9), so that the intercept is included in the model by choosing $X_1\equiv 1$. The linear regression model is typically called a parametric model, but strictly speaking it is a semiparametric model, if we do not make a parametric assumption about the distribution of the error term ϵ and about the distribution of the explanatory variables X. We can derive asymptotic distribution theory for the parameter β, without making parametric assumptions about the distribution of ϵ or of X.

When we have identically distributed observations $(X_1, Y_1), \ldots, (X_n, Y_n)$ from the linear regression model (2.35), then the least squares estimator is

$$\hat{\beta} = (\mathbf{X}'\mathbf{X})^{-1}\mathbf{X}'\mathbf{y}, \tag{2.36}$$

where $\mathbf{X} = (X_1, \ldots, X_n)'$ is the $n \times K$-matrix, and $\mathbf{y} = (Y_1, \ldots, Y_n)'$ is the $n \times 1$ vector.

Homoskedasticity and Independence If $(X_1, Y_1), \ldots, (X_n, Y_n)$ are i.i.d. observations from the model (2.35), $E(X\epsilon) = 0$, $E(XX')$ is invertible, and

$$E(\epsilon^2 XX') = \sigma^2 E(XX'), \tag{2.37}$$

where $E\epsilon^2 = \sigma^2$, then it can be shown that

$$\sqrt{n}\left(\hat{\beta} - \beta\right) \xrightarrow{d} N\left(0, \sigma^2[E(XX')]^{-1}\right), \tag{2.38}$$

as $n \to \infty$. Assumption (2.37) is called a homoskedasticity assumption. The asymptotic distribution follows from

$$\sqrt{n}\left(\hat{\beta} - \beta\right) = \left(\frac{1}{n}\sum_{i=1}^{n} X_i X_i'\right)^{-1} \times n^{-1/2}\sum_{i=1}^{n} X_i \epsilon_i. \tag{2.39}$$

By the law of large numbers we have $n^{-1}\sum_{i=1}^{n} X_i X_i' \xrightarrow{p} E(XX')$, and by the central limit theorem we obtain $n^{-1/2}\sum_{i=1}^{n} X_i\epsilon_i \xrightarrow{d} N(0, E(\epsilon^2 XX'))$, as $n \to \infty$. For more details, see Wooldridge (2005, p. 54).

We have to estimate σ^2 and $[E(X'X)]^{-1}$ in order to be able to apply the asymptotic distribution in (2.38). We can use the variance estimate

$$\hat{\sigma}^2 = \frac{1}{n}\sum_{i=1}^{n}(Y_i - \hat{\beta}' X_i)^2. \tag{2.40}$$

The matrix $A^{-1} = [E(X'X)]^{-1}$ can be estimated by $\hat{A}^{-1}$, where

$$\hat{A} = \frac{1}{n}\mathbf{X}'\mathbf{X} = \frac{1}{n}\sum_{i=1}^{n} X_i X_i'. \tag{2.41}$$

The asymptotic distribution of the least squares estimator given in (2.38) can be used to derive asymptotic confidence intervals and asymptotic distributions of test statistics. We apply the asymptotic distribution in Section 2.1.5.

Heteroskedasticity Let us continue the study of the asymptotic distribution of the least squares estimator (2.36) in the linear regression model (2.35).

Assumption (2.37) is interpreted as a homoskedasticity assumption, and we remove this assumption. If $(X_1, Y_1), \ldots, (X_n, Y_n)$ are i.i.d. observations from the model (2.35), $E(X\epsilon) = 0$, and $E(XX')$ is invertible, then

$$\sqrt{n}\left(\hat{\beta} - \beta\right) \xrightarrow{d} N\left(0, A^{-1}BA^{-1}\right), \tag{2.42}$$

as $n \to \infty$, where $A = E(XX')$, and $B = E(\epsilon^2 XX')$. The asymptotic normality in (2.42) follows in the same way as (2.38), but now we do not use the simplification obtained from the homoskedasticity assumption (2.37). The homoskedasticity assumption (2.37) implies that $A^{-1}B = \sigma^2$, so under this assumption we get the simplified asymptotic distribution $N\left(0, \sigma^2 A^{-1}\right)$.

The matrix B is estimated by

$$\hat{B} = \frac{1}{n}\sum_{i=1}^{n}(Y_i - \hat{\beta}' X_i)^2\, X_i X_i'.$$

The matrix A is estimated similarly as in (2.41). The estimator $\hat{A}^{-1}\hat{B}\hat{A}^{-1}$ was suggested in White (1980). For more details, see Wooldridge (2005, p. 55).

Heteroskedasticity and Autocorrelation We study the asymptotic distribution of the least squares estimator (2.36) in the linear regression model (2.35), but now we remove the assumption of independence and allow weak dependence between the consecutive observations.

If $(X_1, Y_1), \ldots, (X_n, Y_n)$ are identically distributed observations from the model (2.35), $E(X\epsilon) = 0$, $E(XX')$ is invertible, and the conditions of the central limit theorem in (1.107) hold for the vector time series $\epsilon_i X_i$, then it can be shown that

$$\sqrt{n}\left(\hat{\beta} - \beta\right) \xrightarrow{d} N\left(0, A^{-1}CA^{-1}\right), \tag{2.43}$$

as $n \to \infty$, where $A = E(XX')$, and

$$C = \sum_{j=-\infty}^{\infty} E\left(\epsilon_t \epsilon_{t+j} X_t X_{t+j}'\right).$$

The asymptotic distribution follows from (2.39) when we note that by the central limit theorem in (1.107), we have

$$n^{-1/2}\sum_{i=1}^{n} X_i \epsilon_i \xrightarrow{d} N(0, C),$$

when $n \to \infty$. Matrix C can be estimated using the method discussed in Section 1.8.3. We apply the estimator (1.110), which gives

$$\hat{C}_{hac} = \hat{\Gamma}(0) + \sum_{j=1}^{n-1} w(j)\left(\hat{\Gamma}(j) + \hat{\Gamma}(j)'\right), \tag{2.44}$$

where

$$\hat{\Gamma}(j) = \frac{1}{n}\sum_{i=1}^{n-j}\left(\hat{\epsilon}_i \hat{\epsilon}_{i+j} X_i X_{i+j}'\right),$$

for $j = 1, \ldots, n-1$, with

$$\hat{\epsilon}_i = Y_i - \hat{\beta}' X_i. \tag{2.45}$$

The weights are defined by $w(j) = K(j/h)$, where $K : \mathbf{R} \to \mathbf{R}$ is a kernel function satisfying $K(x) = K(-x)$, $K(0) = 1$, $|K(x)| \le 1$ for all x, and $K(x) = 0$ for $|x| > 1$. The estimator $\hat{C}_{hac}$ is heteroskedasticity and autocorrelation robust estimator of the asymptotic covariance matrix (HAC estimator), and it was proposed in Newey & West (1987).

The estimator $\hat{C}_{hac}$ in (2.44) is robust both to heteroskedasticity and to autocorrelation. We can easily make more restrictive estimators. For example, if we assume homoskedasticity and zero autocorrelation for the time series of errors ϵ_i, but want to be robust against autocorrelation in the time series of the explanatory variables X_i, then we can define

$$\hat{C}_{ac} = \hat{\sigma}^2 \sum_{j=-(n-1)}^{n-1} w(j)\, \hat{\Gamma}_X(j),$$

where $\hat{\sigma}^2 = n^{-1} \sum_{i=1}^{n} \hat{\epsilon}^2$ with residuals $\hat{\epsilon}_i$ as in (2.45), and

$$\hat{\Gamma}_X(j) = \frac{1}{n} \sum_{i=1}^{n-j} X_i X'_{i+j},$$

for $j = 0, \ldots, n-1$.

Asymptotic Distribution of the GMM Estimator Let us consider the GMM estimator $\hat{\beta}_{gmm}$, defined in (2.26) as

$$\hat{\beta}_{gmm} = (\mathbf{X}'\mathbf{Z}W\mathbf{Z}'\mathbf{X})^{-1}\mathbf{X}'\mathbf{Z}W\mathbf{Z}'\mathbf{y},$$

where $\mathbf{X} = (X_1, \ldots, X_n)'$ is the $n \times K$ matrix, $\mathbf{Z} = (Z_1, \ldots, Z_n)'$ is the $n \times L$ matrix, W is a $L \times L$ matrix of weights, and $\mathbf{y} = (Y_1, \ldots, Y_n)'$ is the $n \times 1$ vector.

If $(X_1, Y_1, Z_1), \ldots, (X_n, Y_n, Z_n)$ are i.i.d. observations, $E(Z\epsilon) = 0$, and the rank of $E(ZX')$ is K, then it can be shown that

$$\sqrt{n}\left(\hat{\beta}_{gmm} - \beta\right) \xrightarrow{d} N(0, V_W), \tag{2.46}$$

as $n \to \infty$, where

$$V_W = (C'WC)^{-1}C'W\Lambda WC(C'WC)^{-1},$$

$C = E(ZX')$, and $\Lambda = E(\epsilon^2 ZZ')$; see Wooldridge (2005, p. 191).

2.1.5 Tests and Confidence Intervals for Linear Regression

We apply the asymptotic distribution in (2.38) to derive tests and confidence intervals.

Hypothesis Testing Let us consider testing the hypotheses

$$H_0 : \beta_k = 0, \qquad H_1 : \beta_k \ne 0,$$

for some $k = 1, \ldots, K$. We can test this null hypothesis with the test statistics

$$T_k = \frac{\hat{\beta}_k - \beta_k}{\hat{\text{sd}}(\hat{\beta}_k)},$$

where

$$\hat{\text{sd}}(\hat{\beta}_k) = \hat{\sigma} \left([\hat{A}^{-1}]_{kk} \right)^{1/2}.$$

Furthermore, $\hat{\sigma}$ is defined by (2.40) and $[\hat{A}^{-1}]_{kk}$ is the element in the kth row and in the kth column of the matrix $\hat{A}^{-1}$, where $\hat{A}$ is defined in (2.41). Under the assumptions for (2.38), we have

$$T_k \xrightarrow{d} N(0,1), \tag{2.47}$$

as $n \to \infty$. For the observed value $|T_k| = t_{obs}$, we calculate the p-value $P(|T_k| > t_{obs}) \approx 2(1 - \Phi(t_{obs}))$, where Φ is the distribution function of the standard normal distribution.

Let us consider testing a more general parameter restrictions. Let R be a $J \times K$ matrix of rank J, where $1 \le J \le K$ and let q be a $J \times 1$ vector. We want to test the hypothesis

$$H_0 : R\beta = q, \qquad H_1 : R\beta \neq q.$$

Let us define the test statistics

$$F = (R\hat{\beta} - q)' \left[\hat{\sigma}^2 R(\mathbf{X}'\mathbf{X})^{-1} R' \right]^{-1} (R\hat{\beta} - q).$$

Now it holds that under the assumptions for (2.38),

$$F \xrightarrow{d} \chi^2(J),$$

as $n \to \infty$, where $\chi^2(J)$ is χ-distribution with degrees of freedom J. For the observed value $F = f_{obs}$, we calculate the p-value $P(F > f_{obs}) \approx (1 - F(f_{obs}))$, where F is the distribution function of the $\chi^2(J)$ distribution.

For example, we obtain the F-test of the null hypothesis

$$H_0 : \beta_2 = \cdots = \beta_K = 0$$

when we choose $R = [0_d \;\, I_{K-1}]$ and $q = 0_{K-1}$, where 0_{K-1} is the $(K-1) \times 1$ vector of zeros and I_{K-1} is the $(K-1) \times (K-1)$ identity matrix.

Confidence Interval Using the asymptotic distribution in (2.47) we can derive an asymptotic confidence interval for β_k, for some $k = 1, \ldots, K$. We have for $0 < \alpha < 1$ that

$$P\left(\hat{\beta}_k - \hat{\text{sd}}(\hat{\beta}_k)\, z_{1-\alpha/2} \le \beta_k \le \hat{\beta}_k + \hat{\text{sd}}(\hat{\beta}_k)\, z_{1-\alpha/2} \right) \longrightarrow 1 - \alpha,$$

as $n \to \infty$, under the assumptions for (2.38), where we use the notation $z_p = \Phi^{-1}(p)$ for the p-quantile of the standard normal distribution.

Confidence Bands Scheffé's confidence band is derived under the assumption

$$Y_i = \beta' X_i + \epsilon_i, \qquad i = 1, \ldots, n,$$

where $\epsilon_i \sim N(0, \sigma^2)$, and $\mathrm{Cov}(\epsilon_i, \epsilon_j) = 0$, when $i \neq j$. Let us denote

$$f(x) = \beta' x, \qquad \hat{f}(x) = \hat{\beta}' x,$$

where $\hat{\beta}$ is the least squares estimator. It holds for $0 < \alpha < 1$ that

$$P(L(x) \leq f(x) \leq U(x) \text{ for all } x) \geq 1 - \alpha,$$

where

$$L(x) = \hat{f}(x) - c\hat{\sigma}\sqrt{x'(\mathbf{X}'\mathbf{X})^{-1}x}, \qquad U(x) = \hat{f}(x) + c\hat{\sigma}\sqrt{x'(\mathbf{X}'\mathbf{X})^{-1}x},$$

with $c = \sqrt{dF_{K,n-K}(1-\alpha)}$, $F_{K,n-K}(1-\alpha)$ is the $1-\alpha$-quantile of the F-distribution with K and $n-K$ degrees of freedom,[16]

$$\hat{\sigma}^2 = \frac{1}{n-K}\sum_{i=1}^{n}(Y_i - \hat{f}(X_i))^2.$$

This confidence band can be found in Scheffé (1959) and Seber (1977, pp. 128–130). As noted by Wasserman (2005), we have

$$\|l(x)\|^2 = \sum_{i=1}^{n} l_i(x)^2 = x'(\mathbf{X}'\mathbf{X})^{-1}x$$

and

$$\mathrm{Var}\left(\hat{f}(x)\right) = \sigma^2 \|l(x)\|^2,$$

where $l_i(x)$ was defined in (2.12), which motivates the result.

2.1.6 Variable Selection

We consider the case where we have a sequence $X_1, \ldots, X_K$ of explanatory variables and we want to choose $k = 1, \ldots, K$ so that the model of the form

$$Y = \beta_1 X_1 + \cdots + \beta_k X_k + \epsilon$$

gives the best predictions. The setting arises naturally in the case of autoregressive time series models, because in these cases there is a natural ordering of the explanatory variables.

[16]Now $P(X > F_{K,n-K}(1-\alpha)) = \alpha$ when $X \sim F_{K,n-K}$.

Let $(X_1, Y_1), \ldots, (X_n, Y_n)$ be regression data, where $X_i = (X_{i,1}, \ldots, X_{i,K})'$, $i = 1, \ldots, n$. Cross validation was defined in (1.116). We can apply cross validation in variable selection by defining

$$\text{MISE}_n(k) = \frac{1}{n} \sum_{i=1}^{n} \left(\hat{\beta}'_{k,-i} X_i^{(k)} - Y_i \right)^2,$$

where $X_i^{(k)} = (X_{i,1}, \ldots, X_{i,k})'$ and $\hat{\beta}_{k,-i}$ is the least squares estimator using the model with the variables $X_1, \ldots, X_k$, and calculated with the data were the ith observation (X_i, Y_i) is removed. To obtain a model with the best predictions, we choose k minimizing $\text{MISE}_n(k)$ among $k = 1, \ldots, K$.

We define next Mallows's C_p criterion and Akaike's criterion for variable selection. The connection of these criterions to the cross validation of the kernel estimator is discussed in Section 3.2.7.

Mallows's C_p criterion is

$$C(k) = \text{SSR}(k) + 2\hat{\sigma}_K^2 k, \tag{2.48}$$

where

$$\text{SSR}(k) = \sum_{i=1}^{n} \left(Y_i - \hat{\beta}'_k X_i^{(k)} \right)^2,$$

$X_i^{(k)} = (X_{i,1}, \ldots, X_{i,k})'$, and

$$\hat{\sigma}_K^2 = \frac{1}{n-K} \sum_{i=1}^{n} \left(Y_i - \hat{\beta}'_K X_i \right)^2,$$

where $\hat{\beta}_k$ is the least squares estimator using the model with the variables $X_1, \ldots, X_k$. The model $X_1, \ldots, X_k$ is chosen which minimizes $C(k)$ over $k = 1, \ldots, K$. The criterion was defined in Mallows (1973). Minimizing Mallows's C_p criterion is equivalent to minimizing an unbiased estimator of the mean averaged squared error, defined in (1.114). See, for example, Ruppert et al. (2003, Section 5.3.3) for the case of fixed design regression.

Akaike's information criterion is

$$\text{AIC}(k) = \log \text{SSR}(k) + 2k/n, \tag{2.49}$$

which was proposed in Akaike (1973). Both the Mallows's C_p and Akaike's information criterion penalize the residual sum of squares with the number of the parameters in the model. These criterions are computationally more attractive than the cross validation, but we show in Section 3.2.7 that in the case of kernel estimators cross validation is not slower to calculate than Mallows's C_p or Akaike's information criterion.

2.1.7 Applications of Linear Regression

Linear regression can be used to describe assets and portfolios. The beta of an asset describes the volatility of the asset with respect to a benchmark, the beta of a portfolio can be used to describe the risk aversion of the investor, and the alpha of a portfolio can be used to measure the performance of the portfolio.

Beta of an Asset We define the beta of an asset by

$$\beta = \frac{\text{Cov}(R_t^{(a)}, R_t^{(b)})}{\text{Var}(R_t^{(b)})},$$

where $R_t^{(a)} = (P_t^{(a)} - P_{t-1}^{(a)})/P_{t-1}^{(a)}$ is the return of the asset, and $R_t^{(b)}$ is the return of the benchmark portfolio. We can see from (2.30) that beta is the regression coefficient in the regression

$$R_t^{(a)} = \alpha + \beta R_t^{(b)} + \epsilon_t.$$

Thus, we can estimate the beta with the historical returns

$$(R_1^{(b)}, R_1^{(a)}), \ldots, (R_T^{(b)}, R_T^{(a)})$$

of the benchmark portfolio and of the asset. The beta of an asset gives information about the volatility of the stock in relation to the volatility of the benchmark. If $\beta < 0$, the asset tends to move in the opposite direction as the benchmark; if $\beta = 0$, the asset is uncorrelated with the benchmark; if $0 < \beta < 1$, the asset tends to move in the same direction as the benchmark but it tends to move less; and if $\beta > 1$, the asset tends to move in the same direction as the benchmark but it tends to move more.

Figure 2.2 shows a linear regression when the daily return of the Daimler is the response variable and the daily return of the DAX 30 index is the explanatory variable. The returns are available for the time period starting at 2000-01-03 and ending at 2013-05-02, making together 3382 daily observations. The estimated regression coefficient is $\hat{\beta} = 1.14$ and the estimated intercept is $\hat{\alpha} = 0.011\%$.

Beta of a Portfolio In the framework of Markowitz theory of portfolio selection it can be shown that the optimal portfolios in the Markowitz sense are a combination of the market portfolio and the risk-free investment.[17] Thus, the returns of the optimal portfolios for the period $t - 1 \mapsto t$ are

$$R_t = (1 - \beta)R_t^F + \beta R_t^M, \tag{2.50}$$

where R_t^F is the return of the risk-free investment and R_t^M is the return of the market portfolio, both returns being for the investment period ending at time t. The

[17]The Markowitz theory of portfolio selection defines the optimal stock portfolios as portfolios maximizing expected return for a given upper bound on the standard deviation of the portfolio return or, equivalently, as portfolios minimizing the standard deviation of the portfolio return for a given lower bound on the expected return of the portfolio. This defines single period portfolio choice, for a given investment period.

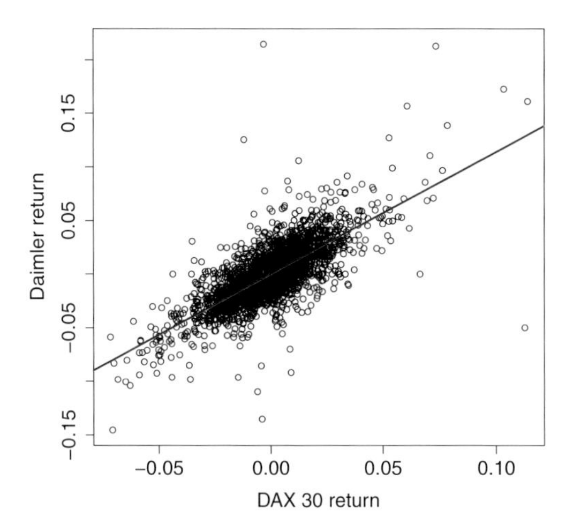

Figure 2.2 *Linear regression of Daimler on DAX 30.* Shown is the estimated linear regression function when the response variable is the return of the Daimler stock and the explanatory variable is the return of the DAX 30 index.

coefficient $\beta \geq 0$ is the proportion invested in the market portfolio. When $0 \leq \beta \leq 1$, then the portfolio is investing available wealth; but if $\beta > 1$, then amount $(\beta - 1)W$ is borrowed and amount $(\beta + 1)W$ is invested in the market portfolio, where W is the investment wealth at the beginning of the period.

The coefficient β is determined by the risk aversion of the investor. For an investor with portfolio returns R_t we do not know the coefficient β, but we obtain from (2.50) that

$$R_t - R_t^F = \beta \left(R_t^M - R_t^F \right).$$

We can collect past returns R_t, $t = 1, \ldots, T$, and use these, together with the past returns R_t^F of the risk-free return and the past returns R_t^M of the market portfolio, to estimate the coefficient β in the linear model

$$R_t - R_t^F = \beta \left(R_t^M - R_t^F \right) + \epsilon_t, \tag{2.51}$$

where ϵ_t is the error term. Now $R_t - R_t^F$ is the response variable and $R_t^M - R_t^F$ is the explanatory variable. The returns R_t^M of the market portfolio are approximated with the returns of a wide market index, like S&P 500 index, Wilshire 5000 index, or DAX 30 index. The risk-free rate R_t^F can be taken to be the rate of return of a government bond.

Alpha of a Portfolio Linear regression can be used to characterize portfolio performance. In (2.51) we have a regression model without a constant term. We extend this model to the model

$$R_t - R_t^F = \alpha + \beta \left(R_t^M - R_t^F \right) + \epsilon_t,$$

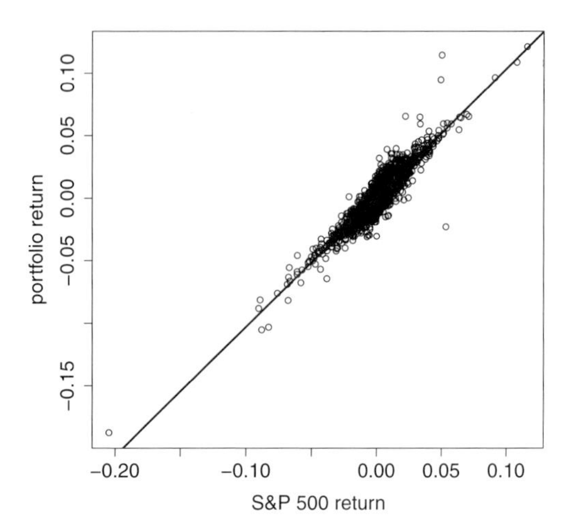

Figure 2.3 *Linear regression of a Markowitz portfolio on S&P 500.* Shown is the estimated linear regression function when the response variable is the return of an actively managed portfolio and the explanatory variable is the return of the S&P 500 index.

where R_t is the return of the actively managed portfolio, R_t^M is the return of the market portfolio, R_t^F is the risk-free rate, and ϵ_t is the error term. The excess return of a market index is chosen as the explanatory variable, and the excess return of the actively managed portfolio is chosen as the response variable. The estimated constant $\hat{\alpha}$ is taken as the measure of the performance, so that larger values of $\hat{\alpha}$ indicate better performance of the portfolio.

Figure 2.3 shows an estimated linear regression function when the daily return of an actively managed portfolio is the response variable and the S&P 500 return is the explanatory variable. The actively managed portfolio is the dynamic Markowitz portfolio whose cumulative wealth is shown in Figure 3.63. The data are the S&P 500 and Nasdaq-100 data described in Section 1.6.2. The estimated constant term is $\hat{\alpha} = 0.034\%$ and the estimated regression coefficient is $\hat{\beta} = 1.03$.

Alpha of a Hedge Fund: Fung–Hsieh Factors In hedge fund performance measurement the alpha is calculated in the linear regression of the hedge fund excess returns on the Fung–Hsieh risk factors. Fung & Hsieh (2004) define seven risk factors: three trend-following risk factors, two equity-oriented risk factors, and two bond-oriented risk factors. The trend-following risk factors are a bond trend-following factor, a currency trend-following factor, and a commodity trend-following factor. The equity-oriented risk factors are the equity market factor[18] and the size

[18]The equity market factor is the S&P 500 index monthly total return.

spread factor.[19] The bond-oriented risk factors are the bond market factor[20] and the credit spread factor.[21] The alpha is now estimated in the multivariate linear regression

$$R_t^H - R_t^F = \alpha + \sum_{k=1}^{7} \beta_k X_t^{(k)} + \epsilon_t,$$

where we denote by R_t^H the hedge fund monthly returns, R_t^F is the monthly risk-free rate, and $X_t^{(k)}$, $k = 1, \ldots, 7$, are the Fung–Hsieh risk factors.

2.2 VARYING COEFFICIENT LINEAR REGRESSION

In the varying coefficient linear regression we approximate the regression function with a linear function of variables X, but we assume that the coefficients are a function of additional variables Z. Thus,

$$E(Y \mid X = x, Z = z) \approx \alpha(z) + \beta(z)'x,$$

where $x \in \mathbf{R}^p$, $z \in \mathbf{R}^d$, $\alpha : \mathbf{R}^d \to \mathbf{R}$, and $\beta : \mathbf{R}^d \to \mathbf{R}^p$. We observe identically distributed observations (X_i, Y_i, Z_i), $i = 1, \ldots, n$, where $X_i \in \mathbf{R}^p$, $Y_i \in \mathbf{R}$, and $Z_i \in \mathbf{R}^d$, and denote the estimator by

$$\hat{f}(x) = \hat{\alpha}(z) + \hat{\beta}(z)'x.$$

For example, it is possible to have $Z_i = X_i$. The varying coefficient model has been studied in Hastie & Tibshirani (1993).

2.2.1 The Weighted Least Squares Estimator

The least squares estimator of functions α and β is defined as

$$(\hat{\alpha}(z), \hat{\beta}(z)) = \text{argmin}_{\alpha \in \mathbf{R}, \beta \in \mathbf{R}^p} \sum_{i=1}^{n} p_i(z) \left(Y_i - \alpha - \beta' X_i\right)^2, \tag{2.52}$$

where weights $p_i(z)$ are such that $p_i(z)$ is large when z is close to Z_i, and $p_i(z)$ is small when z is far from Z_i. We can use the kernel weights

$$p_i(z) = K_h(z - Z_i), \qquad i = 1, \ldots, n,$$

where $K : \mathbf{R}^d \to \mathbf{R}$ is the kernel function, $K_h(z) = K(z/h)/h^d$, and $h > 0$ is the smoothing parameter. Note that we do not have to normalize the weights to sum

[19]The size spread is the Wilshire Small Cap 1750 minus the Wilshire Large Cap 750 monthly return or Russell 2000 index monthly total return minus the S&P 500 monthly total return.

[20]The bond market factor is the monthly change in the 10-year treasury constant maturity yield (month end-to-month end).

[21]The credit spread is the monthly change in the Moody's Baa yield minus the 10-year treasury constant maturity yield (month end-to-month end).

to one, like in the definition of the kernel weights in (3.7). We can also use other weights defined in Chapter 3, like the nearest-neighbor weights.

The least squares estimator can be written as

$$(\hat{\alpha}(z), \hat{\beta}(z)')' = (\mathbf{X}'\mathbf{P}\mathbf{X})^{-1}\mathbf{X}'\mathbf{P}\mathbf{y}, \tag{2.53}$$

where $\mathbf{X}$ is the $n \times (p+1)$-matrix whose ith row is $(1, X_i')$, $\mathbf{y} = (Y_1, \ldots, Y_n)'$ is the $n \times 1$ column vector, and $\mathbf{P}$ is the $n \times n$ diagonal matrix with the diagonal elements $p_i(z)$, $i = 1, \ldots, n$. The solution (2.53) is derived similarly as the solution of the usual least squares regression in (2.10). We denote for shortness $\gamma = (\alpha, \beta')'$, and write the least squares criterion in (2.52) with the matrix notation as

$$(\mathbf{y} - \mathbf{X}\gamma)'\mathbf{P}(\mathbf{y} - \mathbf{X}\gamma) = \mathbf{y}'\mathbf{P}\mathbf{y} - 2\gamma'\mathbf{X}'\mathbf{P}\mathbf{y} + \gamma'\mathbf{X}'\mathbf{P}\mathbf{X}\gamma.$$

Derivating this with respect to γ, and setting the gradient to zero, we get the equations

$$\mathbf{X}'\mathbf{P}\mathbf{X}\,\gamma = \mathbf{X}'\mathbf{P}\mathbf{y},$$

which leads to the solution (2.53). In the case $p = 1$ we get

$$\beta(z) = \frac{\sum_{i=1}^n p_i(z)(X_i - \bar{X})(Y_i - \bar{Y})}{\sum_{i=1}^n p_i(z)(X_i - \bar{X})^2}, \qquad \alpha(z) = \bar{Y} - \beta(z)\bar{X}, \tag{2.54}$$

where

$$\bar{X} = \sum_{i=1}^n p_i(z)\, X_i, \qquad \bar{Y} = \sum_{i=1}^n p_i(z)\, Y_i.$$

We can write the weighted least squares estimator as

$$\hat{f}(x, z) = \sum_{i=1}^n l_i(x, z)\, Y_i,$$

where

$$l_i(x, z) = X_i' \left[\mathbf{X}'\mathbf{P}\mathbf{X}\right]^{-1} \mathbf{x} p_i(z),$$

where $\mathbf{x} = (1, x_1, \ldots, x_p)'$.

Figure 2.4 illustrates the weights $l_i(x, z)$ in the case of one-dimensional explanatory variables $X = Z$. We have used the standard Gaussian kernel and smoothing parameter $h = 0.7$ in the weights $p_i(z)$. Panel (a) shows a perspective plot of the function $(x, X_i) \mapsto l_i(x, x)$, where $X_1, \ldots, X_n$ is a simulated sample of size $n = 200$ from the uniform distribution on $[-1, 1]$. Panel (b) shows the six functions $x \mapsto l_i(x, x)$ for the choices $X_i = -1, -0.5, \ldots, 1$.

2.2.2 Applications of Varying Coefficient Regression

Many applications of varying coefficient regression arise in the cases where we are restricted to make an approximation of the response variable Y using a linear

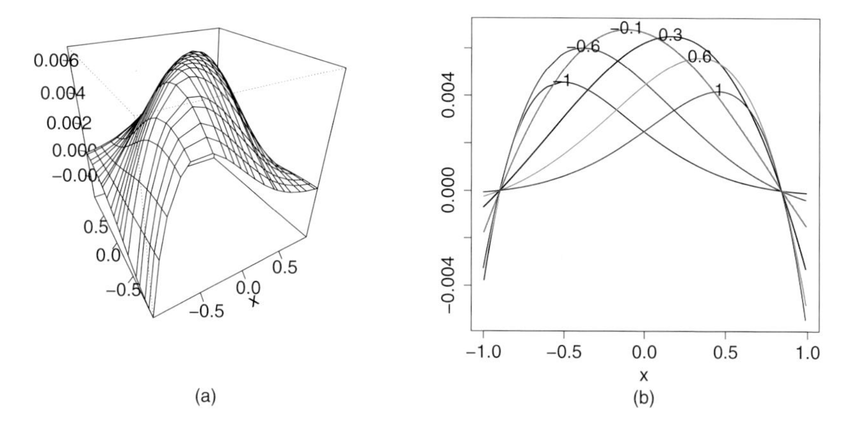

Figure 2.4 *Weights in varying coefficient linear regression.* (a) The function $(x, X_i) \mapsto l_i(x, x)$. (b) The six slices $x \mapsto l_i(x, x)$ for the choices $X_i = -1, -0.5, \ldots, 1$.

combination. In these cases the only way to introduce nonlinearity is through the nonlinearity of the coeffiecients. For example, a portfolio is always a linear combination of tradable assets, but nonlinearity can be introduced by making the coefficients a nonlinear function.

We mention below two applications of varying coefficient regression: hedge fund index replication and performance measurement with the conditional alpha.

Replication of a Hedge Fund Index We have available monthly returns $Y_1, \ldots, Y_T \in \mathbf{R}$ of a hedge fund index and we want to replicate this time series using the returns $X_1, \ldots, X_T \in \mathbf{R}^p$ of tradable assets. This gives us a new hedge fund that is cheaper to create and more liquid than a typical hedge fund. However, the returns of the new hedge fund can be similar as the returns of the hedge fund index. The replication is done conditionally on the information contained in the observations $Z_1, \ldots, Z_T \in \mathbf{R}^d$.

The replication is made using weights b_T chosen at time T, and holding the weights until the returns at time $T+1$ are realized. The replication is

$$\hat{Y}_{T+1} = b_T' X_{T+1},$$

where the portfolio vector $b_T \in \mathbf{R}^p$ satisfies $\sum_{i=1}^p b_{T,i} = 1$. We can also write

$$\hat{Y}_{T+1} = b_T(Z_T)' X_{T+1},$$

where Z_T is the information used to choose the weights. The weights b_T are chosen as

$$b_T = \text{argmin}_{b \in S_N} \sum_{t=t_0}^{T} (Y_t - b' X_t)^2 \, p_{t-1}(Z_T), \tag{2.55}$$

where $S_N = \{s \in \mathbf{R}^p : \sum_{i=1}^p s_i = 1\}$. We use the kernel weights

$$p_{t-1}(Z_T) = K_h(Z_T - Z_{t-1}),$$

where $K_h(x) = K(x/h)/h^d$ is the scaled kernel function, $K : \mathbf{R}^d \to \mathbf{R}$ is the kernel function, and $h > 0$ is the smoothing parameter. Now the weight $p_{t-1}(Z_T)$ is large at those time points $t-1$ that are such that the relevant information Z_{t-1}, available at that time, is close to the current relevant information Z_T.

As an example we use the index of long/short equity hedge funds from the CSFB database. The tradable assets which we use to replicate the CSFB long/short hedge fund index are the S&P 500 stock index, the spread between the Russell 2000 small cap stock index and the Russell 1000 large cap stock index, and the one-month risk-free investment.[22] We take the risk-free rate equal to zero (on the other hand we do not include transaction costs). Thus, $X_t = (X_t^1, X_t^2, X_t^3)$, $p = 3$, and

$$X_t^1 = \frac{\mathrm{SP500}_t}{\mathrm{SP500}_{t-1}}, \quad X_t^2 = 1 + \frac{\mathrm{SC}_t^R}{\mathrm{SC}_{t-1}^R} - \frac{\mathrm{LC}_t^R}{\mathrm{LC}_{t-1}^R}, \quad X_t^3 = 1,$$

where $\mathrm{SP500}_t$ is the price of the S&P 500 index at time t, SC_t^R is the price of the Russell 2000 small cap index at time t, and LC_t^R is the price of the Russell 1000 large cap index at time t. The conditioning variables Z_t are chosen to be a function of lagged values of X_t:

$$Z_t = (X_t, X_{t-1}, \ldots, X_{t-k+1}), \tag{2.56}$$

where $k \geq 1$ is the number of lags. Thus $Z_t \in \mathbf{R}^d$, where $d = pk$. We also used the copula transform, as defined in Section 1.7.2, to normalize the marginals of Z_t to have approximately standard normal distribution. We chose $t_0 = k + 1$ in (2.55) and used the smoothing parameter $h = 1$ and the autoregression parameter $k = 8$.

We use the linear replication as a benchmark. The linear replication is defined by the weights

$$b_T^{lin} = \mathrm{argmin}_{b \in S_N} \sum_{t=1}^{T} (Y_t - b' X_t)^2. \tag{2.57}$$

Figure 2.5 shows the wealth processes of kernel replication (blue), linear replication (green), long/short hedge fund index (red), and S&P 500 index (black), starting with value 1 at 31 January 1994 and ending at 29 January 2010. The wealth process of the replicating time series with weights b_t is defined as

$$W_1 = 1, \qquad W_{t+1} = W_t \cdot (1 + b_t' X_{t+1}).$$

For the kernel strategy, as defined in (2.55), the annualized mean return is 7.0% and the annualized standard deviation of returns is 10.4%.[23] The Sharpe ratio for

[22]We thank Juha Joenväärä for providing the hedge fund data. The stock index data is obtained from Yahoo.

[23]The annualized mean is defined as 12 times the average of the monthly returns, and the annualized standard deviation is defined as $\sqrt{12}$ times the sample standard deviation of the monthly returns.

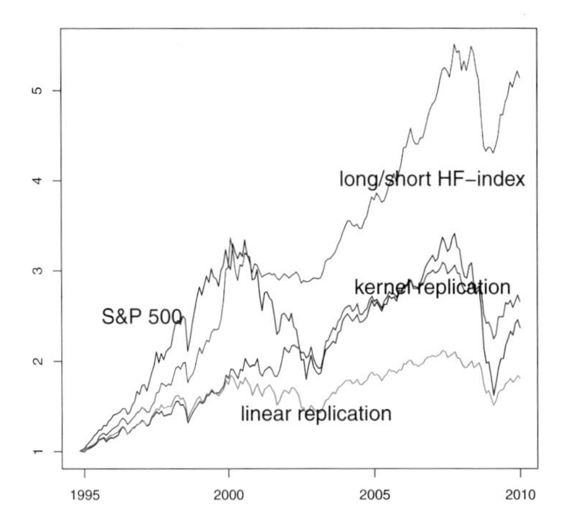

Figure 2.5 *Hedge fund index replication.* Wealth processes of kernel replication (blue), linear replication (green), long/short hedge fund index (red), and S&P 500 index (black), starting with value 1 at 31 January 1994 and ending at 29 January 2010.

the kernel strategy is 0.67. The Sharpe ratio is defined in (1.88) as the ratio of the annualized mean of the returns to the annualized standard deviation of the returns. We do not use the excess returns.

As a comparison, the annualized mean of the monthly S&P 500 index returns during our investigation period is 6.4%, the annualized standard deviation is 15.5%, and the Sharpe ratio is 0.41. For the CSFB long/short hedge fund index the annualized mean return is 10.2%, the annualized standard deviation is 10.0%, and the Sharpe ratio is 1.02. For the dynamic linear replication, as defined in (2.57), the annualized mean return is 4.2%, the annualized standard deviation is 10.2%, and the Sharpe ratio is 0.42.

We conclude that the kernel strategy seems to give a better performance than the linear strategy, but the kernel strategy is worse than the target hedge fund index. On the other hand, the Sharpe ratio of the kernel strategy is better than the Sharpe ratio of the S&P 500 index.

Conditional Alpha In Section 2.1.7 we applied a linear model to the evaluation of portfolio performance. The performance was measured by the estimate $\hat{\alpha}$ of the constant term α of linear regression. We can use varying coefficient regression to estimate conditional alpha. It has been argued that the conditional alpha measures better hedge fund performance, since hedge funds do not use long only strategies but apply short selling and buying and writing of options.

We choose a collection of risk factors $X_t^1, \ldots, X_t^p$ and make a linear regression of hedge fund return Y_t on these risk factors. The unconditional alpha is defined as

$$\hat{\alpha} = \text{argmin}_\alpha \min_{\beta_1, \ldots, \beta_p} \sum_{t=1}^{T} \left(Y_t - \alpha - \beta_1 X_t^1 - \cdots - \beta_p X_t^p\right)^2.$$

The conditional alpha, conditionally on the information $Z_t \in \mathbf{R}^d$ at time t, is defined as

$$\hat{\alpha}\left(Z_{t_0}\right) = \text{argmin}_\alpha \min_{\beta_1, \ldots, \beta_p} \sum_{t=1}^{T} \left(Y_t - \alpha - \beta_1 X_t^1 - \cdots - \beta_p X_t^p\right)^2 p_t(Z_{t_0}),$$

where

$$p_t(Z_{t_0}) = K_h(Z_t - Z_{t_0}),$$

where $K_h(x) = K(x/h)/h^d$ is the scaled kernel function, $K : \mathbf{R}^d \to \mathbf{R}$ is the kernel function, and $h > 0$ is the smoothing parameter.

2.3 GENERALIZED LINEAR AND RELATED MODELS

In a generalized linear model the assumption

$$E(Y \,|\, X = x) = G(\alpha + \beta' x),$$

is made, where $Y \in \mathbf{R}$, $X \in \mathbf{R}^d$, $\alpha \in \mathbf{R}$ and $\beta \in \mathbf{R}^d$ are unknown parameters, and $G : \mathbf{R} \to \mathbf{R}$ is a known link function. A generalized linear model generalizes the linear model $E(Y \,|\, X = x) = \alpha + \beta' X$. The link function G can introduce nonlinearity to the model. In Section 4.1 we consider single-index models, where the link function is unknown and is estimated using the data.

2.3.1 Generalized Linear Models

Let us have regression data $(X_1, Y_1), \ldots (X_n, Y_n)$. We denote the estimator of $f(x) = E(Y \,|\, X = x)$ in the generalized linear model

$$\hat{f}(x) = G(\hat{\alpha} + x'\hat{\beta}), \qquad x \in \mathbf{R}^d, \tag{2.58}$$

where $\hat{\alpha} \in \mathbf{R}$ and $\hat{\beta} \in \mathbf{R}^d$ are estimated using the regression data, and $G : \mathbf{R} \to \mathbf{R}$ is a known function. The function G is often called called a link function, but sometimes its inverse G^{-1} is called the link function, and it is assumed that G is a monotone function.

The term "generalized linear model" was introduced by Nelder & Wedderburn (1972). Generalized linear models are treated extensively in McCullagh & Nelder (1989).

Generalized Linear Models and Exponential Families In (1.68) we defined a one-parameter canonical exponential family, where the density functions are

$$p(y, \theta) = p(y) \exp\{y\theta - b(\theta)\}, \tag{2.59}$$

where $y, \theta \in \mathbf{R}$ and $B : \mathbf{R} \to \mathbf{R}$. In (1.70) we showed that if Y follows $p(\cdot, \theta)$, then

$$EY = b'(\theta),$$

where b' is the derivative of d. Thus, under the assumption that the response variable follows exponential family density $p(\cdot, \theta)$, the natural link function is $G = b'$, and we have the generalized linear model

$$E(Y \mid X = x) = b'(\alpha + \beta' x). \tag{2.60}$$

In (1.71) we showed that if Y follows $p(\cdot, \theta)$, then

$$\text{Var}(Y) = b''(\theta).$$

Thus, the natural link function for variance regression is $G = b''$, and we have the generalized linear model for variance regression

$$\text{Var}(Y \mid X = x) = b''(\alpha + \beta' x).$$

The general exponential family regression was introduced in Section 1.3.2. A special case of that modeling approach is obtained if we model the conditional distribution of Y by

$$Y \mid X = x \sim p(y, G(\alpha + \beta' x)).$$

Estimation As usual, the least squares criterion can be used to estimate the parameters: we minimize

$$\sum_{i=1}^{n} (Y_i - G(\alpha + \beta' X_i))^2 \tag{2.61}$$

over $\alpha \in \mathbf{R}$ and $\beta \in \mathbf{R}^d$.

Alternatively, maximum likelihood estimation can be used. Let us make the assumption

$$f_{Y \mid X=x}(y, v) = p(y) \exp\{y\theta(x) - b(\theta(x))\}, \qquad y \in \mathbf{R}, \tag{2.62}$$

where $x \in \mathbf{R}^d$, and $\theta(x) = \alpha + \beta' x$. If $(Y_1, X_1), \ldots, (Y_n, X_n)$ are independent, then the density of the observations is

$$\prod_{i=1}^{n} f_{Y_i, X_i}(y_i, x_i) = \prod_{i=1}^{n} f_{Y_i \mid X_i = x_i}(y_i) \, f_{X_i}(x_i), \tag{2.63}$$

and, using (2.62), we see that to maximize the likelihood (the density function or the probability mass function of the observations), we need to maximize

$$\sum_{i=1}^{n} [Y_i \theta(X_i) - b(\theta(X_i))] = \sum_{i=1}^{n} [Y_i(\alpha + \beta' X_i) - b(\alpha + \beta' X_i)]$$

over $\alpha \in \mathbf{R}$ and $\beta \in \mathbf{R}^d$.

2.3.2 Binary Response Models

Binary response models were introduced in Section 1.2.1. In a binary response model the response variable Y takes only values 0 and 1, and we write $Y \sim \text{Bernoulli}(p)$, where $0 \le p \le 1$. For a Bernoulli distributed random variable Y, it holds that $EY = P(Y = 1) = p$. We also have

$$E[Y \,|\, X = x] = P(Y = 1 \,|\, X = x).$$

Let us denote

$$p(x) = E[Y \,|\, X = x],$$

where $p : \mathbf{R}^d \to [0, 1]$. In binary response models we want to estimate the function p. We get a generalized linear model when we have

$$p(x) = G(\alpha + \beta' x), \tag{2.64}$$

where $G : \mathbf{R} \to [0, 1]$ is a known link function and $\alpha \in \mathbf{R}$ and $\beta \in \mathbf{R}^d$ are unknown parameters. In the probit model, G is the distribution function of the standard normal distribution and in the logit model the distribution function is the distribution function of the logistic distribution. In the *linear probability model* we take

$$p(x) = \alpha + \beta' x,$$

so that in the linear probability model the probability could be negative or larger than one.

Binary Response Model as an Exponential Model The probability mass function of a Bernoulli distributed random variable Y is

$$f_Y(y) = p^y (1 - p)^{1-y}, \qquad y \in \{0, 1\}.$$

We can write the probability mass function as

$$f_Y(y) = \exp\left\{ y \log\left(\frac{p}{1-p}\right) + \log(1 - p) \right\} = \exp\{y\theta - b(\theta)\},$$

where

$$\theta = \log\left(\frac{p}{1-p}\right) \quad \Leftrightarrow \quad p = \frac{e^\theta}{1 + e^\theta}$$

and

$$b(\theta) = \log\left(1 + e^\theta\right) = -\log(1 - p).$$

We have written the probability mass function as in (2.59). Now,

$$b'(\theta) = \frac{e^\theta}{1 + e^\theta},$$

so that b' is the distribution function of the logistic distribution. Using (2.60), we get the model

$$p(x) = b'(\alpha + \beta' x).$$

Latent Variable Approach We can obtain probit and logit models using the latent variable approach. In the latent variable approach the model for Y is

$$Y = \begin{cases} 1, & \text{if } \alpha + \beta' X + \epsilon > 0, \\ 0, & \text{if } \alpha + \beta' X + \epsilon \le 0, \end{cases} \tag{2.65}$$

where ϵ is an error term. This is called the latent variable approach because $Y^* = \alpha + \beta' X + \epsilon$ is the latent variable that is not observed. Then, the regression function is

$$\begin{aligned} E(Y \mid X = x) &= P(Y = 1 \mid X = x) \\ &= P(\alpha + \beta' x + \epsilon > 0) \\ &= 1 - P(\epsilon \le -(\alpha + x'\beta)) \\ &= 1 - F_\epsilon(-(\alpha + x'\beta)), \end{aligned}$$

where F_ϵ is the distribution function of ϵ. If the distribution of ϵ is symmetric around 0, then $F_\epsilon(t) = 1 - F_\epsilon(-t)$ and we get

$$E(Y \mid X = x) = F_\epsilon(\alpha + x'\beta).$$

Thus, the generalized linear model in (2.64) is obtained with the link function F_ϵ being the distribution function of the error distribution:

$$p(x) = F_\epsilon(\alpha + x'\beta).$$

The typical examples are the case where ϵ has the standard Gaussian distribution and the case where ϵ has the logistic distribution.

1. *Probit Model* In the probit model the distribution function of the error distribution is the standard Gaussian distribution function defined as

$$F_\epsilon(t) = \int_{-\infty}^{t} \phi(u)\, du, \qquad t \in \mathbf{R},$$

where

$$\phi(u) = \frac{1}{\sqrt{2\pi}} \exp\left\{ -\frac{1}{2} u^2 \right\}, \qquad u \in \mathbf{R}.$$

The inverse of the standard Gaussian distribution function is called the probit function. The generalized linear model with binary responses and the standard Gaussian distribution function as the link function is called the *probit model.*

2. *Logit Model* In the logit model the distribution function of the error distribution is the logistic distribution function defined as

$$F_\epsilon(t) = \frac{1}{1 + e^{-t}}, \qquad t \in \mathbf{R}.$$

The variance of the logistic distribution is $\pi^2/3$, so the distribution function of the standardized logistic distribution is $1/(1 + e^{-t/(\pi/\sqrt{3})})$. The standardized logistic distribution is close to the standard Gaussian distribution, but the tails of the logistic distribution are fatter than the tails of the Gaussian distribution. The inverse of the logistic distribution function is called the logit function and we have

$$\text{logit}(p) = \log\left(\frac{p}{1-p}\right), \qquad p \in (0,1).$$

The generalized linear model with binary responses and the logistic distribution function as the link function is called the *logit model*.

We can also write a model for Y as

$$Y = \begin{cases} 1, & \text{if } \beta_1' X_1 + \epsilon_1 > \beta_2' X_2 + \epsilon_2, \\ 0, & \text{otherwise,} \end{cases}$$

where ϵ_1 and ϵ_2 are independent error terms. When ϵ_1 and ϵ_2 are normal random variables, then we get the probit model. When ϵ_1 and ϵ_2 are distributed according to the Gumbel distribution (Type 1 extreme value distribution), then $\epsilon_1 - \epsilon_2$ is distributed according to the logistic distribution and we get the logit model. The distribution function of the Gumbel distribution is $F(y) = \exp(-\exp(-y))$. This modeling approach is called the *random utility approach*.

In the Tobit model

$$Y = \max\{0, \beta' X + \epsilon\} = \begin{cases} \beta' X + \epsilon, & \text{if } \beta' X + \epsilon > 0, \\ 0, & \text{if } \beta' X + \epsilon \le 0, \end{cases} \tag{2.66}$$

where ϵ is an unobserved zero mean error term, assumed to be normally distributed and independent of X.

Estimation in the Binary Response Model Let us consider estimation in the binary choice models. We can use the least squares criterion, defined in (2.61), to estimate the parameters. Let us consider some details of maximum likelihood estimation. Assume that we have an i.i.d. sequence $(Y_1, X_1), \ldots, (Y_n, X_n)$ from a binary choice model. The conditional probability mass function is

$$f_{Y \mid X=x}(y) = p(x)^y (1 - p(x))^{1-y}, \qquad y \in \{0,1\}, \; x \in \mathbf{R}^d,$$

where $p : \mathbf{R}^d \to [0,1]$. Under the generalized linear model, we have $p(x) = G(\alpha + \beta' x)$. Using (2.63), we see that to maximize the likelihood, we need to maximize

$$\prod_{i=1}^{n} G(\alpha + \beta' X_i)^{Y_i} (1 - G(\alpha + \beta' X_i))^{1-Y_i},$$

over $\alpha \in \mathbf{R}$ and $\beta \in \mathbf{R}^d$. In probit and logit models we need to maximize

$$\prod_{i=1}^{n} F_\epsilon\left(w_i(\alpha + \beta' x_i)\right),$$

where $w_i = 2y_i - 1$, so that $w_i = 1$ when $y_i = 1$ and $w_i = -1$ when $y_i = 0$. This holds because $F_\epsilon(t) = 1 - F_\epsilon(-t)$, and thus $1 - G(\alpha + \beta' x_i) = G(-\alpha - \beta' x_i)$.

2.3.3 Growth Models

Consider the model

$$Y = \prod_{i=1}^{d} X_i^{\beta_i} + \epsilon,$$

where ϵ is an additive error term with $E(\epsilon \mid X = x) = 0$. We can write the regression function as

$$E(Y \mid X = x) = \exp\left\{ \sum_{i=1}^{d} \beta_i \log_e(X_i) \right\}.$$

Thus we have a generalized linear model where the link function is $G(t) = e^t$. Note that the model

$$Y = \prod_{i=1}^{d} X_i^{\beta_i} \cdot \epsilon,$$

where ϵ is an multiplicative error with $E(\log_e \epsilon \mid X = x) = 0$, can be transformed to a model on $\log_e Y$, and then the regression function is

$$E(\log_e(Y) \mid X = x) = \sum_{i=1}^{d} \beta_i \log_e(X_i).$$

Thus we have obtained a linear model on the transformed variables.

2.4 SERIES ESTIMATORS

We call a series estimator of the regression function any estimate of the type

$$\hat{f}(x) = \sum_{k=1}^{K} \hat{w}_k g_k(x), \qquad x \in \mathbf{R}^d,$$

where $g_k : \mathbf{R}^d \to \mathbf{R}$ are suitable functions and $\hat{w}_k \in \mathbf{R}$ are weights that are determined using the regression data $(X_1, Y_1), \ldots, (X_n, Y_n)$.

2.4.1 Least Squares Series Estimator

Given a sequence $g_1, \ldots, g_K$ of functions $\mathbf{R}^d \to \mathbf{R}$, we can find a series estimator by solving the weights $\hat{w}_1, \ldots, \hat{w}_K$ as the minimizers of the least squares functional

$$\sum_{i=1}^{n} \left(Y_i - \sum_{k=1}^{K} w_k g_k(X_i) \right)^2, \tag{2.67}$$

where $(X_1, Y_1), \ldots, (X_n, Y_n)$ are regression data. The solution can be found in matrix notation from (2.10): The weights $\hat{w} = (\hat{w}_1, \ldots, \hat{w}_K)'$ can be written as

$$\hat{w} = (\mathbf{G}'\mathbf{G})^{-1}\mathbf{G}'\mathbf{y}, \tag{2.68}$$

where $\mathbf{G}$ is $n \times K$ matrix with elements $[\mathbf{G}]_{ik} = g_k(X_i)$ and $\mathbf{y} = (Y_1, \ldots, Y_n)'$ is a column vector of length n.

Using (2.11), we can write the least squares series estimator as

$$\hat{f}(x) = \sum_{i=1}^{n} l_i(x)\, Y_i,$$

where

$$(l_1(x), \ldots, l_n(x)) = G(x)'(\mathbf{G}'\mathbf{G})^{-1}\mathbf{G}',$$

with $G(x) = (g_1(x), \ldots, g_K(x))'$.

Analogously to the ridge regression, we can replace the least squares criterion in (2.67) with the penalized least squares criterion

$$\sum_{i=1}^{n} \left(Y_i - \sum_{k=1}^{K} w_k g_k(X_i) \right)^2 + \lambda \sum_{k=1}^{K} w_k^2,$$

where $\lambda \geq 0$ is the penalization parameter. The coefficient vector of the series ridge regression estimate, minimizing the penalized least squares criterion, is $\hat{w} = (\hat{w}_1, \ldots, \hat{w}_K)'$, defined by

$$\hat{w} = (\mathbf{G}'\mathbf{G} + \lambda I)^{-1}\mathbf{G}'\mathbf{y},$$

where G and $\mathbf{y}$ are as in (2.68), and I is the $K \times K$ identity matrix.

Section 2.4.2 considers using the basis functions of an orthonormal basis, but we can consider also nonorthogonal basis functions. For example, we can use non-normalized Gaussian density functions

$$g_k(x) = \exp\left\{ -\frac{\|x - \mu_k\|^2}{2\sigma_k^2} \right\},$$

where $\mu_k \in \mathbf{R}^d$ and $\sigma_k > 0$.

2.4.2 Orthonormal Basis Estimator

We assume now that $\{g_k\}_{k=1,2,\ldots}$ is an orthonormal basis and that $g_1(x), \ldots, g_K(x)$ is a finite subset of this basis. The basis can be the Fourier basis or a wavelet basis, for example. We define the estimator of the regression function $f : \mathbf{R}^d \to \mathbf{R}$ in three steps, using regression data $(X_1, Y_1), \ldots, (X_n, Y_n)$. First we define the estimator of the function $g = f \cdot f_X$, where $f_X : \mathbf{R}^d \to \mathbf{R}$ is the density function of X. The estimator of g is

$$\hat{g}(x) = \sum_{k=1}^{K} \hat{w}_k g_k(x), \qquad x \in \mathbf{R}^d, \tag{2.69}$$

where

$$\hat{w}_k = \frac{1}{n} \sum_{i=1}^{n} Y_i\, g_k(X_i).$$

Second we define the estimator of the density function f_X of X as

$$\hat{f}_X(x) = \sum_{k=1}^{K} \hat{\theta}_k g_k(x), \qquad x \in \mathbf{R}^d, \tag{2.70}$$

where

$$\hat{\theta}_k = \frac{1}{n} \sum_{i=1}^{n} g_k(X_i).$$

Finally, the estimator of the regression function f is

$$\hat{f}(x) = \frac{\hat{g}(x)}{\hat{f}_X(x)}, \qquad x \in \mathbf{R}^d. \tag{2.71}$$

Note that the linear projection of $g(x)$ on the subspace spanned by $g_1(x), \ldots, g_K(x)$ is the function

$$\tilde{g}(x) = \sum_{k=1}^{K} w_k g_k(x),$$

where

$$w_k = \int_{\mathbf{R}^d} g(x) g_k(x)\, dx = \int_{\mathbf{R}^d} f(x) g_k(x)\, dP_X(x) = E_X f(X) g_k(X).$$

Thus, the coefficient w_k can naturally be estimated with the arithmetic mean $\hat{w}_k$. Also, the linear projection of $f_X(x)$ on the subspace spanned by $g_1(x), \ldots, g_K(x)$ is the function

$$\tilde{f}_X(x) = \sum_{k=1}^{K} \theta_k g_k(x),$$

where

$$\theta_k = \int_{\mathbf{R}^d} f_X(x) g_k(x)\, dx = E_X g_k(X).$$

Thus, the coefficient θ_k can naturally be estimated with the arithmetic mean $\hat{\theta}_k$.

We have defined the regression function estimator in (2.71) for the case of random design regression. In the case of one-dimensional fixed design regression, there is no need to estimate the density of the design distribution as in (2.70). Instead, we assume that the fixed design points $x_1, \ldots, x_n \in \mathbf{R}$ are obtained with $x_i = F^{-1}(z_i)$, where $z_1, \ldots, z_n \in [0, 1]$ is a regular uniform grid and F is a distribution function of a continuous distribution. Then we replace the estimate $\hat{f}_X$ of the design density with the density F'.

We can write the regression function estimator $\hat{f}$, defined in (2.71), in a linear form (1.2) as

$$\hat{f}(x) = \sum_{i=1}^{n} l_i(x)\, Y_i,$$

where

$$l_i(x) = \frac{1}{n\hat{f}_X(x)} \sum_{k=1}^{K} g_k(X_i) g_k(x).$$

If thresholding is applied to the regression function estimator in (2.71), then linearity can be lost. For example, in hard thresholding the regression estimator $\hat{g}$ in (2.69) is replaced by

$$\hat{g}_{hard}(x) = \sum_{k=1}^{K} I_{\{k:\,|\hat{w}_k| \geq \lambda\}}(k)\, \hat{w}_k g_k(x), \tag{2.72}$$

where $\lambda > 0$ is a threshold.

2.4.3 Splines

We restrict now to the case of one explanatory variable, so that $d = 1$. We define the estimator

$$\hat{f}(x) = \sum_{j=1}^{K} \hat{\beta}_j g_j(x), \tag{2.73}$$

where $\hat{\beta}_j$ minimize the least squares criterion

$$\sum_{i=1}^{n} (Y_i - f(X_i; \beta))^2 + \lambda \int_{-\infty}^{\infty} \left(\frac{\partial^2}{\partial x^2} f(x; \beta) \right)^2 dx, \tag{2.74}$$

where

$$f(x; \beta) = \sum_{j=1}^{K} \beta_j g_j(x),$$

$\beta = (\beta_1, \ldots, \beta_K)'$, and $\lambda \geq 0$ is the penalization parameter. Note that the penalized least squares criterion in (2.74) can be written with the matrix notation as

$$(\mathbf{y} - G\beta)'(\mathbf{y} - G\beta) + \lambda \beta' \Omega \beta, \tag{2.75}$$

where $\mathbf{y}$ is the $n \times 1$ vector with elements $[\mathbf{y}]_i = Y_i$, G is the $n \times K$ matrix with elements $[G]_{ij} = g_j(X_i)$, and Ω is the $K \times K$ matrix with elements $[\Omega]_{jk} = \int g_j''(x) g_k''(x)\, dx$. We can find the minimizer of (2.75) in the similar way as the least squares estimator was derived in (2.10). Indeed,

$$(\mathbf{y} - G\beta)'(\mathbf{y} - G\beta) + \lambda \beta' \Omega \beta = \mathbf{y}'\mathbf{y} - 2\beta' B' \mathbf{y} + \beta'(G'G + \lambda\Omega)\beta.$$

Derivating this with respect to β, and setting the gradient to zero leads to the equations

$$(G'G + \lambda\Omega)\beta = G'\mathbf{y}$$

and we get the solution

$$\hat{\beta} = (G'G + \lambda\Omega)^{-1} G'\mathbf{y}.$$

The penalization estimator minimizing (2.74) can be used when the collection $g_1, \ldots, g_K$ is a spline basis. We define two bases.

The first basis is the truncated power basis. Let $k_0 < \cdots < k_{L-1}$ be given knot points in interval $(0, 1)$. Denote $B_j(x) = x^j$, for $j = 0, \ldots, m-2$, and $B_j(x) = (x - k_{j-m+1})_+^{m-1}$, for $j = m-1, \ldots, m+L-2$. This is collection of $m + L - 1$ functions. When $m = 3$, we call the collection a cubic spline basis. The truncated power basis is a basis for the collection of m-splines. This collection consists of functions on the interval $(0, 1)$, that are continuous, that are $m - 1$ order polynomial on intervals $(k_0, k_1), \ldots, (k_{L-2}, k_{L-1})$, and have continuous $m-2$ order derivatives at the knots.

The second basis is the B-spline basis. Korostelev & Korosteleva (2010, Chapter 11) contains a detailed description of B-splines, when the knots are equally spaced. A standard B-spline of order m is defined recursively.

1. Let $S_1(u) = I_{[0,1]}(u)$.
2. Let $S_m(u) = \int_{-\infty}^{\infty} S_{m-1}(z) I_{[0,1)}(u - z)\, dz$, for $m = 2, 3, \ldots$.

Now S_m is the probability density function of a sum of m independent random variables uniformly distributed on $[0, 1]$. Let $m \geq 2$. The spline of order m is defined by

$$\gamma_k(x) = h^m S_m\left(\frac{x - 2hk}{2h}\right) I_{[0,1]}(x),$$

where $k = -m+1, \ldots, Q-1$, $Q = 1/(2h)$, and $h > 0$ is such that $Q \geq 1$ is integer. Korostelev & Korosteleva (2010, Lemma 11.9) state that the set of functions γ_k, $k = -m+1, \ldots, Q-1$, forms a basis in the linear subspace of the piecewise polynomials of order $m - 1$ that are defined in bins B_q,

$$B_q = [2(q-1)h, 2qh), \qquad q = 1, \ldots, Q,$$

and have continuous derivatives up to order $m - 2$.

Figure 2.6(a) shows B-splines γ_k of order $m = 2$, for $k = -1, \ldots, 4$ and Figure 2.6(b) B-splines γ_k of order $m = 3$, for $k = -2, \ldots, 4$.

2.5 CONDITIONAL VARIANCE AND ARCH MODELS

Section 1.1.4 contains an introduction to variance estimation. We consider now the time series setting and use autoregression with p lags, so that the response variable is $Y = Y_t$, and the explanatory variables are $X = (Y_{t-1}^2, \ldots, Y_{t-p}^2)$. We assume that

$$E(Y \mid X = x) = 0$$

and discuss methods for estimating

$$\mathrm{Var}(Y \mid X = x) = E(Y^2 \mid X = x).$$

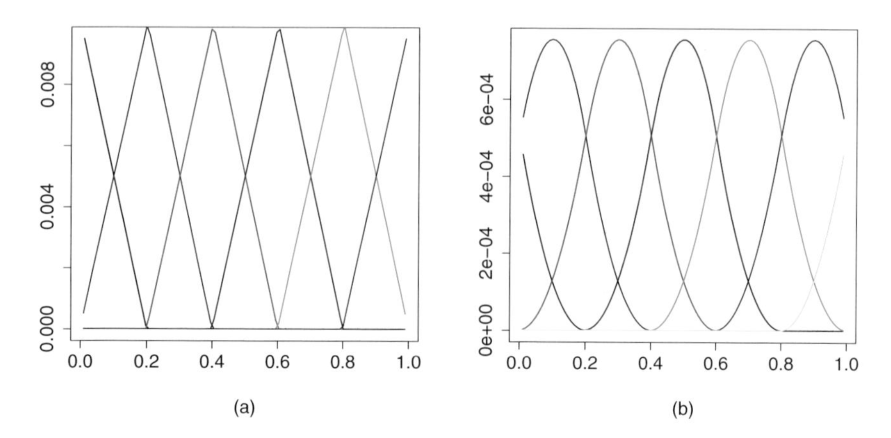

Figure 2.6 *B-splines.* (a) B-splines γ_k of order $m = 2$, for $k = -1, \ldots, 4$. (b) B-splines γ_k of order $m = 3$, for $k = -2, \ldots, 4$.

In the linear variance autoregression the conditional variance is approximated by a linear function of the previous squared observations:

$$\begin{aligned} E(Y^2 \mid X = x) &= E(Y_t^2 \mid Y_{t-1}^2 = y_{t-1}^2, \ldots, Y_{t-p}^2 = y_{t-p}^2) \\ &\approx \alpha + \beta_1 y_{t-1}^2 + \cdots + \beta_p y_{t-p}^2. \end{aligned} \tag{2.76}$$

We define below the least squares estimator and the ARCH estimator. We illustrate the methods with the S&P 500 data. The performance comparison of the estimators is postponed to Section 2.6.

2.5.1 Least Squares Estimator

Given the observed time series $Y_1, \ldots, Y_T$, we can estimate the regression coefficients $\alpha, \beta_1, \ldots, \beta_p$ in (2.76) by minimizing the sum of the squared errors:

$$\sum_{t=p+1}^{T} \left(Y_t^2 - \alpha - \beta_1 Y_{t-1}^2 - \cdots - \beta_p Y_{t-p}^2\right)^2.$$

The least squares estimator is given in (2.4) and (2.5), when the explanatory variables are $(Y_{t-1}^2, \ldots, Y_{t-p}^2)$ and the response variable is Y_t^2. The regression function estimator is

$$\hat{f}(y_{t-1}, \ldots, y_{t-p}) = \hat{\alpha} + \hat{\beta}_1 y_{t-1}^2 + \cdots + \hat{\beta}_p y_{t-p}^2,$$

where $\hat{\alpha}, \hat{\beta}_1, \ldots, \hat{\beta}_p$ are the least squares estimators. We denote below

$$\hat{\sigma}_t^2 = \hat{f}(Y_{t-1}, \ldots, Y_{t-p}).$$

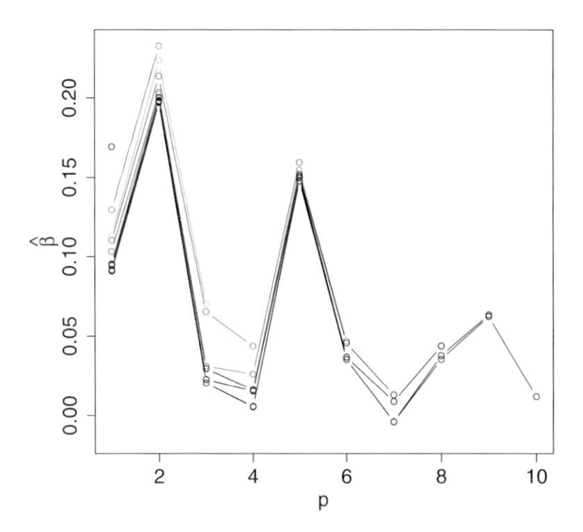

Figure 2.7 *S&P 500 volatility: parameter estimates in the least squares regression.* Shown are the 10 curves $i \mapsto \hat{\beta}_i^{(p)}$, for $p = 1, \ldots, 10$, where $i = 1, \ldots, p$.

We apply the estimator to the S&P 500 returns data described in Section 1.6.1. Observations Y_t are the daily net returns of the S&P 500 index.

Figure 2.7 shows a sequence of the least squares parameter estimates. We use lag numbers $p = 1, \ldots, 10$ and fit a linear model for each lag value. For each lag value p we get the coefficient estimates $\hat{\beta}_i^{(p)}$, $i = 1, \ldots, p$. The 10 curves $i \mapsto \hat{\beta}_i^{(p)}$, $i = 1, \ldots, p$, for $p = 1, \ldots, 10$, are shown in Figure 2.7. We see that the parameter estimates are about the same for each model. The highest value is about 0.2 for the second coefficients $\hat{\beta}_2^{(p)}$. The estimates of the intercept are not shown. For example, for the one-lag model ($p = 1$) we have $\hat{\alpha}_1^{(1)} = 7.9 \times 10^{-5}$ and for the ten-lag model ($p = 10$) we have $\hat{\alpha}_1^{(10)} = 3.8 \times 10^{-5}$.

2.5.2 ARCH Model

In the ARCH model the variance estimator is linear in the squares of the previous observations as in (2.76), but now the time series is modeled and maximum likelihood estimation is used to estimate the parameters.

Definition of ARCH Model The ARCH model (autoregressive conditional heteroskedastic model) is a special case of a conditional heteroskedasticity model, defined in (1.16). It is also a special case of the GARCH model (generalized autoregressive conditional heteroskedastic model), defined in Section 3.9.2.

In the ARCH(p) model it is assumed that

$$Y_t = \sigma_t \epsilon_t, \qquad t = 0, \pm 1, \pm 2, \ldots,$$

where

$$\sigma_t^2 = \alpha + \beta_1 Y_{t-1}^2 + \cdots + \beta_p Y_{t-p}^2 \tag{2.77}$$

with parameters satisfying restrictions $\alpha \geq 0$, $\beta_i \geq 0$. The noise ϵ_t is an i.i.d. process with $E\epsilon_t = 0$ and $\mathrm{Var}(\epsilon_t) = 1$. It is assumed that ϵ_t is independent of $Y_{t-1}, Y_{t-2}, \ldots$. The ARCH model was introduced in Engle (1982) for modeling U.K. inflation rates. The ARCH(p) process is strictly stationary if $\sum_{i=1}^{p} \beta_i < 1$; see Fan & Yao (2005, Theorem 4.3) and Giraitis, Kokoszka & Leipus (2000).

Variance in ARCH Models In the ARCH model, we have

$$\mathrm{Var}(Y_t \mid Y_{t-1}, Y_{t-2}, \ldots) = \sigma_t^2,$$

as shown in (1.17). Thus,

$$\mathrm{Var}(Y_t \mid Y_{t-1}, Y_{t-2}, \ldots) = E(Y_t^2 \mid Y_{t-1}, Y_{t-2}, \ldots) = \alpha + \beta_1 Y_{t-1}^2 + \cdots + \beta_p Y_{t-p}^2$$

and the ARCH model leads to an estimator of the conditional variance which is linear in the squared observations.

Maximum Likelihood Estimation of ARCH Parameters Estimation of the parameters $\alpha, \beta_1, \ldots, \beta_p$ can be done using the method of maximum likelihood, if we make distributional assumptions on ϵ_t. Let us denote the density of ϵ_t by $f_\epsilon : \mathbf{R} \to \mathbf{R}$. Then the conditional density of Y_t, given $Y_{t-1}, \ldots, Y_{t-p}$, is

$$f_{Y_t \mid Y_{t-1}, \ldots, Y_{t-p}}(y) = \frac{1}{\sigma_t} f\left(\frac{y}{\sigma_t}\right).$$

When we have observed $Y_1 = y_1, \ldots, Y_T = y_T$, then the likelihood function is

$$\begin{aligned} L(\alpha, \beta_1, \ldots, \beta_p) &= f_{Y_1, \ldots, Y_p}(y_1, \ldots, y_p) \prod_{t=p+1}^{T} f_{Y_t \mid Y_{t-1}=y_{t-1}, \ldots, Y_1=y_1}(y_t) \\ &= f_{Y_1, \ldots, Y_p}(y_1, \ldots, y_p) \prod_{t=p+1}^{T} f_{Y_t \mid Y_{t-1}=y_{t-1}, \ldots, Y_{t-p}=y_{t-p}}(y_t). \end{aligned}$$

Let us ignore the term $f_{Y_1, \ldots Y_p}(y_1, \ldots, y_p)$ and define the conditional likelihood

$$\begin{aligned} \tilde{L}(\alpha, \beta_1, \ldots, \beta_p) &\stackrel{def}{=} L(\alpha, \beta_1, \ldots, \beta_p \mid Y_p = y_p, \ldots, Y_1 = y_1) \\ &= \prod_{t=p+1}^{T} f_{Y_t \mid Y_{t-1}=y_{t-1}, \ldots, Y_{t-p}=y_{t-p}}(y_t). \end{aligned}$$

The parameters are estimated by maximizing the conditional likelihood, and we get

$$(\hat{\alpha}, \hat{\beta}_1, \ldots, \hat{\beta}_p) = \mathrm{argmax}_{\alpha, \beta_1, \ldots, \beta_p} \log \tilde{L}(\alpha, \beta_1, \ldots, \beta_p).$$

The logarithm of the conditional likelihood is

$$\log \tilde{L}(\alpha, \beta_1, \ldots, \beta_p) = -\frac{1}{2} \sum_{t=p+1}^{T} \log \sigma_t^2 + \sum_{t=p+1}^{T} \log f_\epsilon\left(\frac{y_t}{\sigma_t}\right), \tag{2.78}$$

where σ_t^2 is obtained from (2.77). If we assume that ϵ_t has the standard normal distribution

$$\epsilon_t \sim N(0, 1),$$

then the conditional density of Y_t given $Y_{t-1}, \ldots, Y_{t-p}$ is

$$f_{Y_t \,|\, Y_{t-1}, \ldots, Y_{t-p}}(y) = \frac{1}{\sqrt{2\pi}\sigma_t} \exp\left\{-\frac{1}{2}\frac{y^2}{\sigma_t^2}\right\}. \tag{2.79}$$

In the case of the Gaussian assumption (2.79), we get

$$(\hat{\alpha}, \hat{\beta}_1, \ldots, \hat{\beta}_p) = \text{argmin}_{\alpha, \beta_1, \ldots, \beta_p} \sum_{t=p+1}^{T} \left(\log \sigma_t^2 + \frac{y_t^2}{\sigma_t^2}\right). \tag{2.80}$$

For example, let $p = 1$ and $\alpha = 0$. Then $\sigma_t^2 = \beta Y_{t-1}^2$, and under the Gaussian assumption we have to minimize

$$(T-1)\log \beta + \frac{1}{\beta} \sum_{t=1}^{T-1} \frac{y_{t+1}^2}{y_t^2},$$

and the minimizer is

$$\hat{\beta} = \frac{1}{T-1} \sum_{t=1}^{T-1} \frac{y_{t+1}^2}{y_t^2}.$$

After estimating the parameters of the ARCH model, we get an estimator for the conditional variance:

$$\text{Var}(Y_t \,|\, Y_{t-1}, \ldots, Y_{t-p}) \approx \hat{\sigma}_t^2 = \hat{\alpha} + \hat{\beta}_1 Y_{t-1}^2 + \cdots + \hat{\beta}_p Y_{t-p}^2.$$

Let us analyze the S&P 500 returns data described in Section 1.6.1. Figure 2.8 shows a sequence of ARCH(p) parameter estimates. We take lag numbers $p = 1, \ldots, 25$ and fit the ARCH model for each lag value. For each lag value p we get the coefficient estimates $\hat{\beta}_i^{(p)}$, $i = 1, \ldots, p$. The 25 curves $i \mapsto \hat{\beta}_i^{(p)}$, $i = 1, \ldots, p$, for $p = 1, \ldots, 25$, are shown in Figure 2.8. We can see that the values $\hat{\beta}_i^{(p)}$ of the estimates are monotonically decreasing as a function of p. The values $\hat{\beta}_i^{(p)}$ are also decreasing as a function of i, although they are not decreasing monotonically. The estimates of the intercepts $\hat{\alpha}^{(p)}$ are not shown, but these estimates are decreasing from $\hat{\alpha}_1^{(1)} = 6.6 \times 10^{-5}$ for the one-lag model ($p = 1$) to $\hat{\alpha}_1^{(20)} = 1.1 \times 10^{-5}$ for the 22-lag model ($p = 22$); after that, there are spikes–for example, $\hat{\alpha}_1^{(23)} = 8.4 \times 10^{-5}$.

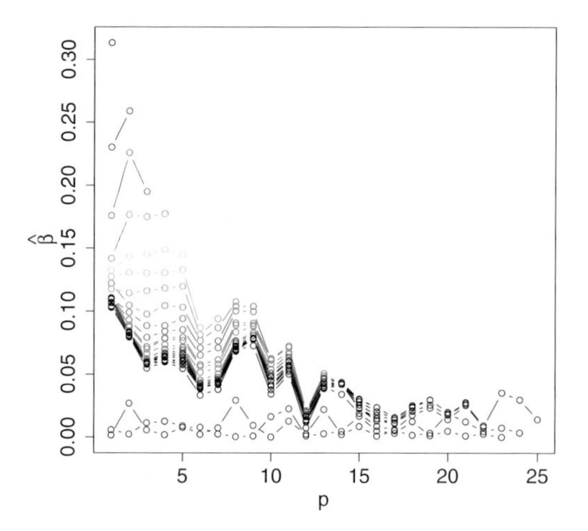

Figure 2.8 *S&P 500 volatility: parameter estimates in ARCH(p) models.* Shown are the 25 curves $i \mapsto \hat{\beta}_i^{(p)}$, for $p = 1, \ldots, 25$, where $i = 1, \ldots, p$.

2.6 APPLICATIONS IN VOLATILITY AND QUANTILE ESTIMATION

We estimate the volatility and quantiles of the S&P 500 returns Y_t using the S&P 500 index data, described in Section 1.6.1. Before studying conditional quantile estimators, we set benchmarks by looking at the performance of some sequentially calculated quantile estimators.

2.6.1 Benchmarks for Quantile Estimation

Figure 1.6 shows the performance of the sequentially calculated empirical quantile estimator. We study now other benchmarks for quantile estimation. Quantile regression was introduced in Section 1.1.6, where a quantile estimator was defined in (1.30), which puts

$$\hat{Q}_p(Y_t \mid Y_{t-1}, Y_{t-2}, \ldots) = \hat{\sigma}_t \, \hat{F}_{\epsilon_t}^{-1}(p),$$

where $\hat{\sigma}_t$ is an estimator of the conditional standard deviation and $\hat{F}_{\epsilon_t}^{-1}(p)$ is an estimator of the p-quantile of the distribution of $\epsilon_t = Y_t/\sigma_t$. In this section, $\hat{\sigma}_t$ is the sequentially calculated sample standard deviation: $\hat{\sigma}_t$ is the standard deviation of $Y_1, \ldots, Y_{t-1}$. The performance is evaluated looking at the differences $p - \hat{p}$ and $\hat{p} - p$, where

$$\hat{p} = \frac{1}{T - t_0} \sum_{t=t_0+1}^{T} I_{(-\infty, \hat{q}_t]}(Y_t),$$

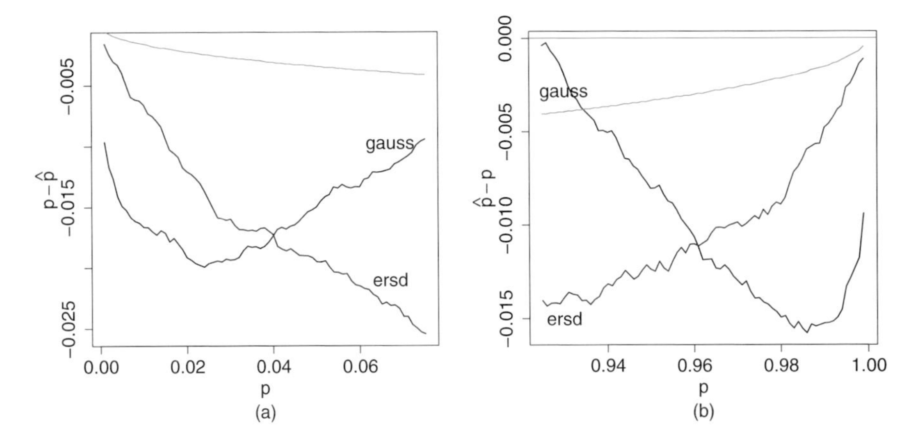

Figure 2.9 *S&P 500 quantiles: Gauss and empirical residuals.* The black line shows the performance of the quantile estimator with the Gaussian residuals and the red line shows the performance of the quantile estimator with the empirical residuals. (a) Functions $p \mapsto p - \hat{p}$ for $p \in [0.001, 0.075]$. (b) Functions $p \mapsto \hat{p} - p$ for $p \in [0.925, 0.999]$. The green lines show level $\alpha = 0.05$ fluctuation bands.

where $\hat{q}_t = \hat{Q}_p(Y_t \mid Y_{t-1}, Y_{t-2}, \ldots)$. The performance measurement is explained in more detail in Section 1.9.4.

Figure 2.9 shows the performance of two quantile estimators. The first quantile estimator uses $\hat{F}_{\epsilon_t}^{-1}(p) = \Phi^{-1}(p)$, where Φ is the distribution function of the standard normal distribution. The second quantile estimator takes $\hat{F}_{\epsilon_t}^{-1}(p)$ to be the empirical quantile of $Y_u/\hat{\sigma}_u$, $u = 1, \ldots, t$. Panel (a) plots $p \mapsto p - \hat{p}$ in the range $p \in [0.001, 0.075]$ and panel (b) plots $p \mapsto \hat{p} - p$ in the range $p \in [0.925, 0.999]$. The black curves are for the estimator with the Gaussian residuals and the red curves are for the estimator with the empirical residuals. A green line is drawn at level 0, and it is accompanied by the level $\alpha = 0.05$ fluctuation bands, defined in (1.130)–(1.131). Figure 2.9 implies that the true distribution has heavier tails than the quantile estimates would indicate. The Gaussian residuals are better for central quantiles and the empirical residuals are better for extreme quantiles. The performance of the empirical quantile, shown in Figure 1.6, is similar to the performance of the empirical residual method for the left tail. For the right tail the empirical residual method is better than the empirical quantile.

Figure 2.10 shows the performance of quantile estimators which use $\hat{F}_{\epsilon_t}^{-1}(p) = \sqrt{(\nu-2)/\nu}\, t_\nu^{-1}(p)$, where t_ν is the distribution function of the t-distribution with ν degrees of freedom, $\nu > 2$. We show the estimators for $\nu = 4,5,6,8,10,20$. Panel (a) plots $p \mapsto p - \hat{p}$ in the range $p \in [0.001, 0.075]$ and panel (b) plots $p \mapsto \hat{p} - p$ in the range $p \in [0.925, 0.999]$. For central quantiles the large values of ν lead to better performance but for extreme quantiles the smaller values of ν lead to better performance. The performance of the estimator with $\nu = 20$ looks similar

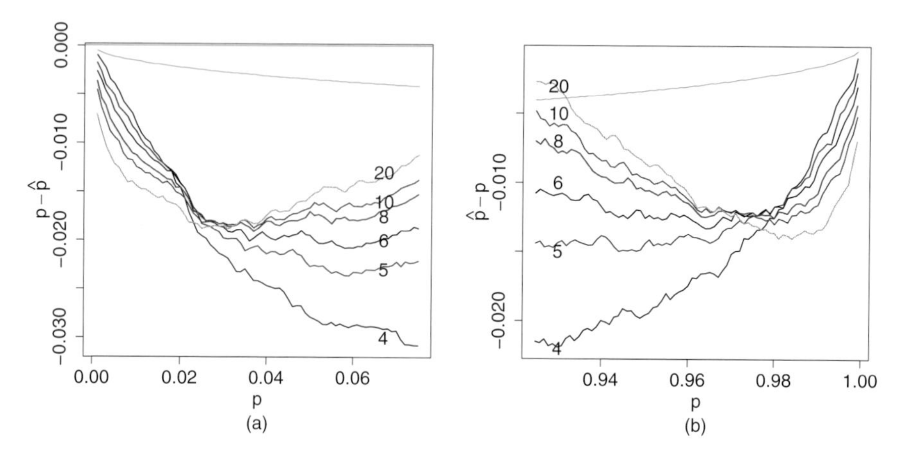

Figure 2.10 *S&P 500 quantiles: Student residuals.* Shown are the performances of quantile estimators when the residual distribution is the standard t-distribution with $\nu = 4, 5, 6, 8, 10, 20$ degrees of freedom. (a) Functions $p \mapsto p - \hat{p}$ for $p \in [0.001, 0.075]$. (b) Functions $p \mapsto \hat{p} - p$ for $p \in [0.925, 0.999]$. The green lines show level $\alpha = 0.05$ fluctuation bands.

to the performance of the estimator with the Gaussian residuals. When $\nu = 5$, the performance is similar to the performance of the estimator with the empirical residuals.

The extreme tails of t-distributions are heavier than the tails of the standard Gaussian distribution. To better interpret the results in Figure 2.10, let us find the points where the quantiles change to be larger for the standard t-distribution. Let $p_\nu \in (0, 1)$ be the solution of the equation

$$\sqrt{(\nu - 2)/\nu}\; t_\nu^{-1}(p) = \Phi^{-1}(p). \tag{2.81}$$

Figure 2.11 plots the functions $\nu \mapsto p_\nu$ for $v \in \{3, 4, \ldots, 25\}$. There are two solutions to (2.81). Panel (a) shows the solution near zero and panel (b) shows the solution near one. For example, when degrees of freedom is $\nu = 12$, then the solutions are $p_\nu = 0.0371$ and $p_\nu = 0.963$. This means that for a level $p < 3.71\%$, the standard t-distribution with $\nu = 12$ degrees of freedom has larger quantiles than the standard normal distribution.

2.6.2 Volatility and Quantiles with the LS Regression

We use the least squares regression to estimate volatility and quantiles with the S&P 500 returns data, described in Section 1.6.1. Observations Y_t are the daily net returns of the S&P 500 index. Figure 2.7 shows a sequence of least squares parameter estimates.

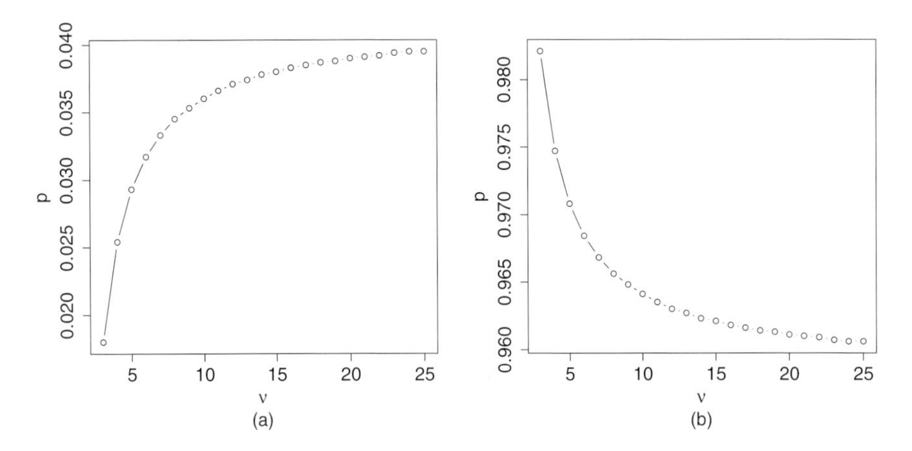

Figure 2.11 *Quantiles of t-distributions compared to quantiles of normal distribution.* Shown is the function $\nu \mapsto p_\nu$, where p_ν is the point where the quantile of the standard t-distribution with degrees of freedom ν is equal to the quantile of the standard normal distribution.

Figure 2.12 shows the performance of the linear least squares estimator in making one step ahead predictions of the squared S&P 500 returns. The performance is shown for the models with lags $p = 1, \ldots, 15$. Panel (a) shows the performance measured with the mean deviation error MDE$^{(1,2)}$, defined in (1.124). Panel (b) shows the performance measured with the mean absolute ratio errors MARE$^{(2)}$, defined in (1.127). The performance is compared to the GARCH(1,1) estimates: We have divided the MADE and the MARE values of the least squares estimator with the MADE and the MARE values of the GARCH(1,1) estimator.[24] We can see that increasing the lag value makes the error smaller. For large lag values the MDE of the least squares regression is about 0.4% larger than the MDE of GARCH(1,1), and the MARE of the least squares regression is about 10% larger than the MARE of GARCH(1,1).

Figure 2.13 shows the annualized volatility estimates for the lag value $p = 5$. We show the sequentially estimated values $\sqrt{250}\,\hat{\sigma}_t$.

We apply the volatility estimates to make estimates of quantiles. The quantile estimators are defined in (1.31)–(1.33), where estimators of the type

$$\hat{Q}_p(Y_t \mid \mathcal{F}_{t-1}) = \hat{\sigma}_t\, F_{\epsilon_t}^{-1}(p)$$

are defined, and F_{ϵ_t} is either the standard normal distribution, the standard t-distribution with the degrees of freedom 5 and 12, or the empirical distribution of the residuals $Y_t/\hat{\sigma}_t$.

[24]GARCH estimation is introduced in Section 3.9.2. For GARCH(1,1) MDE$^{(1,2)}$ is 0.0602 and MARE$^{(2)}$ is 1.087.

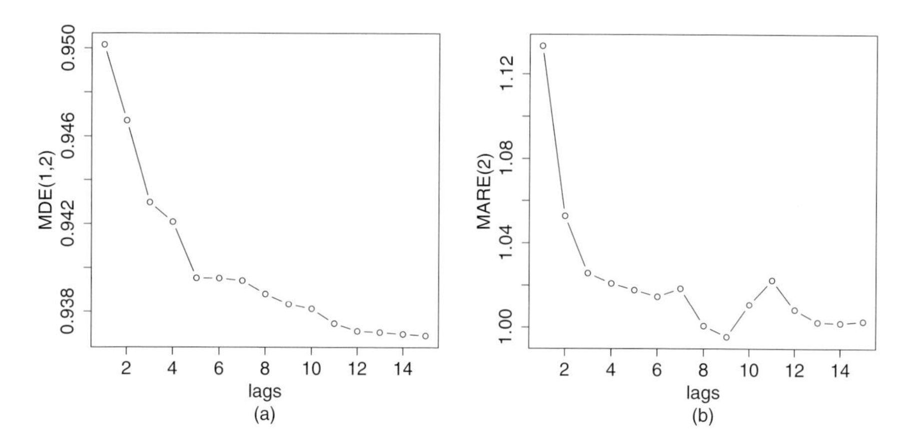

Figure 2.12 *S&P 500 Volatility: The performance of the least squares regression.* Panel (a) shows the mean deviation errors ($\mathrm{MDE}^{(1,2)}$). Panel (b) shows the mean absolute ratio errors with the exponent 2 ($\mathrm{MARE}^{(2)}$). The performance is measured for the lag values $p = 1, \ldots, 15$. The errors are relative to the $\mathrm{MDE}^{(1,2)}$ and $\mathrm{MARE}^{(2)}$ of the GARCH(1,1).

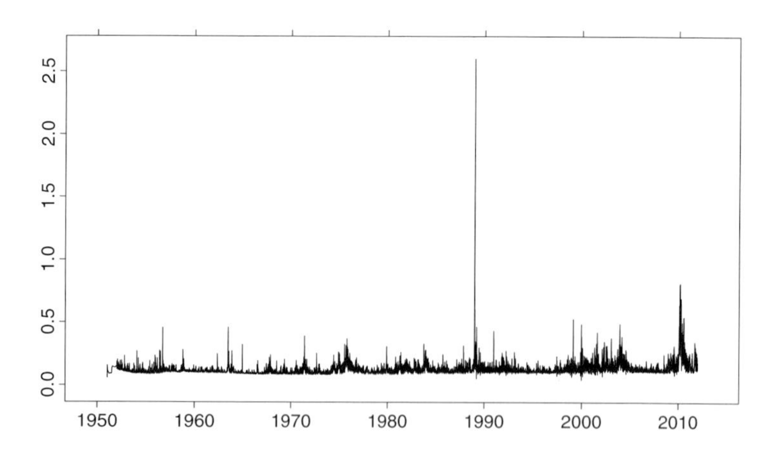

Figure 2.13 *S&P 500 volatility: The volatility estimates with the least squares regression.* The time series $\sqrt{250}\,\hat{\sigma}_t$ of the annualized volatility estimates using the least squares regression with $p = 5$ lags.

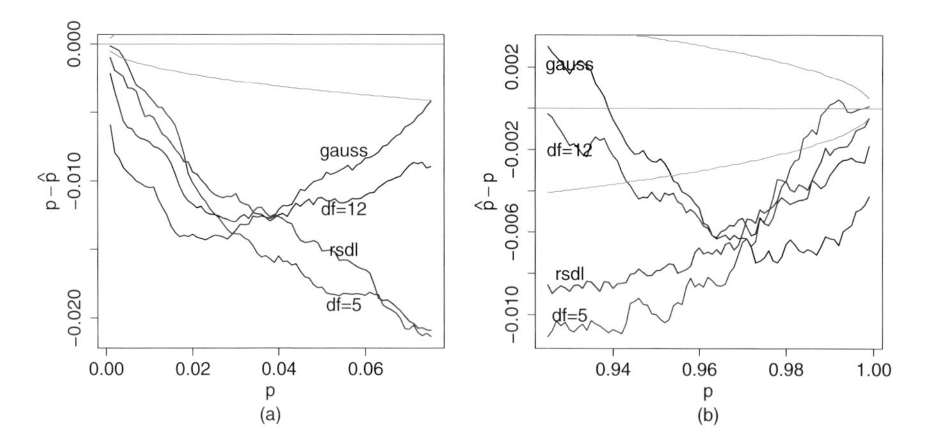

Figure 2.14 *S&P 500 quantiles: The performance of least squares quantile estimators.* Panel (a) shows the curves $p \mapsto p - \hat{p}$ for $p \in [0.001, 0.025]$ and panel (b) shows the curves $p \mapsto \hat{p} - p$ for $0.975 \le p \le 0.999$.

Figure 2.14 shows the performance of the quantile estimators. Let $\hat{p}$ be the number of exceedances, defined in (1.129). Number $\hat{p}$ is the proportion of the next day observations that exceeded the quantile estimate. Panel (a) shows the curves $p \mapsto p - \hat{p}$ for $0.001 \le p \le 0.075$, and panel (b) shows the curves $p \mapsto \hat{p} - p$ for $0.925 \le p \le 0.999$. The black curves show the case of the standard normal innovations, the blue curves show the case of the standard t-distribution with degrees of freedom 12, the red curves show the case of standard t-distribution with degrees of freedom 5, and the dark green curves show the case of empirical residuals. A green line is drawn at level 0, and it is accompanied with the level $\alpha = 0.05$ fluctuation bands, defined in (1.130)–(1.131).

2.6.3 Volatility with the Ridge Regression

We analyze the S&P 500 returns data, described in Section 1.6.1. Figure 2.15 shows the performance of the ridge regression estimator in making one step ahead predictions of the squared S&P 500 returns. The performance is shown for the models with lags $p = 2, 5, 10, 20$ and for the ridge parameters $\lambda = 0, 100, 1000$. The ridge parameter $\lambda = 0$ leads to the least squares regression. Panel (a) shows the performance measured with the mean deviation error $\text{MDE}^{(1,2)}$, defined in (1.124). Panel (b) shows the performance measured with the mean absolute ratio error $\text{MARE}^{(2)}$, defined in (1.127). The performance is compared with the performance of GARCH(1,1) estimates: We have divided the MDE and the MARE values of the least squares estimator with the MDE and the MARE values of the GARCH(1,1) estimator. GARCH estimation is introduced in Section 3.9.2. The black curve with the labels 1 shows the least squares regression, the red curve with the labels 2 shows the ridge regression

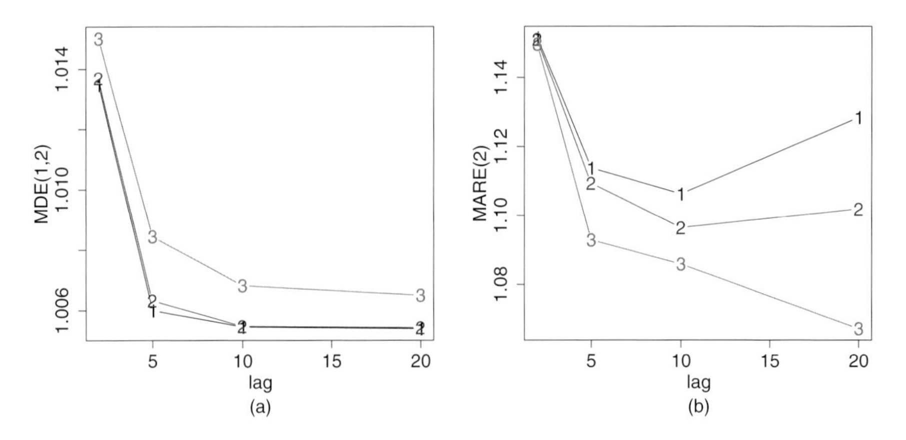

Figure 2.15 *S&P 500 volatility: The performance of the ridge regression.* (a) Mean deviation errors ($\mathrm{MDE}^{(1,2)}$). (b) Mean absolute ratio errors ($\mathrm{MARE}^{(2)}$). The performance is measured for the lag values $p = 2, 5, 10, 20$. The black curve with the labels 1 show the least squares regression, the red curve with the labels 2 show the ridge regression with $\lambda = 100$, and the green curve with the labels 3 show the ridge regression with $\lambda = 1000$. The errors are relative to the MDE and MARE of GARCH(1,1).

with $\lambda = 100$, and the green curve with the labels 3 shows the ridge regression with $\lambda = 1000$. We can see that for the MDE criterion the least squares regression gives the best results but for the MARE criterion the ridge regression with a large parameter λ gives the best results.

2.6.4 Volatility and Quantiles with ARCH

We analyze the S&P 500 returns data described in Section 1.6.1. Figure 2.8 shows a sequence of ARCH(p) parameter estimates.

Figure 2.16 shows the performance of the ARCH estimator in making one-step-ahead predictions of the squared S&P 500 returns. The performance is shown for the models with lags $p = 1, \ldots, 6$. Panel (a) shows the performance measured with the mean absolute deviation error $\mathrm{MDE}^{(1,2)}$, defined in (1.124). Panel (b) shows the performance measured with the mean absolute ratio errors $\mathrm{MARE}^{(2)}$, defined in (1.127). We have divided the MDE and the MARE values of the least squares estimator with the MDE and the MARE values of the GARCH(1,1) estimator. We can see that increasing the lag value makes the error smaller. For large lag values the MDE of the least squares regression is about 0.4% larger than the MDE of GARCH(1,1), and the MARE of the least squares regression is about 10% larger than the MDE of GARCH(1,1). The first year of observations is used in to estimate the parameters, but the first year is not used in the performance measure.

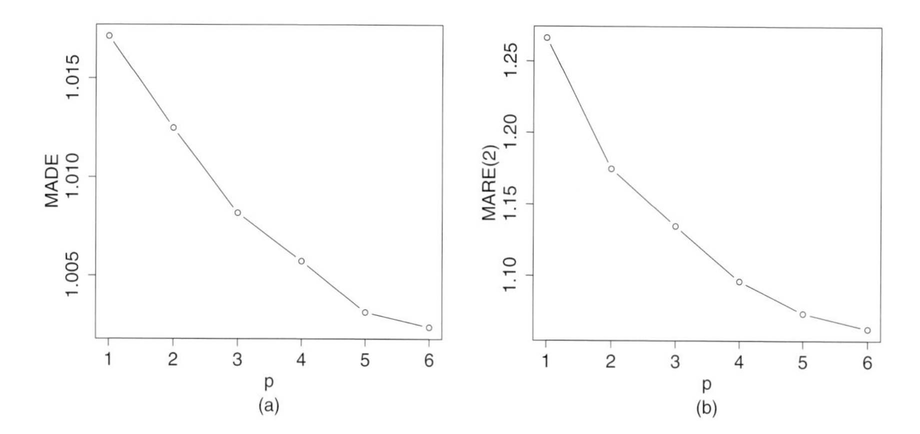

Figure 2.16 *S&P 500 volatility: The performance of ARCH(p).* (a) Mean deviation errors ($\text{MDE}^{(1,2)}$). (b) Mean absolute ratio errors ($\text{MARE}^{(2)}$). The performance is measured for the lag values $p = 1, \ldots, 6$. The errors are relative to the MDE and MARE of GARCH(1,1).

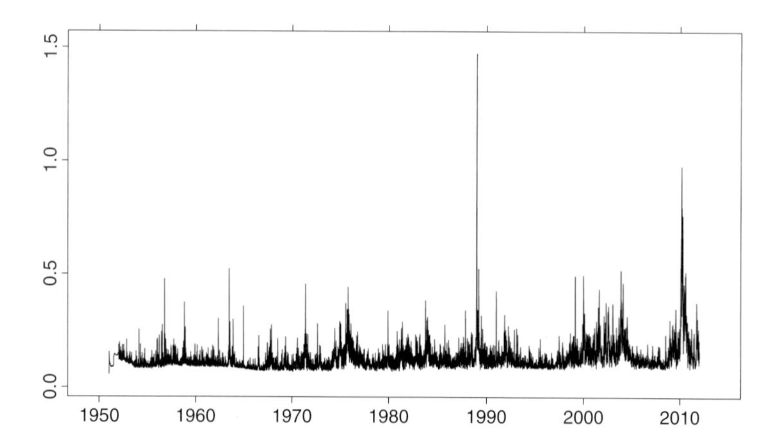

Figure 2.17 *S&P 500 volatility: The volatility estimates with ARCH(5).* The time series $\sqrt{250}\,\hat{\sigma}_t$ of the annualized volatility estimates using ARCH(p) with $p = 5$ lags.

Figure 2.17 shows the annualized volatility estimates for the lag value $p = 5$. The estimation proceeds out-of-sample: $\hat{\sigma}_t$ is estimated using data $Y_1, \ldots, Y_{t-1}$.

Figure 2.18 shows the performance of the quantile estimators with the curves $p \mapsto \hat{p} - p$ and $p \mapsto p - \hat{p}$, where $\hat{p}$ is the proportion of the next-day observations that exceeded the quantile estimate, as defined in (1.129). We apply the three estimators defined in (1.31)–(1.33). Panel (a) shows the cases $0.001 \le p \le 0.075$, and panel (b) shows the cases $0.925 \le p \le 0.999$. The black curves show the case of standard

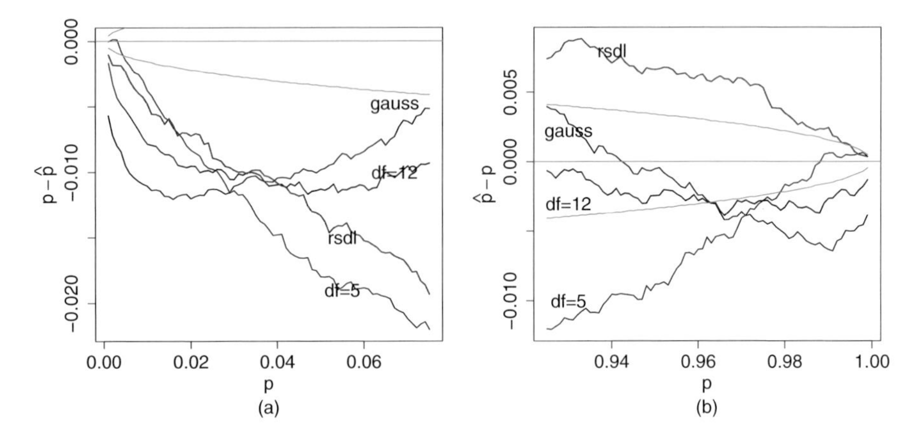

Figure 2.18 *S&P 500 quantiles: The performance of ARCH(5) quantile estimators.* (a) The curves $p \mapsto p - \hat{p}$ for $p \in [0.001, 0.075]$. (b) The curves $p \mapsto \hat{p} - p$ for $p \in [0.925, 0.999]$.

normal innovations, the blue curves show the case of standard t-distributed with degrees of freedom 12, the red curves show the case of standard t-distributed with degrees of freedom 5, and the dark green curves show the case of empirical residuals. The level $\alpha = 0.05$ fluctuation bands are shown with light green. The Gaussian innovations give best results overall. The results are quite similar to the results of least squares regression shown in Figure 2.14.

2.7 LINEAR CLASSIFIERS

In the two-class case we call a classifier $\hat{g} : \mathbf{R}^d \to \{0, 1\}$ linear, if the boundary of the classification sets is linear:

$$\hat{g}(x) = \begin{cases} 0, & \text{if } \alpha + \beta' x \le 0, \\ 1, & \text{if } \alpha + \beta' x > 0, \end{cases}$$

for some $\alpha \in \mathbf{R}$ and $\beta \in \mathbf{R}^d$. In the multiclass case, when $\hat{g} : \mathbf{R}^d \to \{0, \ldots, K-1\}$, we call a classifier linear if the sets $\{x \in \mathbf{R}^d : \hat{g}(x) = k\}$, $k = 0, \ldots, K-1$, are intersections of half spaces.

Linear Regression-Based Classifiers The regression-based classifier $\hat{g}$ was defined in (1.78), where

$$\hat{g}(x) = \text{argmax}_{k=0,\ldots,K-1}\, \hat{p}_k(x),$$

and $\hat{p}_k(x)$ is an estimator of the probability $P(Y = k \mid X = x)$. We can write, as in (1.80),

$$P(Y = k \mid X = x) = E\left(I_{\{k\}}(Y) \mid X = x\right),$$

and use linear regression with data $(X_i, I_{\{k\}}(Y_i))$, $i = 1, \ldots, n$, to estimate the conditional expectations. Now,

$$\hat{g}(x) = \text{argmax}_{k=0,\ldots,K-1} \left[\hat{\alpha}_k + \hat{\beta}_k' x\right],$$

where $\hat{\alpha}_k$ and $\hat{\beta}_k$ are the estimated linear regression coefficients. Thus we have obtained a linear classifier. The linear regression based classifier can suffer from the masking problem, as explained in Hastie et al. (2001, p. 105).

Density-Based Classifiers The density-based classifier was defined in (1.83). The population version of the classifier is

$$g(x) = \text{argmax}_{k=0,\ldots,K-1} p_k f_{X|Y=k}(x),$$

where $f_{X|Y=k}$ is the class density function and $p_k = P(Y = k)$ is the class prior probability.

In the quadratic discriminant analysis it is assumed that the class densities are multivariate normal densities:

$$f_{X\,|\,Y=k}(x) = |\Sigma_k|^{-1/2} \phi\left((x - \mu_k)' \Sigma_k^{-1} (x - \mu_k)\right),$$

for $k = 0, \ldots, K-1$, where $\phi(t) = (2\pi)^{-d/2} \exp\{-t/2\}$, $\mu_k \in \mathbf{R}^d$ is the mean, and Σ_k is the $d \times d$ covariance matrix. The empirical classifier is obtained by estimating μ_k and Σ_k, typically using sample means and sample covariance matrices. Now we have that

$$g(x) = \text{argmax}_{k=0,\ldots,K-1} \delta_k(x), \tag{2.82}$$

where

$$\delta_k(x) = 2 \log p_k - \log\left(|\Sigma_k|\right) - (x - \mu_k)' \Sigma_k^{-1} (x - \mu_k).$$

This discrimination rule leads to quadratic boundaries of classification sets.

In the linear discriminant analysis, it is assumed that the class densities are multivariate normal densities with equal covariance matrices. The classification function is defined by (2.82), but the discriminant function is

$$\delta_k(x) = 2 \log p_k - (x - \mu_k)' \Sigma^{-1} (x - \mu_k),$$

where Σ is the common covariance matrix of the class distributions.

In the two-class case we have

$$\log \frac{p_1 f_{X|Y=1}(x)}{p_0 f_{X|Y=0}(x)} = b'x + c,$$

where

$$b = \Sigma^{-1}(\mu_1 - \mu_0),$$

$$c = -\frac{1}{2}(\mu_1 + \mu_0)' \Sigma^{-1} (\mu_1 - \mu_0) + \log \frac{p_1}{p_0},$$

and Σ is the common covariance matrix. Thus the classification set in the two class case has a linear boundary, since we have

$$\{x \in \mathbf{R}^d : p_1 f_{X|Y=1}(x) \geq p_0 f_{X|Y=0}(x)\} = \{x \in \mathbf{R}^d : b'x + c \geq 0\}.$$

In the multiclass case the decision sets are intersections of half spaces, since all the pairwise decision boundaries are linear.

Empirical Risk Minimization-Based Linear Classifiers A classifier based on empirical risk minimization is defined in (1.84) as

$$\hat{g} = \mathrm{argmin}_{g \in \mathcal{G}} \gamma_n(g), \tag{2.83}$$

where $\mathcal{G}$ is a class of functions $g : \mathbf{R}^d \to \{0, \ldots, K-1\}$ and $\gamma_n(g)$ is the empirical error of classifier g. We get a linear classifier if the class $\mathcal{G}$ of functions is chosen suitably. For example, define

$$\mathcal{G} = \left\{ g(\,\cdot\,, \theta) : \theta \in \mathbf{R}^{K(d+1)} \right\}, \tag{2.84}$$

where

$$\begin{aligned} g(x, \theta) &= \mathrm{argmax}_{k=1,\ldots,K-1} \delta_k(x, \theta), \\ \delta_k(x, \theta) &= \alpha_k - (x - \mu_k)' \hat{\Sigma}^{-1} (x - \mu_k), \\ \theta &= (\alpha_1, \ldots, \alpha_K, \mu_1, \ldots, \mu_K), \end{aligned}$$

and $\hat{\Sigma}$ is the sample covariance matrix, calculated using the complete learning sample. This leads to an optimization problem with the number of parameters $K(d+1)$. We can reduce the number of parameters by defining

$$\begin{aligned} \delta_k(x, \theta) &= \alpha_k - (x - \hat{\mu}_k)' \hat{\Sigma}^{-1} (x - \hat{\mu}_k), \\ \theta &= (\alpha_1, \ldots, \alpha_K), \end{aligned}$$

where $\hat{\mu}_k$ is the sample mean calculated from the class k learning sample. Now there are only K parameters over which the optimization is done in (2.83). The reduction was suggested in Hastie et al. (2001, p. 110).

CHAPTER 3

KERNEL METHODS AND EXTENSIONS

We use the term "local averaging" to refer to a method of regression function estimation where the estimator can be written as

$$\hat{f}(x) = \sum_{i=1}^{n} p_i(x) Y_i, \qquad x \in \mathbf{R}^d, \tag{3.1}$$

where $p_i(x) \geq 0$, $\sum_{i=1}^{n} p_i(x) = 1$, and the weights satisfy the properties that $p_i(x)$ is close to zero when X_i is distant from x and $p_i(x)$ is large when X_i is near x. The estimator $\hat{f}(x)$ of regression function is a weighted average of Y_i, $i = 1, \ldots, n$, where more weight is given to the observations which are such that X_i is close to x. Regressogram, kernel estimator, and nearest-neighbor estimator are special cases of local averaging. Local polynomial estimators can also be considered to be in the class of local averaging estimators, but we discuss local linear estimators in Section 5.2, where also local likelihood estimators are discussed. We can motivate local averaging with the following two observations.

First, a local average can be obtained as a modification of interpolation. Let us first consider the problem where we want to estimate the value of a function $f : \mathbf{R}^d \to \mathbf{R}$ at point $x \in \mathbf{R}^d$, when we have available only the values $f(x_1), \ldots, f(x_n)$, for some collection of points $x_1, \ldots, x_n \in \mathbf{R}^d$. Several interpolation methods, including piecewise constant and polynomial interpolation of the values $f(x_1), \ldots, f(x_n)$, can be used to obtain an approximation of the value $f(x)$ at any point $x \in \mathbf{R}^d$. For example, we can estimate

$$f(x) \approx f(x_{i(x)}),$$

where $i(x)$ is the index of observation closest to x:

$$\|x - x_{i(z)}\| = \min\{\|x - x_i\| : i = 1, \ldots, n\}.$$

In the setting of regression function estimation, we do not observe the exact values of the function, but only values with are corrupted with noise:

$$Y_i = f(x_i) + \epsilon_i,$$

where ϵ_i, $i = 1, \ldots, n$, are random errors. We could choose as the estimate the value Y_i corresponding to the x_i that is closest to x:

$$f(x) \approx Y_{i(x)},$$

where $i(x)$ is as before, but this estimator would contain too much random variation since the value of the estimator depends on one error term $\epsilon_{i(x)}$. It is a better idea to take a local average over several observations as in (3.1), so that the random variation is diminished by averaging over several error terms.

Second, the estimator (3.1) can be obtained as an extension of the least squares method. Namely, (3.1) is the solution of the locally weighted least squares minimization problem:

$$\hat{f}(x) = \text{argmin}_{\theta \in \mathbf{R}} \sum_{i=1}^{n} p_i(x)(Y_i - \theta)^2. \tag{3.2}$$

Note that the taking $p_i(x) \equiv 1/n$ leads to the arithmetic average

$$\hat{\theta} = \frac{1}{n} \sum_{i=1}^{n} Y_i.$$

Local empirical risk is studied in Section 5.2.1, where the estimator defined in (3.2) is called local constant estimator and extended to local linear or quadratic estimators.

Section 3.1 defines a regressogram. Section 3.2 covers kernel estimators. Sections 3.3–3.10 discuss the nearest-neighbor estimator, classification with local averaging, median smoothing, conditional density and distribution function estimation, conditional quantile, variance, and covariance estimation.

Applications to risk management are given in Section 3.11, where the conditional variance, covariance, and quantile estimation with the S&P 500 return data is considered. Note that we already studied in Section 3.9.2 the GARCH(1,1) fitting with the S&P 500 return data and in Section 3.9.3 the moving average estimator for the S&P 500 return data. Applications to portfolio selection are given in Section 3.12, where regression function estimation with utility maximization, classification, and regression function estimation with the Makowitz criterion are studied.

3.1 REGRESSOGRAM

Regressograms are one of the simplest nonparametric estimators of a regression function. A regressogram is a piecewise constant regression function estimator. The X-observation space is covered by disjoint bins, and the value of a regressogram in a bin is the average of the Y-values for the X-values inside that bin. The bins are typically rectangles but they can also be hexagons, for example. The name "regressogram" was coined by Tukey (1961). The name is related to "histogram," which denotes a piecewise constant estimator of a density function, analogous to a regressogram.

A regressogram, based on data $(X_1, Y_1), \ldots, (X_n, Y_n)$, is determined by a collection $A_1, \ldots, A_N \subset \mathbf{R}^d$ of sets such that they are disjoint and their union covers the observed explanatory variables:

1. $A_i \cap A_j = \emptyset$, when $i \neq j$,
2. $\{X_1, \ldots, X_n\} \subset U_{j=1}^N A_j$.

The regressogram is defined as

$$\hat{f}_n(x) = \hat{Y}_{A_j}, \qquad \text{if } x \in A_j,$$

where $\hat{Y}_{A_j}$ is the average of those response variables whose corresponding explanatory variable is in A_j. We can write, using the notation $I_A(x) = 1$ if $x \in A$ and $I_A(x) = 0$ if $x \notin A$,

$$\hat{Y}_A = \frac{1}{n_A} \sum_{i=1}^n Y_i \, I_A(X_i), \tag{3.3}$$

where n_A is the number of explanatory variables inside A:

$$n_A = \sum_{i=1}^n I_A(X_i).$$

The definition of a regressogram can be written as

$$\hat{f}_n(x, \mathcal{P}) = \sum_{j=1}^N \hat{Y}_{A_j} I_{A_j}(x) = \sum_{A \in \mathcal{P}} \hat{Y}_A I_A(x), \qquad x \in \mathbf{R}^d, \tag{3.4}$$

where we have also made the dependence of the regressogram on the partition $\mathcal{P} = \{A_1, \ldots, A_N\}$ explicit. Changing the order of summation in (3.4), we get

$$\hat{f}_n(x) = \sum_{j=1}^N \left(\frac{1}{n_{A_j}} \sum_{i=1}^n Y_i \, I_{A_j}(X_i) \right) I_{A_j}(x) = \sum_{i=1}^n p_i(x) \, Y_i,$$

where

$$p_i(x) = \sum_{j=1}^N \frac{1}{n_{A_j}} I_{A_j}(X_i) I_{A_j}(x) = \frac{1}{n_{A_x}} I_{A_x}(X_i), \tag{3.5}$$

and $A_x \in \{A_1, \ldots, A_N\}$ is such that $x \in A_x$. (By symmetry we can as well write $p_i(x) = I_{A_{X_i}}(x)/n_{A_{X_i}}$.) Thus regressogram can be written as a local average as in (3.1) that have the form

$$\hat{f}_n(x) = \sum_{i=1}^{n} p_i(x)\, Y_i,$$

where $p_i(x) = p_i(x, X_1, \ldots, X_n) \geq 0$ and $\sum_{i=1}^{n} p_i(x) = 1$. The weights $p_i(x)$ satisfy the property that the weight $p_i(x)$ is large when X_i is close to x and the weight $p_i(x)$ is small when X_i is far away from x.

A regressogram is completely determined by defining a partition of the space of explanatory variables. We discuss only partitions made of rectangles. We distinguish between regular and irregular partitions. In the one-dimensional case a regular partition is a collection of intervals of length h and an irregular partition is a collection of intervals of differing lengths. In the multivariate case we can distinguish between isotropic and anisotropic regular partitions. An isotropic regular partition is a partition where all rectangles have the same side lengths h and thus the partition is a collection of cubes of volume h^d (cubic partition). An anisotropic regular partition is a partition where the side lengths of the rectangles are the same in one direction but differ across dimensions, having side lengths $h_1, \ldots, h_d$ and volumes $h_1 \cdots h_d$. In the multivariate case an irregular partition consists of rectangles, where each rectangle can have a different volume and shape.

Regular partitions depend on the data through the smoothing parameter h, or in the anisotropic case through smoothing parameters $h_1, \ldots, h_d$. The smoothing parameters can be chosen by cross-validation or a plug-in method, for example. Irregular partitions depend more heavily on the data, because the shapes and volumes of the sets of the partition are chosen using data. The methods for irregular partition selection are discussed in Section 5.5.

3.2 KERNEL ESTIMATOR

We define the kernel regression estimate, compare it to the regressogram, define Gasser–Müller and Priestley–Chao estimators, define moving averages, consider kernel estimation with locally stationary data, mention the curse of dimensionality, discuss the smoothing parameter selection, discuss the possible definitions of the effective sample size used by a kernel estimator, define the kernel estimator of partial derivatives, and give pointwise confidence intervals for the kernel estimator.

3.2.1 Definition of the Kernel Regression Estimator

The kernel estimator of the regression function is defined as

$$\hat{f}(x) = \sum_{i=1}^{n} p_i(x)\, Y_i, \tag{3.6}$$

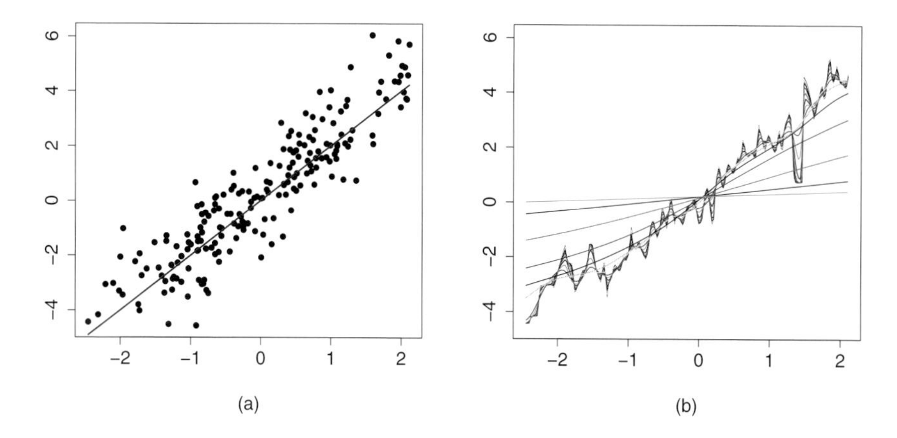

Figure 3.1 *Kernel estimates of a linear function.* (a) The data and the true regression function (red). (b) Kernel estimators of the regression function with a sequence $h = 0.02, \ldots, 5$ of smoothing parameters.

where

$$p_i(x) = \frac{K_h(x - X_i)}{\sum_{i=1}^{n} K_h(x - X_i)}, \qquad i = 1, \ldots, n, \tag{3.7}$$

$K : \mathbf{R}^d \to \mathbf{R}$ is the kernel function, $K_h(x) = K(x/h)/h^d$, and $h > 0$ is the smoothing parameter. The kernel estimator is called also a Nadaraya–Watson estimator, since it was defined by Nadaraya (1964) and Watson (1964).

Figure 3.1 illustrates the effect of the smoothing parameter in estimating a one-dimensional linear regression function $f(x) = 2x$. The data $(X_1, Y_1), \ldots, (X_n, Y_n)$ are i.i.d. with $n = 200$, and $Y_i = f(X_i) + \epsilon_i$, where $\epsilon_i \sim N(0, 1)$, $X_i \sim N(0, 1)$, and X_i and ϵ_i are independent. Panel (a) shows the true regression function with the red line and shows the data with black bullets. Panel (b) shows the kernel estimators of f with smoothing parameters $h = 0.02, 0.02004, \ldots, 2.7, 5.$[25] The kernel is the standard Gaussian density function. When $h \to \infty$, the estimate is converging to the constant function whose value is always the arithmetic mean $\bar{Y}$: $\hat{f}(x) \equiv \bar{Y}$. For small h the estimate is a nonsmooth function.

Figure 3.2 illustrates the effect of the smoothing parameter in estimating a quadratic one-dimensional function. The true regression function is $f(x) = x^2$, $Y_i = f(X_i) + \epsilon_i$, $i = 1, \ldots, n = 200$, where $\epsilon_i \sim N(0, 1)$, and $X_i \sim N(0, 1)$, where X_i and ϵ_i are independent. Panel (a) shows the true regression function and the data. Panel (b) shows the kernel estimators of f for the same sequence of smooth-

[25]We have defined the sequence of N smoothing parameters on the interval $[h_1, h_2]$ by first defining grid points $g_i = h_1 + i\delta$ with equal step sizes $\delta = (h_2 - h_1)/(N - 1)$, $i = 0, \ldots, N - 1$. Then $h_i = a10^{g_i} + b$, where $a = (h_2 - h_1)/(10^{h_2} - 10^{h_1})$ and $b = h_1 - a10^{h_1}$.

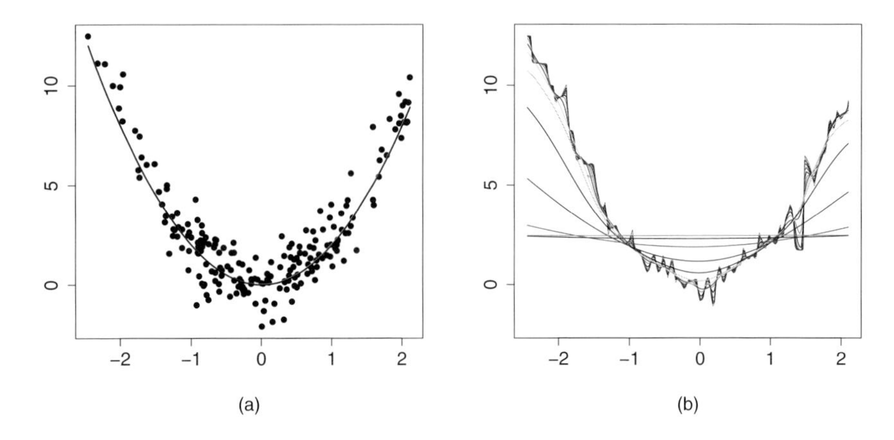

Figure 3.2 *Kernel estimates of a quadratic function.* (a) The data and the true regression function. (b) Kernel estimators of the regression function with a sequence $h = 0.02, \ldots, 5$ of smoothing parameters.

ing parameters as in Figure 3.1, when the kernel is the standard Gaussian density function.

Figure 3.3 illustrates the weights $p_i(x)$ in the case of one-dimensional explanatory variable X. We have used the standard Gaussian kernel and smoothing parameter $h = 0.2$. Panel (a) shows a perspective plot of the function $(x, X_i) \mapsto p_i(x)$, where $X_1, \ldots, X_n$ is a simulated sample of size $n = 200$ from the uniform distribution on $[-1, 1]$. Panel (b) shows the six functions $x \mapsto p_i(x)$ for the choices $X_i = -1, -0.5, \ldots, 1$. We can compare the kernel weights to the linear weights in Figure 2.1. We can see that the kernel weights are local: The weight $p_i(x)$ is positive only for x in a neighborhood of X_i. In contrast, the linear weights $l_i(x)$ are nonzero for almost all x.

3.2.2 Comparison to the Regressogram

When $K = I_{[-1,1]^d}$, we obtain

$$K_h(x - X_i) = h^{-d} I_{R_h(x)}(X_i),$$

where

$$R_h(x) = [x - h, x + h] = [x_1 - h, x_1 + h] \times \cdots \times [x_d - h, x_d + h].$$

Thus,

$$p_i(x) = \frac{1}{n_{R_h(x)}} I_{R_h(x)}(X_i), \tag{3.8}$$

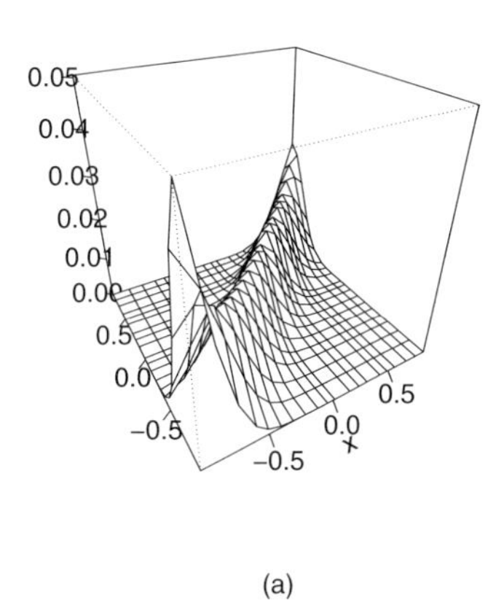

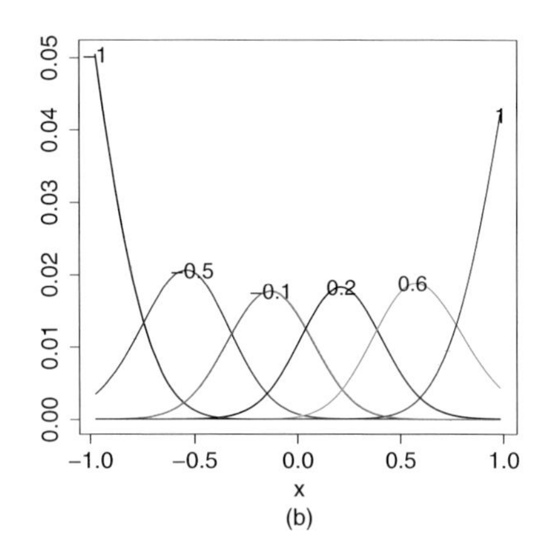

Figure 3.3 *Weights in kernel regression.* (a) The function $(x, X_i) \mapsto p_i(x)$. (b) The six slices $x \mapsto p_i(x)$ for the choices $X_i = -1, -0.5, \ldots, 1$.

where

$$n_{R_h(x)} = \sum_{i=1}^{n} I_{R_h(x)}(X_i)$$

is the number of observations X_i in $R_h(x)$. The weights in (3.8) look similar to the weights of regressogram in (3.5),. However, there is an important difference.

The kernel estimator with the uniform kernel can be written as

$$\hat{f}(x) = \hat{Y}_{R_h(x)}, \tag{3.9}$$

where

$$\hat{Y}_R = \frac{1}{n_R} \sum_{i=1}^{n} Y_i \, I_R(X_i), \qquad n_R = \#\{X_i \in R\}.$$

In contrast, regressogram is defined by

$$\hat{f}(x) = \hat{Y}_{R_x}, \tag{3.10}$$

where R_x is the bin containing x. Thus the difference between the kernel estimator with the uniform kernel and a regressogram is that in the regressogram the average is taken over the rectangle R_x which is the rectangle R in the partition such that $x \in R$, whereas in the kernel estimator the average is taken over the rectangle $R_h(x)$ which is defined as the rectangle whose center is x. Thus kernel estimator is a moving average, and regressogram takes averages over a fixed partition. In the case of a regressogram, it can happen that x is close to the boundary of R_x, and it is better to use a moving average over those values of the response variable where the corresponding points of the explanatory variables are in a symmetric neighborhood of x, as is done when using a kernel estimator.

3.2.3 Gasser–Müller and Priestley–Chao Estimators

Gasser–Müller and Priestley–Chao weights are an alternative in the one-dimensional case to the Nadaraya–Watson weights in (3.7). Now we assume that $X_i \in \mathbf{R}$. These estimators are traditionally used in the case of fixed design regression.

In the Gasser–Müller estimator we use the definition (3.6) but define the weights as

$$p_i(x) = \int_{s_{i-1}}^{s_i} K_h(x-u)\,du,$$

where we assume that the observations are sorted so that $X_1 \le \cdots \le X_n$ and denote $s_i = (X_i + X_{i+1})/2$ for $i = 1, \ldots, n-1$. If it is known that $X_1, \ldots, X_n \in [a, b]$, then $s_0 = a$ and $s_n = b$, but we can also take $s_1 = -\infty$, $s_n = \infty$. This estimator was defined in Gasser & Müller (1979). See also Gasser & Müller (1984).

In the Priestley–Chao estimator, we take the definition (3.6) but define the weights as

$$p_i(x) = (X_i - X_{i-1})K_h(x - X_i), \tag{3.11}$$

where we assume that the observations are sorted so that $X_1 \le \cdots \le X_n$ and $X_0 = a$ when it is assumed that $X_1, \ldots, X_n \in [a, b]$. This estimator was defined in Priestley & Chao (1972).

3.2.4 Moving Averages

Moving averages can be used in time–space smoothing or in time–space prediction. Time–space smoothing and prediction are done using regression techniques with the time parameter as the explanatory variable, as explained in (1.48). The underlying model is typically a signal with noise model of the type $Y_t = \mu_t + \sigma_t \epsilon_t$.

Two-sided moving averages are used in time–space smoothing, and one-sided moving averages are used in time–space prediction. One-sided moving averages can also be used to derive an explanatory variable to be used in state–space smoothing; see, for example, Franke et al. (2004, Section 18.4).

Two-Sided Moving Average Let us consider time series observations $Y_1, \ldots, Y_T$. We choose the explanatory variable $X_t = t$. When the kernel is $K(x) = I_{[-1,1]}(x)$, then the kernel estimator is the moving average

$$\hat{f}(t) = \frac{1}{2h+1} \sum_{i=-h}^{h} Y_{t+i},$$

where $h = 0, 1, 2, \ldots$. We get a wider class of moving averages by choosing a general kernel function $K : \mathbf{R} \to \mathbf{R}$ and smoothing parameter $h > 0$. The moving average[26] is

$$\hat{f}(t) = \sum_{i=1}^{T} p_i(t)\, Y_i, \tag{3.12}$$

where

$$p_i(t) = \frac{K((t-i)/h)}{\sum_{j=1}^{T} K((t-j)/h)}. \tag{3.13}$$

The smoothing parameter $h > 0$ controls the length of the smoothing neighborhood.

One-Sided Moving Average In the time series setting, we need to use one-sided moving averages to make predictions. This can be obtained by choosing the kernel $I_{[0,1]}(x)$. Now,

$$\hat{f}(t) = \frac{1}{h+1} \sum_{i=t-h}^{t} Y_i,$$

where $h = 0, 1, 2, \ldots$. To get a more flexible class of moving averages, we use a general kernel function $K : [0, \infty) \to \mathbf{R}$ and smoothing parameter $h > 0$. We can take $K(x) = \exp(-x) I_{[0,\infty)}(x)$, for example.[27] The one-sided moving average[28] is

$$\hat{f}(t) = \sum_{i=1}^{t} p_i(t)\, Y_i, \tag{3.14}$$

where

$$p_i(t) = \frac{K((t-i)/h)}{\sum_{j=1}^{t} K((t-j)/h)}. \tag{3.15}$$

Exponential Moving Average The exponential moving average is a one-sided moving average obtained by taking $K(x) = \exp(-x)\, I_{[0,\infty)}(x)$ and

$$h = -\frac{1}{\log \gamma},$$

where $0 < \gamma < 1$. Now the estimator (3.14) is equal to

$$\hat{f}(t) = \sum_{i=1}^{t} p_i(t)\, Y_i = \frac{1-\gamma}{1-\gamma^t} \sum_{i=1}^{t} \gamma^{t-i} Y_i. \tag{3.16}$$

[26]The $2k+1$-period moving average with the period number $k = 0, 1, \ldots$ is defined as $\hat{f}(t) = \sum_{i=t-k}^{t+k} p_i(t)\, Y_i$, where $p_i(t) = K((t-i)/h) / \sum_{j=t-k}^{t+k} K((t-j)/h)$. The step number k is taken to be large enough so that the moving average is taken over a large enough number of observations but we do not use k as a smoothing parameter. For example, if we have observed time series $Y_1, \ldots, Y_T$, and we want to calculate $\hat{f}(t)$, we can take $k = \min\{t-1, T-t\}$.

[27]Note that Gijbels, Pope & Wand (1999) use half-kernels, which are kernel functions that are zero in their positive arguments, like $K(x) = \exp(x) I_{(-\infty,0]}(x)$.

[28]The k-period moving average with the period number $k = 0, 1, \ldots$ is defined by $\hat{f}(t) = \sum_{i=t-k}^{t} p_i(t)\, Y_i$, where $p_i(t) = K((t-i)/h) / \sum_{j=t-k}^{t} K((t-j)/h)$. For example, if we have observed time series $Z_1, \ldots, Z_t$, and we want to calculate $\hat{f}(t)$, it is natural to take $k = t-1$.

Indeed, now $\gamma = \exp(-1/h)$ and

$$\exp\left(-\frac{t-i}{h}\right) = \gamma^{t-i}.$$

Using the summation formula of geometric series,[29] we obtain

$$\sum_{j=1}^{t} \gamma^{t-i} = \frac{1-\gamma^t}{1-\gamma}.$$

We get a slightly different exponential moving average by making the recursive definition

$$\mathrm{ma}(t) = (1-\gamma)Y_t + \gamma \mathrm{ma}(t-1), \tag{3.17}$$

where $0 \le \gamma \le 1$. This leads to

$$\mathrm{ma}(t) = (1-\gamma)\sum_{i=1}^{t} \gamma^{t-i} Y_i,$$

when the moving average is calculated from $Y_t, \ldots, Y_1$, and we choose the initial value $\mathrm{ma}(1) = (1-\gamma)Y_1$.

Smoothing parameter selection for exponential smoothing has been considered in Gijbels et al. (1999).

3.2.5 Locally Stationary Data

Let us consider time series (X_t, Y_t), $t = 0, \pm 1, \pm 2, \ldots$. Strong stationarity means that sequence $(X_t, Y_t), \ldots, (X_{t+h}, Y_{t+h})$ is identically distributed as the sequence $(X_u, Y_u), \ldots, (X_{u+h}, Y_{u+h})$, for each $t, u, h \in \mathbf{Z}$. A locally stationary time series is such that the distribution of $(X_t, Y_t), \ldots, (X_{t+h}, Y_{t+h})$ is close to the distribution of $(X_u, Y_u), \ldots, (X_{u+h}, Y_{u+h})$, if t and u are close to each other.

Locally Stationary AR(1) Model Let $Y_1, \ldots, Y_T$ be an observed sequence from the model

$$Y_t = \beta_t Y_{t-1} + \epsilon_t, \qquad t = 1, \ldots, T, \tag{3.18}$$

where $Y_0 = 0$, $\beta_t = \beta(t/T)$, $\beta : [0,1] \to \mathbf{R}$, and $\epsilon_1, \ldots, \epsilon_T$ are i.i.d. $N(0,1)$. This differs from the usual autoregressive model so that the coefficient β_t is changing through time. We choose $\beta(x) = x^{1/2}$. This type of nonstationary time series has been considered, for example, in Dahlhaus (1997).

Figure 3.4 illustrates estimation with the data from model (3.18). Panel (a) shows the simulated time series with $T = 1000$ observations. Panel (b) shows the sequence β_t of the true coefficients with the black curve. The sequential estimates

$$\hat{\beta}_t^{seq} = \frac{\sum_{i=2}^{t} Y_i Y_{i-1}}{\sum_{i=2}^{t} Y_{i-1}^2}$$

[29]For $0 < r < 1$, we have $\sum_{j=0}^{t-1} r^j = (1-r^t)/(1-r)$.

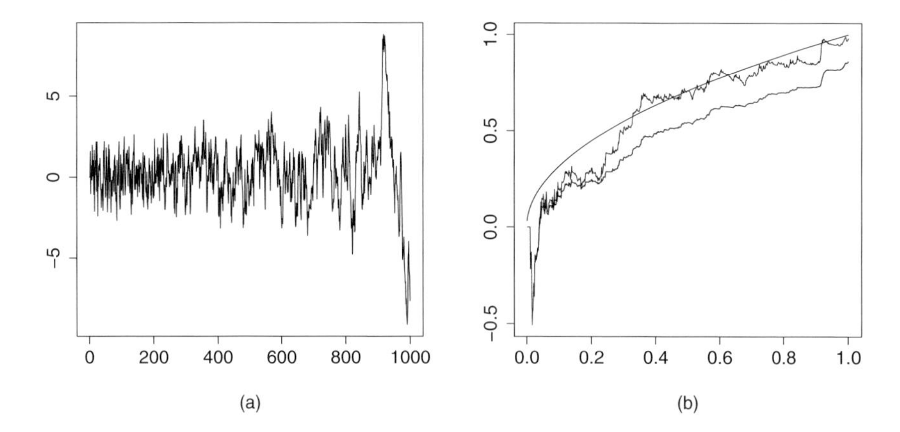

Figure 3.4 *Locally stationary data.* (a) A realization of a locally stationary AR(1) time series with the model (3.18). (b) The true coefficients are shown with the black curve, the sequentially estimated coefficients are shown with the blue curve, and the moving average estimates are shown with the red curve.

are shown with the blue curve. The sequential estimates are the least squares estimators defined in (2.10), but now with the data (Y_{i-1}, Y_i), $i = 2, \ldots, t$. The moving average estimates

$$\hat{\beta}_t^{ma} = \frac{\sum_{i=2}^{t} p_i(t)\, Y_i Y_{i-1}}{\sum_{i=2}^{t} p_i(t)\, Y_{i-1}^2}$$

are shown with the red curve, where the weights $p_i(t)$ are defined in (3.15) with the kernel function $K(x) = \exp(-x) I_{[0,\infty)}(x)$.

Figure 3.5 shows the sequence of true regression curves and the moving average estimates of the regression curves. Panel (a) shows the true regression curves $f_t(x) = \beta_t x$, $x \in [-9, 9]$. The curves are shown for the times $t \in \{100, 150, 200, \ldots, 950\}$. Panel (b) shows the estimated regression curves $\hat{f}_t(x) = \hat{\beta}_t^{ma} x$, when the moving average method is used to estimate the sequence of coefficients.

Combined Time–Space and State–Space Smoothing Let

$$Y_t = f_t(X_t) + \epsilon_t, \qquad t = 1, \ldots, T, \tag{3.19}$$

where $f_t : \mathbf{R}^d \to \mathbf{R}$ are functions that are changing smoothly with time. We have now data $(X_1, Y_1), \ldots, (X_T, Y_T)$. It makes sense to combine time space smoothing and state space smoothing. We will give more weight to the closest in time observations. The estimator is

$$\hat{f}_t(x) = \sum_{i=1}^{t} w_i(x, t) Y_i, \tag{3.20}$$

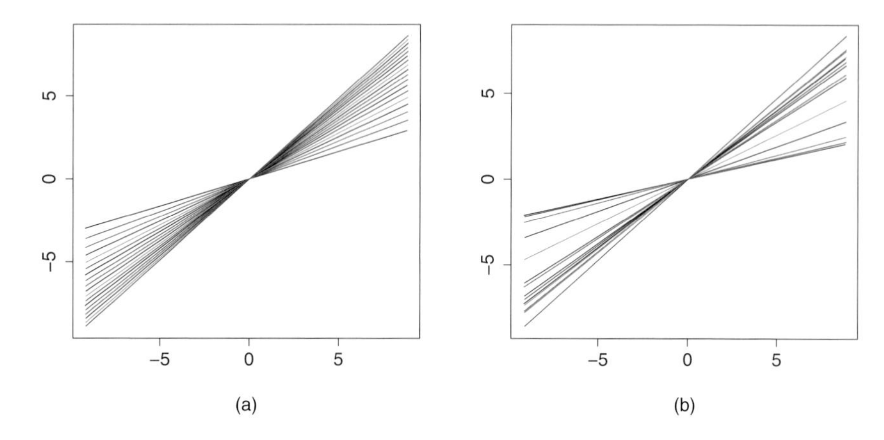

Figure 3.5 *Locally stationary estimation.* (a) The sequence of true regression functions in the locally stationary AR(1) model (3.18). (b) The sequence of estimated regression functions $\hat{f}_t^{ar}$.

where the weights have the form

$$w_i(x,t) = \frac{K((x-X_i)/h)\,L((t-i)/g)}{\sum_{j=1}^n K((x-X_j)/h)\,L((t-j)/g)}, \qquad i = 1,\ldots,t,$$

where $K : \mathbf{R}^d \to \mathbf{R}$, $L : \mathbf{R} \to \mathbf{R}$ are kernel functions and $h > 0$, $g > 0$ are smoothing parameters. Reasonable choices for kernel function $L : \mathbf{R} \to \mathbf{R}$ can be $L(t) = I_{[0,1]}(t)$, $L(t) = (1-t^k)\,I_{[0,1]}(t)$, or $L(t) = \exp(-t^k)\,I_{[0,\infty)}(t)$, where $k = 1, 2, \ldots$.

Let us illustrate the locally stationary nonlinear regression model (3.19). Note that the locally stationary AR(1) model (3.18) is obtained as a special case by choosing $X_t = Y_{t-1}$ and $f_t(x) = \beta(t/T)\,x$. We choose now

$$f_t(x) = 0.5\,\phi\left(x - \mu_t^{(1)}\right) + 0.5\,\phi\left(x - \mu_t^{(2)}\right),$$

where $\mu_t^{(1)} = -2t/T$, $\mu_t^{(2)} = 2t/T$, and ϕ is the density function of the standard normal distribution. The design variables X_t are i.i.d. $N(0,1)$ and the errors ϵ_t are i.i.d. $N(0, 0.1^2)$.

Figure 3.6(a) shows data simulated from the model (3.19) with the sample size $T = 1000$. Panel (b) shows the sequence of true regression functions f_t for the times $t \in \{100, 150, 200, \ldots, 950\}$. The highest unimodal curve is at time $t = 100$ and the lower two modal curve is at time $t = 950$.

Figure 3.7 shows the estimated regression functions in the locally stationary nonlinear regression model (3.19). Panel (a) shows the sequential estimates

$$\hat{f}_t^{seq}(x) = \sum_{i=1}^{t} p_i(x) Y_i,$$

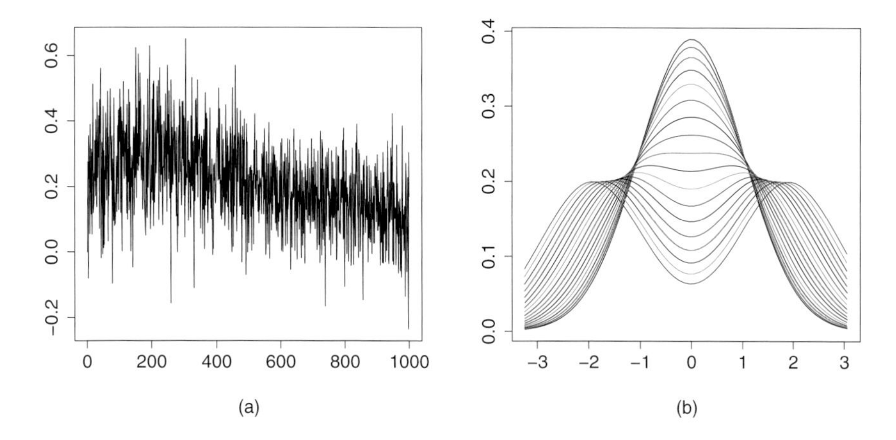

Figure 3.6 *Locally stationary data from a nonlinear model.* (a) A realization from a locally stationary nonlinear regression model (3.19). (b) The sequence of true regression functions f_t.

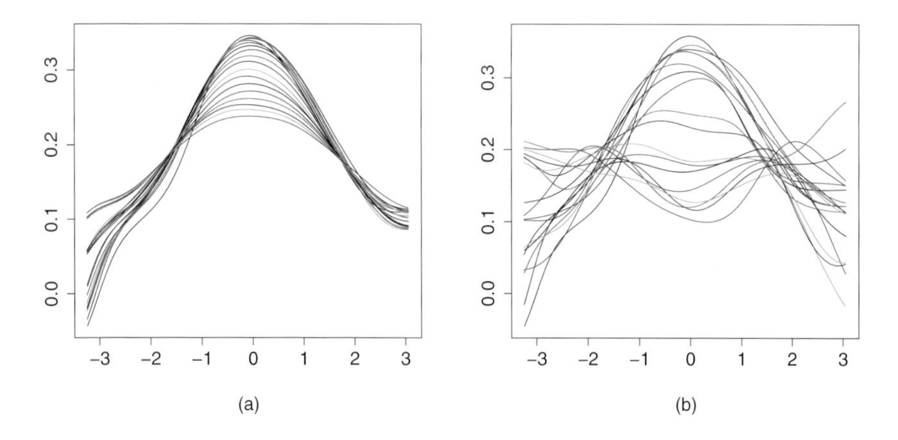

Figure 3.7 *Locally stationary estimation in a nonlinear model.* (a) The sequence of sequential state–space smoothing estimates $\hat{f}_t^{seq}$. (b) The sequence of state–space smoothing and time–space smoothing combining estimates $\hat{f}_t$.

where $p_i(x)$ are the kernel weights (3.7) with the standard Gaussian kernel and the smoothing parameter h is 0.5. Panel (b) shows the state–space smoothing and time–space smoothing combining estimates $\hat{f}_t$ defined in (3.20), when K is the standard Gaussian kernel, L is the exponential kernel, $h = 0.5$, and $g = 50$.

3.2.6 Curse of Dimensionality

Let us consider the case where the kernel function is chosen as $K(x) = I_{[-1/2,1/2]^d}(x)$. Then the support of the scaled kernel function K_h is $[-h/2, h/2]^d$, and this support has volume h^d. When the explanatory variable has uniform distribution on $[0,1]^d$, then the number of observations in the support of K_h is typically $n \cdot h^d$. For example, when $n = 1000$, $h = 0.1$, and $d = 3$, then there is typically one observation in the support of K_h. In general, local neighborhoods are almost empty in high-dimensional spaces, and thus kernel estimators are not efficient in high-dimensional spaces.

Bellman (1961) coined the phrase "curse of dimensionality" while discussing the computational complexity in optimization over many dimensions. Simonoff (1996) gives a detailed discussion of the concept of the curse of dimensionality.

3.2.7 Smoothing Parameter Selection

The kernel estimator $\hat{f} = \hat{f}_h$ depends on the smoothing parameter $h > 0$. We can use data-based methods for choosing the smoothing parameter.

Cross-Validation Cross-validation was defined in (1.116). In cross-validation the smoothing parameter $h > 0$ is chosen which minimizes the empirical mean integrated squared error

$$\text{MISE}_n(h) = \frac{1}{n} \sum_{i=1}^{n} \left(Y_i - \hat{f}_{h,-i}(X_i) \right)^2,$$

where $\hat{f}_{h,-i}$ is the leave-one-out kernel estimator. The leave-one-out kernel estimator is otherwise similar to the kernel estimator, but it is calculated with the data where the ith observation is removed. The leave-one-out kernel estimator is defined as

$$\hat{f}_{h,-i}(x) = \sum_{j=1, j \neq i}^{n} p_{j,-i}(x)\, Y_j,$$

where

$$p_{j,-i}(x) = \frac{p_j(x)}{\sum_{j=1, j\neq i}^{n} p_j(x)}, \qquad j = 1, \ldots, n, \;\; j \neq i.$$

Note that the smoothing parameter cannot be chosen as a minimizer of the sum of the squared residuals $n^{-1} \sum_{i=1}^{n} \left(Y_i - \hat{f}_h(X_i) \right)^2$, because this quantity can be made arbitrarily small by letting $h \downarrow 0$. The mean integrated squared error was defined in (1.113) as

$$\text{MISE}(h) = E \int_{\mathbf{R}^d} \left(\hat{f}_h(x) - f(x) \right)^2 f_X(x)\, dx.$$

The quantity $\text{MISE}_n(h)$ is an unbiased estimator of $\text{MISE}(h)$.

We can write the empirical mean integrated squared error as

$$\text{MISE}_n(h) = \frac{1}{n} \sum_{i=1}^{n} \left(Y_i - \hat{f}_h(X_i) \right)^2 (1 - p_i(X_i))^{-2}. \tag{3.21}$$

Indeed,

$$\text{MISE}_n(h) = \frac{1}{n}\sum_{i=1}^{n}\left(Y_i - \hat{f}_h(X_i)\right)^2 \left(\frac{Y_i - \hat{f}_{h,-i}(X_i)}{Y_i - \hat{f}_h(X_i)}\right)^2$$

and

$$\begin{aligned}\frac{Y_i - \hat{f}_h(X_i)}{Y_i - \hat{f}_{h,-i}(X_i)} &= \frac{Y_i \sum_j p_j(X_i) - \sum_j p_j(X_i)Y_j}{Y_i \sum_{j\neq i} p_{j,-i}(X_i) - \sum_{j\neq i} p_{j,-i}(X_i)Y_j} \\ &= \frac{Y_i \sum_j p_j(X_i) - \sum_j p_j(X_i)Y_j}{Y_i \sum_{j\neq i} p_j(X_i) - \sum_{j\neq i} p_j(X_i)Y_j} \times \sum_{j\neq i} p_j(X_i) \\ &= 1 - p_i(X_i),\end{aligned}$$

because $\sum_{j\neq i} p_j(X_i) = 1 - p_i(X_i)$.

Generalized Cross-Validation Let us write the cross-validation criterion (3.21) as

$$\text{MISE}_n(h) = \frac{1}{n}\sum_{i=1}^{n}\left(Y_i - \hat{f}_h(X_i)\right)^2 \text{Pen}(p_i(X_i)),$$

where $\text{Pen}(u) = (1-u)^{-2}$. The generalized cross-validation criterion is obtained by replacing $p_i(X_i)$ with the average $n^{-1}\sum_{i=1}^{n} p_i(X_i)$. The generalized cross-validation criterion is

$$\text{GCV}(h) = \text{SSR}(h) \times \text{Pen}\left(n^{-1}D\right),$$

where $\text{Pen}(u) = (1-u)^{-2}$,

$$\text{SSR}(h) = \frac{1}{n}\sum_{i=1}^{n}\left(Y_i - \hat{f}_h(X_i)\right)^2,$$

and

$$D = \sum_{i=1}^{n} p_i(X_i). \tag{3.22}$$

Ruppert et al. (2003, Section 5.3.2) give an example which suggests that cross-validation and generalized cross-validation are typically close to each other.

We can interpret $D = \sum_{i=1}^{n} p_i(X_i)$ as analogous to the number K of the parameters in the linear model, as was noted in Ruppert et al. (2003, Section 2.5.2). We observed in (2.11) that when $\hat{f}_{lin}$ is the linear regression estimator, then

$$\hat{f}_{lin}(x) = \sum_{i=1}^{n} l_i(x)\, Y_i,$$

where $l_i(x) = X_i'(\mathbf{X}'\mathbf{X})^{-1}x$ and $\mathbf{X} = (X_1, \ldots, X_n)'$ is the $n \times K$ matrix of the observed values of the explanatory variables. It holds that

$$\sum_{i=1}^{n} l_i(X_i) = K. \tag{3.23}$$

Indeed, we can show (3.23) as follows. We have that $l_i(X_i) = X_i'(\mathbf{X}'\mathbf{X})^{-1}X_i$. Then,

$$\sum_{i=1}^{n} l_i(X_i) = \mathrm{tr}\big(\mathbf{X}(\mathbf{X}'\mathbf{X})^{-1}\mathbf{X}'\big) = \mathrm{tr}\big(\mathbf{X}'\mathbf{X}(\mathbf{X}'\mathbf{X})^{-1}\big) = \mathrm{tr}(I_K) = K.$$

We can consider other penalizing functions $\mathrm{Pen} : (0, \infty) \to \mathbf{R}$ than $\mathrm{Pen}(u) = (1 - u)^{-2}$. Note that for small u, we have

$$(1 - u)^{-2} \approx 1 + 2u \approx \exp(2u).$$

The choice $\mathrm{Pen}(u) = 1 + 2u$ is related to Mallows's C_p criterion, defined in (2.48). When we choose $\mathrm{Pen}(u) = 1 + 2u$, then

$$\mathrm{GCV}(h) = \mathrm{SSR}(h)\,(1 + 2n^{-1}D) = \mathrm{SSR}(h) + 2\hat{\sigma}^2 D, \tag{3.24}$$

where we denote $\hat{\sigma}^2 = n^{-1}\mathrm{SSR}(h)$. The generalized cross-validation criterion in (3.24) and the Mallows's C_p criterion in (2.48) are similar expressions. The estimators $\hat{\sigma}^2$ and $\hat{\sigma}_P^2$ both estimate the error variance, although the estimators are different. By comparing (3.22) and (3.23), we can claim that D and K both express the number of parameters in the model.

The choice $\mathrm{Pen}(u) = \exp(2u)$ is related to Akaike's information criterion, defined in (2.49). When we choose $\mathrm{Pen}(u) = \exp(2u)$, then

$$\log \mathrm{GCV}(h) = \log \mathrm{SSR}(h) + 2n^{-1}D.$$

3.2.8 Effective Sample Size

In the case of a regressogram, as defined in (3.4), and in the case of a nearest neighborhood estimator, as defined in (3.29), it is easy to determine how many observations are used in the local average. In these cases the weights $p_i(x)$ are either zero or a positive constant, the same for all observations $i = 1, \ldots, n$. Thus we know how many observations have an influence on the local average. In the case of a kernel estimator, it is not straightforward to characterize how many observations have an influence on the local average. For the kernel estimator with a nonnegative kernel, the only constraints that the weights satisfy are $0 \le p_i(x) \le 1$ and $\sum_{i=1}^{n} p_i(x) = 1$. When the kernel function has an infinite support, as in the case of the Gaussian kernel, the weights are positive for all observations. However, a Gaussian kernel has tails that are so light that the effective number of observations used in the estimator is much smaller than the total number of observations.

We mention three heuristic methods to measure the effective sample size. Variance and entropy base measures in (3.25) and (3.26) can be used for any local averaging estimator, but the equivalent kernel based measure in (3.27) can be used only for the kernel regression estimator.

Variance and Entropy If Y_i are i.i.d. with $\mathrm{Var}(Y_i) = \sigma^2$, then

$$\mathrm{Var}\left(\sum_{i=1}^{n} p_i(x)\, Y_i\right) = \sigma^2 \sum_{i=1}^{n} p_i^2(x).$$

Thus it is natural to measure the effective number of observations using the Euclidean norm of the vector $(p_1(x), \ldots, p_n(x))$:

$$n_{var}(x) = \left(\sum_{i=1}^{n} p_i^2(x)\right)^{-1}. \tag{3.25}$$

We have $1 \le n(x) \le n$. When $p_i(x) = 1$ only for one $i \in \{1, \ldots, n\}$, then $n_{var}(x) = 1$ and the weights are maximally concentrated (the effective number of observations is the smallest possible). When $p_i(x) = n^{-1}$ for all $i = 1, \ldots, n$, then $n_{var}(x) = n$ and the weights are maximally diffused (the effective number of observations is the largest possible).

We can also use the entropy and define

$$n_{ent}(x) = \exp\left\{-\sum_{i=1}^{n} p_i(x) \log_e p_i(x)\right\}, \tag{3.26}$$

where $\sum_{i=1}^{n} p_i(x) \log_e p_i(x)$ is the entropy. When $p_i(x) = 1$ only for one $i \in \{1, \ldots, n\}$, then the entropy is 0 and $n_{ent}(x) = 1$. When $p_i(x) = n^{-1}$ for all $i = 1, \ldots, n$, then the entropy is $\log_e n$ and $n_{ent}(x) = n$.

Equivalent Kernel Theory We present the equivalent kernel theory for the one-dimensional kernel estimators. Fan & Yao (2005, Section 5.4), suggest that the kernel K_2 using the bandwidth h_2 performs nearly the same as the kernel K_1 using the bandwidth

$$h_1 = \frac{\alpha(K_1)}{\alpha(K_2)}\, h_2,$$

where

$$\alpha(K) = \left(\int_{-\infty}^{\infty} u^2 K(u)\, du\right)^{-2/5} \|K\|_2^{2/5}.$$

This follows because the asymptotically optimal bandwidth, minimizing the mean integrated squared error, is

$$h_{opt} = \alpha(K)\, \|f''\|_2^{-2/5}\, n^{-1/5},$$

when the unknown regression function f is twice differentiable.

The kernel estimator with kernel

$$K_1(x) = I_{[-1/2,1/2]}(x)$$

and bandwidth h_1 uses as many observations as occur in the interval $[x - h_1/2, x + h_1/2]$. The mapping $N_x : (0, \infty) \to \{1, \ldots, n\}$, defined by

$$N_x(h_1) = \sum_{i=1}^{n} I_{[x-h_1/2, x+h_1/2]}(X_i),$$

gives the number of observations in the interval $[x - h_1/2, x + h_1/2]$. Thus kernel estimator with kernel K_2 and bandwidth h_2 uses

$$n_{ker}(x) = N_x\left(\frac{3^{2/5}}{\alpha(K_2)}\, h_2\right) \tag{3.27}$$

observations, because $\alpha(K_1) = 3^{2/5} = 1.551.$[30]

Effective Sample Size for Time Series Data Let us study the three ways to define the effective sample size in the case of time–space smoothing. Let us assume to have time series $Z_1, \ldots, Z_T$ and calculate the effective sample size when we make one-sided exponential moving average at time T. The three methods of calculating the effective sample size are defined in (3.25), (3.26), and (3.27). The exponential moving average is defined in (3.16).

The smoothing parameter $h > 0$ is replaced by the coefficient $0 < \gamma < 1$, which are related as $h = -1/\log(\gamma)$. The variance-based effective sample size (3.25) gives[31]

$$n_{var}(T) = \frac{1+\gamma}{1-\gamma}\,\frac{1-\gamma^T}{1+\gamma^T}.$$

The entropy-based effective sample size $n_{ent}(T)$, defined in (3.26), does not have such closed form expression but we can calculate it numerically.

Let us consider time series data and the one-sided moving average defined in (3.14). Let us consider the kernel

$$K_1(x) = I_{[0,1]}(x).$$

We have, as before, that $\alpha(K_1) = 3^{2/5} = 1.551846$. For the time series data $Z_1, \ldots, Z_T$ we have that

$$N_T(h_1) = \sum_{t=1}^{T} I_{[T-h_1, T]}(t) \approx h_1,$$

for each $t = 1, \ldots, T$. Thus, for the series data the effective sample size is

$$n_{ker} = \frac{3^{2/5}}{\alpha(K_2)}\, h_2.$$

[30] For $K = I_{[-1/2,1/2]}(x)$, $\int_{-\infty}^{\infty} K^2 = 1$, and $\int_{-\infty}^{\infty} u^2 K(u)\, du = 1/3$.

[31] We have that $\sum_{i=1}^{T} \gamma^{2(t-i)} = (1 - \gamma^{2T})/(1 - \gamma^2)$.

For example, let us choose $K_2(x) = \exp(-x)I_{[0,\infty)}(x)$. Now $\alpha(K_2) = 2^{-3/5} = 0.659754$.[32]

The following table gives the effective sample sizes n_{var} n_{ent}, and n_{ker} for several values of γ and $h = -1/\log(\gamma)$. Note that n_{var} and n_{ent} depend on the sample size T, whereas n_{ker} does not depend on T. For a small sample size it can happen that $n_{ker} > T$. We have used $T = 1000$ in the table. Fan & Gu (2003, Table 1) shows a similar table for γ and n_{ker}.

γ	0.90	0.91	0.92	0.93	0.94	0.95	0.96	0.97	0.98	0.99
h	9.5	10.6	12.0	13.8	16.2	19.5	24.5	32.8	49.5	99.5
n_{var}	19	21	24	28	32	39	49	66	99	199
n_{ent}	26	29	33	37	44	53	67	89	135	270
n_{ker}	22	25	28	32	38	46	58	77	116	234

3.2.9 Kernel Estimator of Partial Derivatives

Let $f(x) = E(Y \mid X = x)$, $f : \mathbf{R}^d \to \mathbf{R}$, be the conditional expectation and let us denote its partial derivatives by

$$D_k f(x) = \frac{\partial}{\partial x_k} f(x), \qquad x \in \mathbf{R}^d, \ k \in \{1, \ldots, d\}.$$

We can estimate the partial derivatives by taking partial derivatives of a kernel estimator. Thus we define the estimator of a partial derivative of a regression function by

$$\widehat{D_k f}(x) = \frac{\partial}{\partial x_k} \left(\sum_{i=1}^{n} p_i(x) Y_i \right) = \sum_{i=1}^{n} q_i(x) Y_i, \tag{3.28}$$

where $q_i(x) = \partial p_i(x)/\partial x_k$. We defined $p_i(x) = K_h(x - X_i)/\sum_{i=1}^n K_h(x - X_i)$, and thus

$$q_i(x) = \frac{1}{\sum_{i=1}^n K_h(x - X_i)} \left(\frac{\partial}{\partial x_k} K_h(x - X_i) - p_i(x) \sum_{i=1}^{n} \frac{\partial}{\partial x_k} K_h(x - X_i) \right),$$

where

$$\frac{\partial}{\partial x_k} K_h(x - X_i) = \frac{1}{h^{d+1}} (D_k K) \left(\frac{x - X_i}{h} \right).$$

For example, let K be the standard Gaussian density function:

$$K(x) = (2\pi)^{-d/2} \exp\left\{ -\tfrac{1}{2} \|x\|^2 \right\}.$$

Then,

$$D_k K(x) = -x_k K(x).$$

[32] For $K(x) = \exp(-x)I_{[0,\infty)}(x)$, $\int_{-\infty}^{\infty} K^2 = 1/2$, and $\int_{-\infty}^{\infty} u^2 K^2(u)\, du = 2$.

In the one-dimensional case we can use also Gasser–Müller or Priestley–Chao estimators, defined in Section 3.2.3. Priestley–Chao weights are $p_i(x) = (X_i - X_{i-1})K_h(x - X_i)$, and thus for this choice we have

$$q_i(x) = (X_i - X_{i-1})\, h^{-2} K'\left(\frac{x - X_i}{h}\right),$$

where K' is the derivative of K.

3.2.10 Confidence Intervals in Kernel Regression

We have defined pointwise confidence intervals in Section 1.10.1. The following result is from Härdle (1990, Theorem 4.2.1). We restrict ourselves to the one-dimensional case $d = 1$, where there is only one explanatory variable. Let $f(x) = E(Y \mid X = x)$ be the regression function, $\sigma^2(x) = \mathrm{Var}(Y \mid X = x)$ be the variance function, and f_X the density function of distribution of X. We assume that f and f_X are two times continuously differentiable, $E(|Y|^{2+\epsilon} \mid X = x) < \infty$ for some $\epsilon > 0$, σ^2 is continuous at x, and $f_X(x) > 0$. Smoothing parameter is chosen as $h = cn^{-1/5}$ for $c > 0$ and kernel function is chosen such that $\int |K|^{2+\epsilon} < \infty$ for some $\epsilon > 0$. Then,

$$n^{2/5}\left(\hat{f}(x) - f(x)\right) \xrightarrow{d} N\big(b(x), v^2(x)\big),$$

as $n \to \infty$, where

$$b(x) = c^2 \mu_2(K)\left(\frac{f''(x)}{2} + \frac{f'(x) f_X'(x)}{f_X(x)}\right),$$

with $\mu_2(K) = \int t^2 K(t)\, dt$ and

$$v^2(x) = \frac{\sigma^2(x) \|K\|_2^2}{c f_X(x)}.$$

To calculate a confidence interval we can take $h = cn^{-1/5} / \log n$. Then,

$$(nh)^{1/2}\left(\hat{f}(x) - f(x)\right) \xrightarrow{d} N\left(0, \frac{\sigma^2(x)\|K\|_2^2}{f_X(x)}\right),$$

as $n \to \infty$. We get the confidence interval

$$\left[\hat{f}(x) - a(x), \hat{f}(x) + a(x)\right],$$

where

$$a(x) = z_{1-\alpha/2} \sqrt{\frac{\|K\|_2^2 \hat{\sigma}^2(x)}{nh \hat{f}_X(x)}},$$

$\hat{f}_X$ is a kernel density estimator of f_X, $\hat{\sigma}^2(x)$ is a kernel estimator of the variance function, and $z_{1-\alpha/2}$ is such that $P(-z_{1-/\alpha/2} \le Z \le z_{1-\alpha/2}) = 1-\alpha$, where $Z \sim N(0,1)$.

A second possibility, suggested in Wasserman (2005, Section 5.7), is to make a confidence interval for $E\hat{f}(x)$ by using

$$(nh)^{1/2}\left(\hat{f}(x) - E\hat{f}(x)\right) \xrightarrow{d} N\left(0, \frac{\sigma^2(x)\|K\|_2^2}{f_X(x)}\right),$$

as $n \to \infty$, when $h = cn^{-1/5}$.

A third possibility is to make a bootstrap confidence interval, which can be constructed in the following way. Generate B bootstrap samples from the original sample $(X_1, Y_1), \ldots, (X_n, Y_n)$. Based on a bootstrap sample $(X_1^*, Y_1^*), \ldots, (X_n^*, Y_n^*)$ a regression function estimate $\hat{f}^*$ is constructed. We obtain a sequence $\hat{f}_1^*(x), \ldots, \hat{f}_B^*(x)$ of estimates. Let $q_{\alpha/2}(x)$ and $q_{1-\alpha/2}(x)$ be the empirical quantiles of the sequence of estimates. Then the confidence interval is

$$\left[q_{\alpha/2}(x), q_{1-\alpha/2}(x)\right].$$

In Section 1.10.2 we have defined the concept of a confidence band, which is different from a confidence interval. Härdle (1990, Section 4.3) gives confidence bands for kernel regression and Sun & Loader (1994) give confidence bands for linear regression and smoothing. See also Wasserman (2005, Section 5.7)

3.3 NEAREST-NEIGHBOR ESTIMATOR

The kernel regression estimator with the uniform kernel is defined in (3.9). This estimator at point $x \in \mathbf{R}^d$ is an average over those Y-values, where the corresponding X-values are in the rectangle whose center is x and whose side length is $2h$, where $h > 0$ is the smoothing parameter. The nearest-neighbor regression estimator changes this estimator in two ways. First, the rectangle is changed to a ball centered at x. Second, the radius of the ball is not a constant value, the same at every point $x \in \mathbf{R}^d$, but the radius is changing in such a way that the ball contains always exactly k X-observations, where $k = 1, 2, \ldots$ is an integer that plays the role of the smoothing parameter. The changing radius brings the advantage that also in the tail areas of the X-distribution, where the observations are sparse, it is guaranteed that the average is over k Y-values, whereas in the case of the kernel estimator the average could be over only few Y-values, or the average could be even over an empty set of Y-values, in which case the kernel estimator would not be defined.

We define the nearest-neighbor estimator for the regression function as

$$\hat{f}(x) = \sum_{i=1}^{n} p_i(x)\, Y_i, \qquad x \in \mathbf{R}^d, \tag{3.29}$$

where

$$p_i(x) = \frac{1}{k} I_{B_{r_{k,x}}(x)}(X_i) \tag{3.30}$$

and

$$r_{k,x} = \min\{r > 0 : \#\{X_i \in B_r(x)\} = k\}, \tag{3.31}$$

where $B_r(x)$ is the ball centered at x with radius r: $B_r(x) = \{y \in \mathbf{R}^d : \|x-y\| \leq r\}$. The radius $r_{k,x}$ is the minimum radius such that the ball with this radius, centered at x, contains exactly k observations. Thus,

$$\sum_{i=1}^{n} I_{B_{r_{k,x}}(x)}(X_i) = \#\{X_i \in B_{r_{k,x}}(x)\} = k.$$

Equivalently, a nearest-neighbor estimator can be defined as

$$\hat{f}(x) = \hat{Y}_{B_{r_{k,x}}(x)}, \tag{3.32}$$

where

$$\hat{Y}_R = \frac{1}{n_R} \sum_{i=1}^{n} Y_i \, I_R(X_i)$$

and

$$n_R = \#\{X_i \in R\}.$$

3.4 CLASSIFICATION WITH LOCAL AVERAGING

We have introduced classification in Section 1.4. In classification the possible values of Y are $\{0, \ldots, K-1\}$, and $X \in \mathbf{R}^d$ is an associated predictive variable. We want to find a classification function $g : \mathbf{R}^d \to \{0, \ldots, K-1\}$, which predicts the class label Y with the predictive variable X. The classification function is constructed from the observations $(X_1, Y_1), \ldots, (X_n, Y_n)$, which are identically distributed with the distribution of (X, Y).

3.4.1 Kernel Classification

We define kernel density classifier and kernel regression classifier and note that they are equivalent.

Classification Based on Kernel Density Estimation The classification function can be constructed using density estimators. A rule for the construction of a classification function with the help of density estimators was given in (1.83), which puts

$$\hat{g}(x) = \mathrm{argmax}_{k=0,\ldots,K-1} \, \hat{p}_k \hat{f}_{X|Y=k}(x), \tag{3.33}$$

where $\hat{f}_{X|Y=k}$ is a density estimator for the class k density and $\hat{p}_k$ is an estimator of the prior probability of class k. We can choose

$$\hat{p}_k = \frac{1}{n} \#\{i = 1, \ldots, n : Y_i = k\}. \tag{3.34}$$

We use now a kernel density estimator as the estimator $\hat{f}_{X|Y=k}$ of the class k density. We define the kernel density estimator in (3.39); and applying this formula to the estimation of the class densities, we get

$$\hat{f}_{X|Y=k}(x) = \frac{1}{n_k} \sum_{i=1}^{n} K_h(x - X_i) I_{\{k\}}(Y_i), \tag{3.35}$$

where $n_k = \#\{i = 1, \ldots, n : Y_i = k\}$.

Classification Based on Kernel Regression Function Estimation The classification function can be constructed using regression function estimators. A rule for the construction of a classification function with the help of regression function estimators was given in (1.78), which puts

$$\hat{g}(x) = \text{argmax}_{k=0,\ldots,K-1}\, \hat{p}_k(x), \tag{3.36}$$

where $\hat{p}_k(x)$ are estimators of $P(Y = k \,|\, X = x)$, for $k = 0, \ldots, K-1$. We estimate $\hat{p}_k(x)$ by defining indicator variables, as in (1.79):

$$Y_i^{(k)} = I_{\{k\}}(Y_i), \qquad i = 1, \ldots, n, \;\; k = 0, \ldots, K-1.$$

Let $\hat{p}_k(x)$ be a kernel regression function estimator, defined in (3.6), constructed using regression data $(X_1, Y_1^{(k)}), \ldots, (X_n, Y_n^{(k)})$, for $k = 0, \ldots, K-1$.

We are led to the same rule as using kernel density function estimation, defined in (3.33)–(3.35). Indeed, we can write the kernel regression function estimator as

$$\begin{aligned} \hat{p}_k(x) &= \frac{1}{\hat{f}_X(x)\, n} \sum_{i=1}^{n} K_h(x - X_i)\, Y_i^{(k)} \\ &= \frac{1}{\hat{f}_X(x)} \frac{n_k}{n} \hat{f}_{X\,|\,Y=k}(x) = \frac{\hat{p}_k \hat{f}_{X\,|\,Y=k}(x)}{\hat{f}_X(x)}, \end{aligned} \tag{3.37}$$

where $\hat{f}_{X\,|\,Y=k}(x)$ is defined in (3.35) and $\hat{f}_X$ is the density estimator

$$\hat{f}_X(x) = \frac{1}{n} \sum_{i=1}^{n} K_h(x - X_i).$$

3.4.2 Nearest-Neighbor Classification

We define the nearest-neighbor density classifier and the nearest-neighbor regression classifier. These are not equivalent, as in the case of kernel estimators, but we note that the nearest-neighbor regression classifier is equivalent to a prototype classifier.

Classification Based on Nearest-Neighbor Density Estimation First we have to define the nearest-neighbor density estimator. The nearest-neighbor density estimator, based on identically distributed data $X_1, \ldots, X_n \in \mathbf{R}$, is

$$\hat{f}_X(x) = \frac{k/n}{\text{volume}(B_{r_{k,x}}(x))},$$

where $k = 1, 2, \ldots$ is the smoothing parameter and $r_{k,x}$ is the smallest radius r so that the ball $B_r(x)$ contains exactly k observations: $r_{k,x} = \min\{r > 0 : \#\{X_i \in B_r(x)\} = k\}$.[33] The volume of ball $B_r(x) \subset \mathbf{R}^d$ with radius r is

$$\text{volume}(B_r(x)) = \frac{\pi^{d/2}}{\Gamma(d/2+1)} r^d.$$

Second we use the density rule (1.83) to define a classification function. Let $(X_1, Y_1), \ldots, (X_n, Y_n)$ be classification data and let

$$\hat{g}(x) = \text{argmax}_{y=0,\ldots,K-1} \hat{p}_y \hat{f}_{X|Y=y}(x),$$

where $\hat{f}_{X|Y=y}$ is a nearest-neighbor density estimator for the class y density and

$$\hat{p}_y = \#\{i = 1, \ldots, n : Y_i = y\}/n$$

is the prior probability of class y.[34] The nearest-neighbor density estimator of the class density is

$$\hat{f}_{X|Y=y}(x) = \frac{k/n}{\text{volume}(B_{r_{k,x,y}}(x))},$$

where

$$r_{k,x,y} = \min\{r > 0 : \#\{X_i \in B_r(x) : Y_i = y\} = k\}.$$

Finally we get the nearest-neighbor regression-based classification rule:

$$\hat{g}(x) = \text{argmax}_{y=0,\ldots,K-1} \frac{\hat{p}_y}{\text{volume}(B_{r_{n,k}(x,y)}(x))},$$

where the multiplier k/n has been left out because it is the same for each class y.

Classification Based on Nearest-Neighbor Regression Function Estimation We apply the regression rule (1.78), which defines the empirical classification function as

$$\hat{g}(x) = \text{argmax}_{y=0,\ldots,K-1} \hat{p}_y(x),$$

where $\hat{p}_y(x)$ are estimators of $P(Y = y \,|\, X = x)$, for $y = 0, \ldots, K-1$. We apply the nearest-neighbor regression estimator, defined in (3.29), when the class label indicators $I_{\{y\}}(Y_i)$, $i = 1, \ldots, n$, are the response variables, for $y = 0, \ldots, K-1$. We obtain the estimators

$$\hat{p}_y(x) = \sum_{i=1}^{n} p_i(x)\, I_{\{y\}}(Y_i), \qquad x \in \mathbf{R}^d, \tag{3.38}$$

[33] Here $B_r(x)$ is the ball centered at x with radius r: $B_r(x) = \{y \in \mathbf{R}^d : \|x - y\| \le r\}$.
[34] We use the class label variable $y \in \{0, \ldots, K-1\}$ instead of the previously used k, because k is traditionally used to denote the smoothing parameter of the k-nearest-neighbor estimators.

where the weights were defined in (3.30) as

$$p_i(x) = \frac{1}{k} I_{B_{r_{k,x}}(x)}(X_i),$$

where $r_{k,x}$ is the smallest radius r such that the ball $B_r(x)$ contains exactly k observations from the observed vectors $X_1, \ldots, X_n$. The classification based on the nearest-neighbor regression is not equivalent to the classification based on the nearest-neighbor density estimation. However, we show next that the classification based on the nearest-neighbor regression is an prototype classifier classifying the new observation to the class whose observed values are most similar to the new observation. The name "prototype methods" was used in Hastie et al. (2001, Section 13).

Classification Based on the Nearest-Neighbor Rule The nearest-neighbor classification rule is

$$\hat{g}(x) = y \Leftrightarrow \text{ the most class labels } Y_i \text{ are } y \text{ in the } k\text{-neighborhood of } x,$$

where $y = 0, \ldots, K-1$. The k-neighborhood of x is defined as the ball $B_{r_{k,x}}(x)$, where $r_{k,x}$ is the smallest radius r such that the ball $B_r(x)$ contains exactly k observations from the observed vectors $X_1, \ldots, X_n$, and $k = 1, 2, \ldots$. Let

$$n_y(x) = \#\{i = 1, \ldots, n : Y_i = y, X_i \in B_{r_{k,x}}(x)\}$$

be the number of Y-observations in the neighborhood $B_{r_{k,x}}(x)$ with the label y. Then,

$$\hat{g}(x) = \text{argmax}_{y=0,\ldots,K-1} n_y(x).$$

We have defined the same classification rule that was obtained with the help of nearest-neighbor regression functions, because

$$n_y(x) = k\, \hat{p}_y(x),$$

where $\hat{p}_y(x)$ is defined in (3.38).

3.5 MEDIAN SMOOTHING

We have defined the regressogram, the kernel regression estimator with the uniform kernel, and the nearest-neighbor estimator in (3.10), (3.9), and (3.32) as averages of Y-values over those X-values that are in in a bin, in a local neighborhood, or in a nearest neighborhood. These estimators of conditional expectation can be changed to estimators of conditional median when we change the sample average to the sample median.

1. The median regressogram is

$$\hat{f}(x) = \text{median}(\{X_i \in R_x\}),$$

where R_x is the bin containing x.

2. A kernel estimator of a conditional median is

$$\hat{f}(x) = \text{median}(\{X_i \in R_h(x)\}),$$

where

$$R_h(x) = [x - h, x + h] = [x_1 - h, x_1 + h] \times \cdots \times [x_d - h, x_d + h]$$

is the rectangle centered at x, with side length $2h$.

3. The nearest-neighbor estimator of the conditional median is

$$\hat{f}(x) = \text{median}(\{X_i \in B_{r_{k,x}}(x)\}),$$

where $B_r(x)$ is the ball centered at x with radius r and

$$r_{k,x} = \min\{r > 0 : \#\{X_i \in B_r(x)\} = k\}$$

is the minimum radius such that the ball with this radius, centered at x, contains exactly k observations.

Above we have used the definition of the sample median, given in (1.10). We can define the median regression estimator for the general kernel weights $p_i(x)$, $i = 1, \ldots, n$, by using the definition (1.8) of the population median. Let us define the random variable $Y_n(x)$, with the discrete distribution

$$P(Y_n(x) = y_i) = p_i(x), \qquad i = 1, \ldots, n,$$

where $y_1 \ldots, y_n$ are the observed values of $Y_1, \ldots, Y_n$. Then we can define the median regression estimator as

$$\hat{f}(x) = \text{median}(Y_n(x)).$$

3.6 CONDITIONAL DENSITY ESTIMATION

We define kernel, histogram, and nearest-neighbor estimators of the conditional density. The conditional density can be defined either by conditioning on the state variables, or it can be defined in the time series setting by conditioning on the information at time t. This leads either to the state–space smoothing or to the time–space smoothing. We will also combine the state–space smoothing and the time–space smoothing to obtain an estimator for locally stationary data.

3.6.1 Kernel Estimator of Conditional Density

We start by defining the unconditional kernel density estimator. After that, we define the conditional kernel density estimators. Finally we note that the conditional density estimator in the state–space sense and the kernel regression estimator can be derived from the unconditional density estimator.

Unconditional Kernel Density Estimator The kernel density estimator $\hat{f}_X(x)$ of the density function $f_X : \mathbf{R}^d \to \mathbf{R}$ of random vector $X \in \mathbf{R}^d$, based on identically distributed data $X_1, \ldots, X_n \in \mathbf{R}^d$, is defined by

$$\hat{f}_X(x) = \frac{1}{n} \sum_{i=1}^{n} K_h(x - X_i), \qquad x \in \mathbf{R}^d, \tag{3.39}$$

where $K : \mathbf{R}^d \to \mathbf{R}$ is the kernel function, $K_h(x) = K(x/h)/h^d$, and $h > 0$ is the smoothing parameter.

We can explain the definition of the kernel density estimator in the following informal way. A density function $f_X : \mathbf{R}^d \to \mathbf{R}$ of a probability distribution is a function which satisfies

$$P(A) = \int_A f_X(x)\, dx,$$

for all measurable $A \subset \mathbf{R}^d$. We can approximate the density at point x by choosing a small set centered at x. For example, let

$$U_{x,h} = \{z \in \mathbf{R}^d : \|z - x\| \le h\}$$

be the ball centered at $x \in \mathbf{R}^d$ with radius $h > 0$. Since

$$\int_{U_{x,h}} f_X \approx f_X(x)\, \lambda(U_{x,h}),$$

where λ is the Lebesgue measure, we get the approximation for small h,

$$f_X(x) \approx \frac{P(U_{x,h})}{\lambda(U_{x,h})}. \tag{3.40}$$

By the law of large numbers, probabilities can be approximated by frequencies, and we have

$$P(U_{x,h}) \approx \frac{\#\{X_i \in U_{x,h}\}}{n} = \frac{1}{n} \sum_{i=1}^{n} I_{U_{x,h}}(X_i),$$

for large n. We can write

$$I_{U_{x,h}}(X_i) = I_{U_{0,1}}\left(\frac{X_i - x}{h}\right)$$

and we have that

$$\lambda(U_{x,h}) = h^d\, \lambda(U_{0,1}).$$

Thus, (3.40) can be written as

$$f_X(x) \approx \frac{1}{nh^d} \sum_{i=1}^{n} I_{U_{0,1}}\left(\frac{X_i - x}{h}\right) = \frac{1}{n} \sum_{i=1}^{n} K_h(x - x_i),$$

where $K(x) = I_{U_{0,1}}(x)$. We get the class of kernel density estimators by letting the kernel function K be any integrable function $K : \mathbf{R}^d \to \mathbf{R}$.

Conditional Kernel Density Estimator Unconditional kernel density estimator was defined in (3.39). In the univariate case a kernel estimator of the density of $Y \in \mathbf{R}$, based on data $Y_1, \ldots, Y_n$, is

$$\hat{f}_Y(y) = \frac{1}{n} \sum_{i=1}^{n} L_g(y - Y_i), \qquad y \in \mathbf{R}, \tag{3.41}$$

where $L : \mathbf{R} \to \mathbf{R}$ is the kernel function, $L_g(y) = L(y/g)/g$, and $g > 0$ is the smoothing parameter.

A kernel estimator of the conditional density of Y given X, based on data $(X_1, Y_1), \ldots, (X_n, Y_n)$, associated with the kernel regression function estimator, is defined as

$$\hat{f}_{Y|X=x}(y) = \sum_{i=1}^{n} p_i(x)\, L_g(y - Y_i), \qquad y \in \mathbf{R},\ x \in \mathbf{R}^d, \tag{3.42}$$

where the weights $p_i(x)$ are the kernel weights defined in (3.7). We can also allow the regressogram weights defined in (3.5) or the nearest-neighbor weights defined in (3.30).

Time–Space Smoothing Let $Y_1, \ldots, Y_T$ be an observed time series. Moving averages were defined in Section 3.2.4. We can use a two-sided moving average to define the estimator of the density function of Y_t as

$$\hat{f}_{Y_t}(y) = \sum_{i=1}^{T} p_i(t)\, L_g(y - Y_i), \qquad y \in \mathbf{R},\ t = 1, \ldots, T, \tag{3.43}$$

where the weights $p_i(t)$ are defined in (3.13). In prediction we use a one-sided moving average and define the estimator of the density function of Y_t as

$$\hat{f}_{Y_t}(y) = \sum_{i=1}^{t} p_i(t)\, L_g(y - Y_i), \qquad y \in \mathbf{R}, \tag{3.44}$$

where the weights $p_i(t)$ are defined in (3.15). A particular case of the one-sided moving average is the exponential moving average. The exponential moving average estimator of the density function is defined analogously to (3.16) as

$$\hat{f}_{Y_t}(y) = \frac{1 - \gamma}{1 - \gamma^t} \sum_{i=1}^{t} \gamma^{t-i}\, L_g(y - Y_i),$$

where $0 < \gamma < 1$, $\gamma = \exp(-1/h)$.

Time– and State–Space Smoothing We can combine time–space smoothing and state–space smoothing. Time–space smoothing can be either two-sided smoothing or one-sided smoothing. The one-sided smoothing is more typical, because it can be used in prediction. Combining the state–space smoothing and the one-sided moving average gives an estimator of the conditional density of Y_t:

$$\hat{f}_{Y_t|X=x}(y) = \sum_{i=1}^{t} w_i(x,t)\, L_g(y - Y_i), \qquad y \in \mathbf{R},\ x \in \mathbf{R}^d,$$

where

$$w_i(x,t) = \frac{p_i(x)\,\pi_i(t)}{\sum_{j=1}^{t} p_j(x)\,\pi_j(t)}, \tag{3.45}$$

with

$$p_i(x) = K((x - X_i)/h), \qquad \pi_i(t) = M((t-i)/a),$$

where $K : \mathbf{R}^d \to \mathbf{R}$, $M : [0,\infty) \to \mathbf{R}$ are the kernel functions, and $h > 0$, $a > 0$ are the smoothing parameters. We can use also the regressogram weights or the nearest-neighbor weights to replace the kernel weights. The regressogram weights are $p_i(x) = I_{A_x}(X_i)/n_{A_x}$, where A_x is the bin which contains x and n_{A_x} is the number of X-observations in A_x. The nearest-neighbor weights are $p_i(x) = I_{B_{r_{k,x}}(x)}(X_i)/k$, where $r_{k,x}$ is the smallest radius such that the ball with this radius, centered at x, contains exactly k X-observations.

Conditional Density Estimator Derived from Density Estimator We show that the kernel estimator of conditional density in the state–space sense, defined in (3.42), can be derived from the kernel density estimator. From definition (3.39) we obtain the kernel density estimator $\hat{f}_{X,Y}(x,y)$ of the density of (X,Y), based on data (X_i, Y_i), $i = 1,\ldots,n$. Let us modify the definition to allow a different smoothing parameter for the x coordinates than for the y coordinate. This density estimator is

$$\hat{f}_{X,Y}(x,y) = \frac{1}{n}\sum_{i=1}^{n} M_{h,g}(x - X_i, y - Y_i), \qquad x \in \mathbf{R}^d,\ y \in \mathbf{R}, \tag{3.46}$$

where $M : \mathbf{R}^{d+1} \to \mathbf{R}$ is the kernel function, $M_{h,g}(x,y) = M(x/h, y/g)/(h^d g)$, and $h > 0$, $g > 0$ are the smoothing parameters. The kernel density estimator $\hat{f}_X(x)$ of the density of X is equal to

$$\hat{f}_X(x) = \frac{1}{n}\sum_{i=1}^{n} K_h(x - X_i), \qquad x \in \mathbf{R}^d,$$

where $K : \mathbf{R}^d \to \mathbf{R}$ and $K_h(x) = K(x/h)/h^d$. An estimator of the conditional density can be defined as

$$\hat{f}_{Y|X=x}(y) = \frac{\hat{f}_{X,Y}(x,y)}{\hat{f}_X(x)}, \qquad x \in \mathbf{R}^d,\ y \in \mathbf{R}. \tag{3.47}$$

Let us assume that

$$M(x,y) = K(x) \cdot L(y),$$

where $L : \mathbf{R} \to \mathbf{R}$. Then,

$$\hat{f}_{X,Y}(x, y) = \frac{1}{n} \sum_{i=1}^{n} K_h(x - X_i) \, L_h(y - Y_i) \tag{3.48}$$

and (3.47) is equal to (3.42) with $p_i(x) = K_h(x - X_i) / \sum_{i=1}^{n} K_h(x - X_i)$.

Regression Estimator Derived from Density Estimator We show that the kernel estimator of the regression function, as defined in (3.6), is under some conditions equal to the conditional mean of a kernel density estimator. We define the regression function estimator $\hat{f}(x)$ as the mean of the estimator of the conditional distribution:

$$\hat{f}(x) = \int_{\mathbf{R}} y \, \hat{f}_{Y|X=x}(y) \, dy, \qquad x \in \mathbf{R}^d,$$

where $\hat{f}_{Y|X=x}(y)$ is defined in (3.47). The estimator $\hat{f}_{X,Y}(x, y)$ of the joint density is defined in (3.48). We have

$$\begin{aligned}
\int_{\mathbf{R}} y \, \hat{f}_{X,Y}(x, y) \, dy &= \frac{1}{n} \sum_{i=1}^{n} K_h(x - X_i) \int_{\mathbf{R}} y \, L_g(y - Y_i) \, dy \\
&= \frac{1}{n} \sum_{i=1}^{n} K_h(x - X_i) \int_{\mathbf{R}} (t + Y_i) \, L(t) \, dt \\
&= \frac{1}{n} \sum_{i=1}^{n} Y_i \, K_h(x - X_i),
\end{aligned}$$

when $\int_{\mathbf{R}} t \, L(t) \, dt = 0$ and $\int_{\mathbf{R}} L = 1$. Thus, $\hat{f}(x)$ has the same definition $\hat{f}(x) = \sum_{i=1}^{n} p_i(x) \, Y_i$, as in (3.6).

3.6.2 Histogram Estimator of Conditional Density

We define first the histogram estimator of the unconditional density and after that we define the histogram estimator of the conditional density.

Unconditional Histogram Density Estimator A histogram estimator of the density of $X \in \mathbf{R}^d$, based on identically distributed observations $X_1, \ldots, X_n$, is defined as

$$\hat{f}_X(y) = \sum_{R \in \mathcal{P}} \frac{n_R / n}{\text{volume}(R)} \, I_R(x), \qquad x \in \mathbf{R}^d, \tag{3.49}$$

where $\mathcal{P}$ is a partition on $\mathbf{R}^d$ and

$$n_R = \#\{i : X_i \in R, \; i = 1, \ldots, n\}$$

is the number of observations in R. Partition $\mathcal{P}$ is a collection $A_1, \ldots, A_N$ of sets that are disjoint and they cover the space of the observed X-values.

Conditional Histogram Density Estimator Unconditional histogram density estimator was defined in (3.49). In the univariate case a histogram estimator of the density of Y, based on data $Y_1, \ldots, Y_n$, is defined as

$$\hat{f}_Y(y) = \sum_{R \in \mathcal{P}} \frac{n_R/n}{\text{volume}(R)} I_R(y), \qquad y \in \mathbf{R},$$

where $\mathcal{P}$ is a partition of $\mathbf{R}$ and

$$n_R = \#\{i : Y_i \in R, \ i = 1, \ldots, n\}$$

is the number of observations in R.

A histogram estimator of the conditional density of Y given X, based on data $Y_1, \ldots, Y_n$, associated with a kernel regression function estimator, is defined as

$$\hat{f}_{Y|X=x}(y) = \sum_{R \in \mathcal{P}} \frac{n_R(x)/n}{\text{volume}(R)} I_R(y), \qquad y \in \mathbf{R}, \, x \in \mathbf{R}^d,$$

where

$$n_R(x) = n \cdot \sum_{i:Y_i \in R} p_i(x),$$

and $p_i(x)$ are defined in (3.7).

Let $Y_1, \ldots, Y_T$ be an observed time series. We can define a histogram estimator of the conditional density using time–space smoothing, analogously to the kernel estimator using time–space smoothing in (3.43) and (3.44). For example, in the case of one-sided moving average we replace $n_R(x)$ by

$$n_R(t) = n \cdot \sum_{i=1}^{t} p_i(t),$$

where the weights $p_i(t)$ are defined in (3.15). We can combine time–space smoothing and state–space smoothing using the weights defined in (3.45).

3.6.3 Nearest-Neighbor Estimator of Conditional Density

We define first the nearest-neighbor estimator of the unconditional density and after that we define the nearest-neighbor estimator of the conditional density.

Unconditional Nearest-Neighbor Density Estimator We define the density estimator using identically distributed observations $X_1, \ldots, X_n \in \mathbf{R}^d$. Let $1 \le k < n$ be an integer. Let us define, as in (3.31),

$$r_k(x) = \min\{r > 0 : \#\{X_i \in B_r(x)\} = k\},$$

where $B_r(x)$ is the ball centered at x with radius r: $B_r(x) = \{y \in \mathbf{R}^d : \|x - y\| \le r\}$. The radius $r_k(x)$ is the minimum radius such that the ball with this radius, centered at x, contains exactly k observations. The nearest-neighbor density estimator is defined as

$$\hat{f}_X(x) = \frac{k/n}{\text{volume}(B_{r_k(x)}(x))}, \qquad x \in \mathbf{R}^d. \tag{3.50}$$

Conditional Nearest-Neighbor Density Estimator The conditional nearest-neighbor density estimator does not seem to be popular, because in the univariate case a kernel density estimator with a local smoothing parameter selection method is typically preferred to the nearest-neighbor density estimator.

In the univariate case the nearest-neighbor estimator of the density of Y, based on data $Y_1, \ldots, Y_n$, is defined as

$$\hat{f}_Y(y) = \frac{k/n}{2r_k(y)}, \qquad y \in \mathbf{R},$$

where $r_k(y) = \min\{r > 0 : \#\{Y_i \in [y - r, y + r]\} = k\}$.

The conditional nearest-neighbor density estimator, based on regression data $(X_1, Y_1), \ldots, (X_n, Y_n)$, is defined as

$$\hat{f}_{Y|X=x}(y) = \frac{k(x)/n}{2r_k(y)}, \qquad y \in \mathbf{R},\ x \in \mathbf{R}^d,$$

where

$$k(x) = n \cdot \sum_{i:Y_i \in [y - r_k(x), y + r_k(x)]} p_i(x),$$

and $p_i(x)$ are the kernel weights defined in (3.7), the regressogram weights defined in (3.5), or the nearest-neighbor weights defined in (3.30).

Let $Y_1, \ldots, Y_T$ be an observed time series. We can define a nearest-neighbor estimator of the conditional density using time–space smoothing. For example, in the case of one-sided moving average we replace $k(x)$ by

$$k(t) = n \cdot \sum_{i=1}^{t} p_i(t),$$

where the weights $p_i(t)$ are defined in (3.15).

3.7 CONDITIONAL DISTRIBUTION FUNCTION ESTIMATION

The unconditional distribution function $F_Y(y) = P(Y \le y)$ can be estimated by the empirical distribution function

$$\hat{F}_Y(y) = \frac{1}{n} \sum_{i=1}^{n} I_{(-\infty, y]}(Y_i) = n^{-1} \#\{i : Y_i \le y, i = 1, \ldots, n\}. \tag{3.51}$$

In the empirical distribution function the probabilities of the half lines $(-\infty, y]$ are estimated with the empirical frequencies.

The conditional distribution function of $Y \in \mathbf{R}$ given $X \in \mathbf{R}^d$ is defined as

$$F_{Y|X=x}(y) = P(Y \le y \,|\, X = x), \qquad y \in \mathbf{R},\ \ x \in \mathbf{R}^d.$$

First we define a local averaging estimator of the conditional distribution function, which uses state–space smoothing. Second we define a time–space smoothing estimator of the conditional distribution function.

3.7.1 Local Averaging Estimator

The estimation of the conditional distribution function can be considered as a regression problem, where the conditional expectation of the random variable $I_{(-\infty,y]}(Y)$ is estimated, as was noted in (1.42). Thus we can define a local averaging estimator of the conditional distribution function as

$$\hat{F}_{Y|X=x}(y) = \sum_{i=1}^{n} p_i(x)\, I_{(-\infty,y]}(Y_i), \tag{3.52}$$

where $p_i(x)$ are the kernel weights defined in (3.7), the regressogram weights defined in (3.5), or the nearest-neighbor weights defined in (3.30).

The local averaging estimator in (3.52) is approximately the same as

$$\hat{F}_{Y|X=x}(y) = \int_{-\infty}^{y} \hat{f}_{Y|X=x}(u)\, du, \qquad y \in \mathbf{R},\ x \in \mathbf{R}^d,$$

where $\hat{f}_{Y|X=x}(u)$ is defined similarly as in (3.42):

$$\hat{f}_{Y|X=x}(u) = \sum_{i=1}^{n} p_i(x) L_g(u - Y_i), \qquad u \in \mathbf{R}.$$

Indeed,

$$\lim_{g \to 0} \int_{-\infty}^{y} L_g(u - Y_i)\, du = I_{(-\infty,y]}(Y_i)$$

for each $y \in \mathbf{R}$, if kernel function $L : \mathbf{R} \to \mathbf{R}$ is such that $\lim_{x \to \infty} L(x) = 0$ and $\lim_{x \to -\infty} L(x) = 0$.

3.7.2 Time–Space Smoothing

The local averaging estimator of the conditional distribution function, as defined in (3.52), can be used in the case of state–space prediction of time series data. The state–space prediction was introduced in (1.45). Local averaging can also be used when the observations $Y_1, \ldots, Y_T$ are a time series of observations that are not identically distributed but only locally identically distributed. Then we can use time–space smoothing.

Moving averages were defined in Section 3.2.4. We can use a two-sided moving average to define the estimator of the distribution function of Y_t as

$$\hat{F}_{Y_t}(y) = \sum_{i=1}^{T} p_i(t)\, I_{(-\infty,y]}(Y_i), \qquad t = 1, \ldots, n,$$

where the weights $p_i(t)$ are defined in (3.13). To use the estimate of the conditional distribution function in prediction, we use a one-sided moving average and define the estimator of the distribution function of Y_t as

$$\hat{F}_{Y_t}(y) = \sum_{i=1}^{t} p_i(t)\, I_{(-\infty,y]}(Y_i), \qquad t = 1, \ldots, n, \tag{3.53}$$

where the weights $p_i(t)$ are defined in (3.15). A particular case of the one-sided moving average is the exponential moving average. The exponential moving average estimator of the distribution function is defined analogously to (3.16) as

$$\hat{F}_{Y_t}(y) = \frac{1-\gamma}{1-\gamma^t} \sum_{i=1}^{t} \gamma^{t-i} \, I_{(-\infty,y]}(Y_i),$$

where $0 < \gamma < 1$, $\gamma = \exp(-1/h)$.

3.8 CONDITIONAL QUANTILE ESTIMATION

Quantile regression was introduced in Section 1.1.6. Let $Y_1, \ldots, Y_n \in \mathbf{R}$ be identically distributed observations. An estimator of a quantile of Y can be defined with the help of the empirical distribution function $\hat{F}_Y(y)$, defined in (3.51). We get the quantile estimator by taking the generalized inverse of the empirical distribution function, as in (1.26):

$$\hat{Q}_p(Y) = \inf\{y : \hat{F}_Y(y) \geq p\},$$

where $0 < p < 1$.

Let $(X_1, Y_1), \ldots, (X_n, Y_n)$ be identically distributed regression data. An estimator of a conditional quantile of Y can be defined with the help of the estimator of the conditional distribution function $\hat{F}_{Y \mid X=x}(y)$, defined in (3.52). We get the conditional quantile estimator by taking the generalized inverse of the estimator of the conditional distribution function:

$$\hat{Q}_p(Y \mid X = x) = \inf\{y : \hat{F}_{Y \mid X=x}(y) \geq p\}. \tag{3.54}$$

The estimator (3.54) can be called a local averaging estimator of the conditional quantile. It holds that

$$\hat{Q}_p(Y \mid X = x) = \begin{cases} Y_{(1)}, & 0 < p \leq p_1(x), \\ Y_{(2)}, & p_1(x) < p \leq p_1(x) + p_2(x), \\ \vdots & \\ Y_{(n-1)}, & \sum_{i=1}^{n-2} p_i(x) < p \leq \sum_{i=1}^{n-1} p_i(x), \\ Y_{(n)}, & \sum_{i=1}^{n-1} p_i(x) < p < 1, \end{cases} \tag{3.55}$$

where the ordered sample is denoted by $Y_{(1)} \leq Y_{(2)} \leq \cdots \leq Y_{(n)}$ and $p_i(x)$ are the kernel weights defined in (3.7), the regressogram weights defined in (3.5), or the nearest-neighbor weights defined in (3.30).

Figure 3.8 shows conditional quantile estimates when kernel weights are used. Panel (a) shows estimates for the levels $p = 0, 1, 0, 2, \ldots, 0.9$, when the smoothing parameter is $h = 0.7$. Panel (b) shows estimates for the level $p = 0.1$ when the smoothing parameters are $h = 0.3, 0.5, 0.7, 0.9$. The standard normal kernel is used in both panels. The data are the same as in Figure 1.1: The data consist of the daily

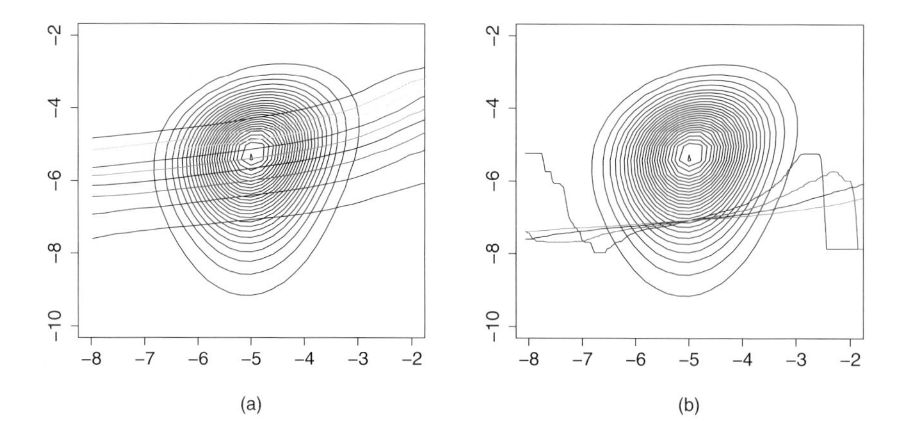

Figure 3.8 *Kernel estimates of conditional quantiles.* (a) Conditional quantile estimates for the levels $p = 0, 1, 0, 2, \ldots, 0.9$, when the smoothing parameter is $h = 0.7$. (b) Estimates for the level $p = 0.1$ when the smoothing parameters are $h = 0.3, 0.5, 0.7, 0.9$. A contour plot of a kernel estimate of the density of (X_t, Y_t) is also shown.

S&P 500 returns $R_t = (S_t - S_{t-1})/S_{t-1}$, where S_t is the price of the index. The explanatory and the response variables as

$$X_t = \log_e \sqrt{\frac{1}{k}\sum_{i=1}^{k} R_{t-i}^2}, \qquad Y_t = \log_e |R_t|.$$

The S&P 500 index data are described more precisely in Section 1.6.1. We show also a contour plot of a kernel estimate of the density of (X_t, Y_t).

Let $Y_1, \ldots, Y_t$ be stationary time series data. We can define one-sided moving average estimator of the conditional quantile by inverting the one sided moving average estimator $\hat{F}_{Y_t}$ of the distribution function, defined in (3.53). This gives

$$\hat{Q}_p(Y_t \mid Y_{t-1}, \ldots) = \begin{cases} Y_{(1)}, & 0 < p \le p_1(t), \\ Y_{(2)}, & p_1(t) < p \le p_1(t) + p_2(t), \\ \vdots & \\ Y_{(t-1)}, & \sum_{i=1}^{t-2} p_i(t) < p \le \sum_{i=1}^{t-1} p_i(t), \\ Y_{(t)}, & \sum_{i=1}^{t-1} p_i(t) < p < 1, \end{cases}$$

where the ordered sample is denoted by $Y_{(1)} \le Y_{(2)} \le \cdots \le Y_{(t)}$ and $p_i(t)$ are the one-sided weights, defined in (3.15).

3.9 CONDITIONAL VARIANCE ESTIMATION

In Section 3.9.1 we define the state–space smoothing of the conditional variance using local averaging. Section 3.9.2 considers GARCH estimation of conditional variance for time series data. Section 3.9.3 considers the moving average estimator of conditional variance for time series data and compares the GARCH(1,1) estimator to the exponential moving average estimator. We study the methods by fitting the S&P 500 return data, but the application to the estimation of S&P 500 return volatility is postponed to Section 3.11.1.

3.9.1 State–Space Smoothing and Variance Estimation

Let $(X_1, Y_1), \ldots, (X_n, Y_n)$ be identically distributed regression data from the distribution of (X, Y), and let us consider the estimation of the conditional variance

$$f(x) = \mathrm{Var}(Y \mid X = x), \qquad x \in \mathbf{R}^d.$$

We can write

$$\mathrm{Var}(Y \mid X = x) = E\left[(Y - f_{reg}(X))^2 \mid X = x \right],$$

where $f_{reg}(x) = E(Y \mid X = x)$ is the regression function. Thus we can estimate the conditional variance by

$$\hat{f}(x) = \sum_{i=1}^{n} p_i(x) \left(Y_i - \hat{f}_{reg}(X_i) \right)^2,$$

where $\hat{f}_{reg}(x)$ is an estimator of the regression function, and $p_i(x)$ are the kernel weights defined in (3.7), the regressogram weights defined in (3.5), or the nearest-neighbor weights defined in (3.30). We can also write

$$\mathrm{Var}(Y \mid X = x) = E\left[(Y^2 \mid X = x \right] - f_{reg}(x)^2.$$

Thus we can estimate the conditional variance by

$$\hat{f}(x) = \sum_{i=1}^{n} p_i(x)\, Y_i^2 - \hat{f}_{reg}(x)^2.$$

We can use the same local averaging to estimate f and f_{reg}, and in this case we have

$$\begin{aligned} \hat{f}(x) &= \sum_{i=1}^{n} p_i(x) \left(Y_i - \sum_{j=1}^{n} p_j(x)\, Y_j \right)^2 \\ &= \sum_{i=1}^{n} p_i(x)\, Y_i^2 - \left(\sum_{i=1}^{n} p_i(x)\, Y_i \right)^2. \end{aligned}$$

We apply state–space smoothing to volatility estimation in Section 3.11.1, see (3.86).

3.9.2 GARCH and Variance Estimation

We define the GARCH model, study the volatility formula of the GARCH model, describe the maximum likelihood estimator, find the multistep prediction formula of GARCH(1,1), compare GARCH(1,1) to ARCH(∞) model, and finally fit the GARCH(1,1) model to S&P 500 returns.

GARCH Model The GARCH model is a special case of a conditional heteroskedasticity model, defined in (1.16). The expression "GARCH model" is a shorthand for "generalized autoregressive conditional heteroskedasticity model." The GARCH model generalizes the ARCH model discussed in Section 2.5.2.

In the GARCH(p, q) model it is assumed that

$$Y_t = \sigma_t \epsilon_t, \qquad t = 0, \pm 1, \pm 2, \ldots$$

and

$$\sigma_t^2 = \alpha_0 + \sum_{i=1}^{q} \alpha_i Y_{t-1}^2 + \sum_{j=1}^{p} \beta_j \sigma_{t-j}^2,$$

where $q \geq 1$, $p \geq 0$, $\alpha_i \geq 0$, $\beta_j \geq 0$, ϵ_t are i.i.d. with $E\epsilon_t = 0$, Var$(\epsilon_t) = 1$, and ϵ_t is independent from $\{Y_{t-1}, Y_{t-2}, \ldots\}$. The GARCH model was introduced in Bollerslev (1986). The GARCH(p, q) process is strictly stationary if

$$\sum_{i=1}^{q} \alpha_i + \sum_{j=1}^{p} \beta_j < 1, \tag{3.56}$$

see Fan & Yao (2005, Theorem 4.4) and Bougerol & Picard (1992).

The version (3.17) of the exponential moving average uses the recursive formula

$$\sigma_t^2 = (1 - \gamma) Y_{t-1}^2 + \gamma \sigma_{t-1}^2. \tag{3.57}$$

We obtain model (3.57) from GARCH(1,1) by choosing $\alpha_0 = 0$, $\alpha_1 = 1 - \gamma$, and $\beta = \gamma$. Thus, this version of the exponential moving average is a special case of the GARCH(1,1). Model (3.57) is called IGARCH(1,1) model, because now $\alpha_1 + \beta = 1$, and this is not a stationary process.

If $\alpha_1 + \beta < 1$ in the GARCH(1,1) model, which implies stationarity by (3.56), then

$$\text{Var}(Y_t) = EY_t^2 = \frac{\alpha_0}{1 - \alpha_1 - \beta}.$$

Indeed, using (3.59) below we have that

$$\begin{aligned} EY_t^2 &= E\sigma_t^2 E\epsilon_t^2 = E\sigma_t^2 = \frac{\alpha_0}{1-\beta} + EY_t^2 \alpha_1 \sum_{k=1}^{\infty} \beta^{k-1} \\ &= \frac{\alpha_0}{1-\beta} + EY_t^2 \frac{\alpha_1}{1-\beta}, \end{aligned}$$

and solving for EY_t^2 gives the result.

Let us assume the condition $\alpha_1 + \beta < 1$, and denote the unconditional variance by $\bar{\sigma}^2 = \mathrm{Var}(Y_t) = \alpha_0/(1 - \alpha_1 - \beta)$. Now we can write the GARCH(1,1) model as

$$\sigma_t^2 = \lambda \bar{\sigma}^2 + \alpha_1 Y_{t-1}^2 + \beta \sigma_{t-1}^2,$$

where

$$\lambda = 1 - \alpha_1 + \beta.$$

Accordingly, GARCH(1,1) can be interpreted as a model, where the first term incorporates the deviation from the long-term unconditional variance; see Hull (2010, Section 9.7, p. 188).

Volatility in GARCH Models In the GARCH model we have

$$\mathrm{Var}\,(Y_t \mid \mathcal{F}_{t-1}) = \sigma_t^2, \tag{3.58}$$

where $\mathcal{F}_{t-1}$ is the sigma-algebra generated by $Y_{t-1}, Y_{t-2}, \ldots$. This was shown in (1.17) for the general conditional heteroskedasticity model. Furthermore, in a stationary GARCH(1,1) model we have

$$\sigma_t^2 = \frac{\alpha_0}{1 - \beta} + \alpha_1 \sum_{k=1}^{\infty} \beta^{k-1} Y_{t-k}^2. \tag{3.59}$$

Equation (3.59) follows by noting that in the GARCH(1,1) model we have

$$\sigma_t^2 = \alpha_0 + \alpha_1 Y_{t-1}^2 + \beta \sigma_{t-1}^2$$

and

$$\sigma_{t-1}^2 = \alpha_0 + \alpha_1 Y_{t-2}^2 + \beta \sigma_{t-2}^2.$$

Continuing this way, for each $k \geq 1$ we get

$$\sigma_t^2 = \alpha_0 \sum_{i=0}^{k-1} \beta^i + \alpha_1 \sum_{j=1}^{k} \beta^{j-1} Y_{t-j}^2 + \beta^k \sigma_{t-k}^2.$$

We assume that $\beta + \alpha_1 < 1$, which ensures stationarity, and so $0 < \beta < 1$, which implies that

$$\beta^k \sigma_{t-k}^2 \xrightarrow{P} 0,$$

and

$$\sum_{j=1}^{k} \beta^{j-1} X_{t-j}^2 \xrightarrow{P} \sum_{j=1}^{\infty} \beta^{j-1} X_{t-j}^2,$$

as $k \to \infty$. Finally,

$$\sum_{i=0}^{\infty} \beta^i = \frac{1}{1 - \beta}.$$

We have proved (3.59) when the covergence is defined as the convergence in probability.

More generally, for the GARCH(p, q) model we have

$$\sigma_t^2 = \frac{\alpha_0}{1 - \sum_{j=1}^p \beta_j} + \sum_{k=1}^\infty d_k Y_{t-k}^2, \tag{3.60}$$

where d_k are obtained from the equation

$$\sum_{k=1}^\infty d_k z^i = \frac{\sum_{i=1}^q \alpha_i z^i}{1 - \sum_{j=1}^p \beta_j z^j},$$

for $|z| \leq 1$; see Fan & Yao (2005, Theorem 4.4). We get the expression

$$\begin{aligned}\sigma_t^2 &= \frac{\alpha_0}{1 - \sum_{j=1}^p \beta_j} + \sum_{i=1}^q \alpha_i Y_{t-i}^2 \\ &+ \sum_{i=1}^q \alpha_i \sum_{k=1}^\infty \sum_{j_1=1}^p \cdots \sum_{j_k=1}^p \beta_{j_1} \cdots \beta_{j_k} Y_{t-i-j_1-\cdots-j_k}^2.\end{aligned}$$

Maximum Likelihood Estimation of the GARCH Parameters Let us have the observations $Y_1, \ldots, Y_T$. Estimation in the GARCH(p, q) model can be done using the method of maximum likelihood. Let us denote the density of ϵ_t by $f_\epsilon : \mathbf{R} \to \mathbf{R}$. In Section 2.5.2 we derived the likelihood function for the ARCH(p) model using the fact that

$$f_{Y_t \mid Y_{t-1}, \ldots, Y_{t-p}}(y) = \frac{1}{\sigma_t} f_\epsilon\left(\frac{y}{\sigma_t}\right).$$

However, in the GARCH(p, q) model we have the expression (3.60) for σ_t^2, and any finite history of Y_t does not fix the value of σ_t^2. Unlike in the ARCH(p) model, σ_t^2 is a sum of infinitely many terms, and we need to truncate the infinite sum in order to be able to calculate the conditional likelihood. Let us denote

$$\tilde{\sigma}_t^2 = \frac{\alpha_0}{1 - \sum_{j=1}^p \beta_j} + \sum_{k=1}^{t-1} d_k Y_{t-k}^2,$$

where d_k are the coefficients in (3.60), so that $\tilde{\sigma}_t^2$ is a function of $Y_1^2, \ldots, Y_{t-1}^2$. The logarithm of the likelihood is obtained similarly as in the ARCH(p) model in (2.78). The conditional likelihood, given the observations $Y_1, \ldots Y_r$, is

$$\log_e L_r(\alpha_0, \ldots, \alpha_q, \beta_1, \ldots, \beta_p) = -\frac{1}{2} \sum_{t=r+1}^T \log_e \sigma_t^2 + \sum_{t=r+1}^T \log f_\epsilon\left(\frac{y_t}{\sigma_t}\right).$$

We modify this and replace σ_t with $\tilde{\sigma}_t$, and define the maximum likelihood estimator as the maximizer of

$$\begin{aligned}&\log_e \tilde{L}_r(\alpha_0, \ldots, \alpha_q, \beta_1, \ldots, \beta_p) \\ &= -\frac{1}{2} \sum_{t=r+1}^T \log_e \tilde{\sigma}_t^2 + \sum_{t=r+1}^T \log f_\epsilon\left(\frac{y_t}{\tilde{\sigma}_t}\right),\end{aligned} \tag{3.61}$$

where $r \geq \max\{p, q\}$. In the case of the Gaussian assumption $\epsilon_t \sim N(0, 1)$ we get

$$\begin{aligned} & (\hat{\alpha}_0, \ldots, \hat{\alpha}_q, \hat{\beta}_1, \ldots, \hat{\beta}_p) \\ = \quad & \operatorname{argmin}_{\alpha_0,\ldots,\alpha_q,\beta_1,\ldots,\beta_p} \left\{ \sum_{t=r+1}^{T} \left(\log_e \tilde{\sigma}_t^2 + \frac{y_t^2}{\tilde{\sigma}_t^2} \right) \right\}. \end{aligned} \tag{3.62}$$

Multistep Prediction with GARCH(1,1) Let us consider the prediction of the h-step ahead squared observation Y_{t+h}^2. In the GARCH(1,1) model with $\alpha_1 + \beta < 1$ the optimal forecast is

$$E\left(Y_{t+h}^2 \,|\, \mathcal{F}_t\right) = \bar{\sigma}^2 + (\alpha_1 + \beta)^{h-1}\left(\sigma_{t+1}^2 - \bar{\sigma}^2\right), \qquad h \geq 1, \tag{3.63}$$

where $\bar{\sigma}^2 = EY_t^2 = \alpha_0/(1 - \alpha_1 - \beta)$ is the unconditional variance. Now the optimality is taken in the sense of the mean squared error. Note that in the case $h = 1$ (3.63) was written in (3.58).

Let us show (3.63) for $h \geq 2$. Let us denote $E(\cdot \,|\, \mathcal{F}_t) = E_t$. Now,

$$\sigma_{t+h}^2 - \bar{\sigma}^2 = \alpha_1\left(Y_{t+h-1}^2 - \bar{\sigma}^2\right) + \beta\left(\sigma_{t+h-1}^2 - \bar{\sigma}^2\right).$$

We have that $E_t Y_{t+h-1}^2 = E_t E_{t+h-2} Y_{t+h-1}^2 = E_t \sigma_{t+h-1}^2$ when $h \geq 2$, and $E_t \sigma_{t+1}^2 = \sigma_{t+1}^2$. Thus,

$$\begin{aligned} E_t\left(\sigma_{t+h}^2 - \bar{\sigma}^2\right) &= (\alpha_1 + \beta) E_t\left(\sigma_{t+h-1}^2 - \bar{\sigma}^2\right) \\ &= (\alpha_1 + \beta)^{h-1} E_t\left(\sigma_{t+1}^2 - \bar{\sigma}^2\right) \\ &= (\alpha_1 + \beta)^{h-1}\left(\sigma_{t+1}^2 - \bar{\sigma}^2\right). \end{aligned}$$

We have shown (3.63).

Let us next consider the prediction of the h-step realized volatility

$$V_{t,h} \stackrel{def}{=} Y_{t+1}^2 + \cdots + Y_{t+h}^2,$$

where $h \geq 1$. Let us denote[35]

$$\sigma_{t,h}^2 \stackrel{def}{=} E\left(Y_{t+1}^2 + \cdots + Y_{t+h}^2 \,|\, \mathcal{F}_t\right).$$

Using (3.63), we have the expression

$$\sigma_{t,h}^2 = h\bar{\sigma}^2 + \left(\sigma_{t+1}^2 - \bar{\sigma}^2\right) \sum_{k=1}^{h} (\alpha_1 + \beta)^{k-1}$$

in the GARCH(1,1) model, where we can write $\sum_{k=1}^{h} (\alpha_1 + \beta)^{k-1} = (1 - (\alpha_1 + \beta)^h)/(1 - \alpha_1 - \beta)$.

[35]Note that in the GARCH(1,1) model the Y_t are conditionally uncorrelated and thus it holds also that $\sigma_{t,h}^2 = E\left[(Y_{t+1} + \cdots + Y_{t+h})^2 \,|\, \mathcal{F}_t\right]$.

Comparison of GARCH(1,1) to Other Models We have defined ARCH(p) in (2.77) so that

$$\sigma_t^2 = \alpha_0 + \alpha_1 Y_{t-1}^2 + \cdots + \alpha_p Y_{t-p}^2.$$

When we compare this to the volatility expression (3.59), we see that the GARCH(1,1) has only three parameters to be estimated in the expression for the conditional variance, whereas the ARCH(p) model has $p+1$ parameters. Squared financial returns have a long-range dependence so that to fit the ARCH model the number of lags p has to be chosen large, but a large number of parameters makes the model more difficult to fit.

GARCH(1,1) can be considered a special case of the ARCH(∞) model, since (3.59) can be written as

$$\sigma_t^2 = \alpha + \sum_{k=1}^{\infty} \beta_k Y_{t-k}^2,$$

where $\beta_k = \alpha_1 \beta^{k-1}$ and $\alpha = \alpha_0/(1-\beta)$. We can obtain a more general ARCH(∞) model by defining

$$\sigma_t^2 = \alpha + \sum_{k=1}^{\infty} \psi_k(\theta) m(Y_{t-k}), \tag{3.64}$$

where $\alpha \in \mathbf{R}$, $\theta \in \mathbf{R}^p$, and $m : \mathbf{R} \to \mathbf{R}$ is called a news impact curve. More generally, following Linton (2009), the news impact curve can be defined as the relationship between σ_t^2 and $y_{t-1} = y$ holding past values σ_{t-1}^2 constant at some level σ^2. In the GARCH(1,1) model the news impact curve is

$$m(y, \sigma^2) = \alpha_0 + \alpha_1 y^2 + \beta \sigma^2.$$

The ARCH(∞) model in (3.64) has been studied in Linton & Mammen (2005), where it was noted that the estimated news impact curve is asymmetric for S&P 500 return data.

S&P 500 Returns with GARCH(1,1) We study fitting the GARCH(1,1) model to the S&P 500 returns data described in Section 1.6.1. Section 3.11.1 studies more specifically the estimation of the conditional variance $\sigma_t^2 = E\left(Y_t^2 \mid Y_{t-1}, Y_{t-2}, \ldots\right)$. GARCH(1,1) fitting has been studied for example in Spokoiny (2000) and in Fan & Yao (2005, Section 4.2.8).

The observed historical net returns are denoted by $Y_1, \ldots, Y_T$, where $Y_t = (P_t - P_{t-1})/P_{t-1}$, and P_t is the value of the index. The GARCH(1,1) model is

$$Y_t = \sigma_t \epsilon_t, \qquad \sigma_t^2 = \alpha_0 + \alpha_1 Y_{t-1}^2 + \beta \sigma_{t-1}^2. \tag{3.65}$$

The maximum likelihood estimators for the parameters, when $\epsilon_t \sim N(0,1)$, are[36]

$$\hat{\alpha}_0 = 7.3 \times 10^{-7}, \qquad \hat{\alpha}_1 = 0.077, \qquad \hat{\beta} = 0.92. \tag{3.66}$$

[36]We have used the R-package "tseries" for the maximum likelihood estimation. When we use the log-returns, then $\hat{\alpha}_0 = 7.6 \times 10^{-7}$, $\hat{\alpha}_1 = 0.079$, and $\hat{\beta} = 0.91$.

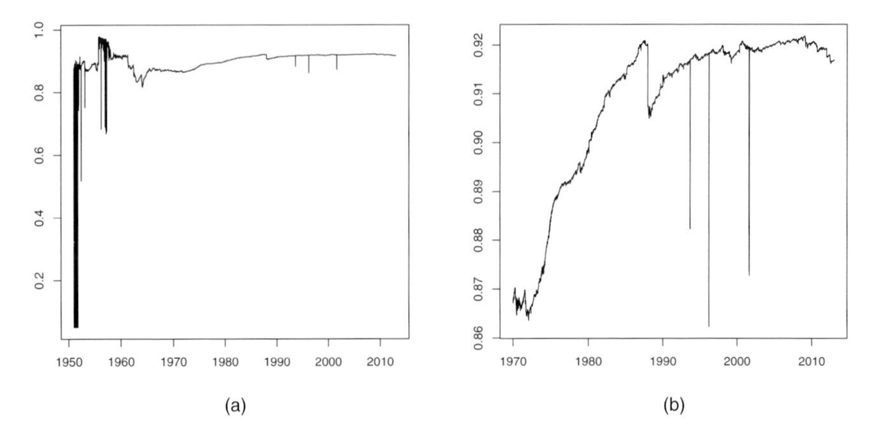

Figure 3.9 *Sequential estimates of β in GARCH(1,1).* We show the fluctuation of the sequential estimates $\hat{\beta}$. Panel (a) shows the estimates after one year (250 days) of observations, and panel (b) shows the estimates starting at 1970-01-02.

Figure 3.9 shows the fluctuation of the sequential estimates $\hat{\beta}_t$. At a day t we have used only the data available at that time to fit the GARCH(1,1) model. Panel (a) shows the estimates after one year (250 days) of observations, and panel (b) shows the estimates starting at 1970-01-02. Panel (a) shows that at the beginning of the estimation period the estimates are fluctuating considerably but the fluctuation of the estimates is steadily decreasing. Panel (b) shows that there is a sudden drop in the estimate values at 1987-10-20, when the index value dropped about 20% in one day. We conjecture that the four one-day downside spikes observed after 1987-10-20 are due to the numerical instability of the maximum likelihood estimation.

Figure 3.10 shows the fluctuation of the ratio

$$\frac{\hat{\sigma}_t^{out}}{\hat{\sigma}_t^{in}},$$

where $\hat{\sigma}_t^{out}$ are the sequential out-of-sample estimates of σ_t and $\hat{\sigma}_t^{in}$ are the in-sample estimates of σ_t. We have calculated σ_t^{out} by using parameter estimates $\hat{\beta}_t$, $\hat{\alpha}_{0,t}$, and $\hat{\alpha}_{1,t}$ that are calculated with data available at time t. We have calculated σ_t^{in} using the parameter estimates $\hat{\beta}$, $\hat{\alpha}_0$, and $\hat{\alpha}_1$ obtained by using the complete sample. In both cases the recursive formula for calculating σ_t in (3.65) has been started after 250 days, and the initial value for the recursion was the sample variance of the 250 first returns.

Let us denote the residuals by

$$\hat{\epsilon}_t = \frac{Y_t}{\hat{\sigma}_t},$$

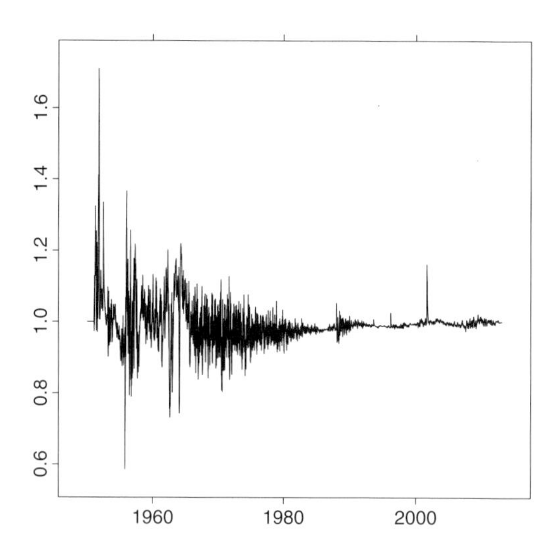

Figure 3.10 *Stability of volatility estimates in GARCH(1,1).* We show the fluctuation of the ratios $\sigma_t^{out}/\sigma_t^{in}$, where σ_t^{out} is the out-of-sample estimate of σ_t and σ_t^{in} is the in-sample estimate of σ_t.

where $\hat{\sigma}_t = \sigma_t^{out}$ is the the out-of-sample estimate of σ_t calculated recursively with the formula $\hat{\sigma}_t^2 = \hat{\alpha}_{t,0} + \hat{\alpha}_{t,1} Y_{t-1}^2 + \hat{\beta}_t \sigma_{t-1}^2$. We diagnose two things. First, we study the distribution of the residuals. Second, we study whether the squared residuals are uncorrelated.

Figure 3.11 shows a left tail plot and a right tail plot of the GARCH(1,1) residuals. Left and right tail plots are defined in Section 6.1.2. Panel (a) shows a left tail plot of the GARCH(1,1) residuals, and panel (b) shows a right tail plot of the GARCH(1,1) residuals. The black circles show the residuals, the red circles show simulated data from the standard normal distribution, and the blue circles show simulated data from the standard Student distribution with 12 degrees of freedom.[37]

Figure 3.12 shows QQ plots of the residuals. QQ plots are explained in Section 6.1.2. Panel (a) shows a QQ plot of the residuals when the comparison is to the normal distribution with the variance equal to the sample variance of the residuals, which is 1.01. Panel (b) shows a QQ plot of the residuals when the comparison is to the Student distribution with 12 degrees of freedom and the variance equal to the sample variance of the residuals.

The tail plots and the QQ plots show that the tails of the residuals are heavier than the tails of the normal distribution. The Student distribution with 12 degrees of freedom gives a better fit, although the extreme left tail of the residuals is not

[37] If Y follows a Student distribution (t-distribution) with degrees of freedom $\nu > 2$, then $\sqrt{(\nu-2)/\nu}\, Y$ follows standard Student distribution with degrees of freedom ν.

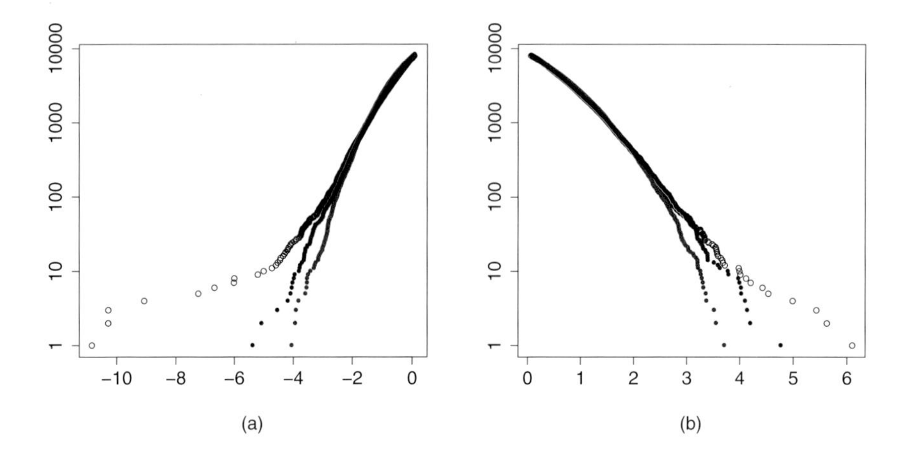

Figure 3.11 *Tail plots of the GARCH(1,1) residuals.* Panel (a) shows a left tail plot of the GARCH(1,1) residuals, and panel (b) shows a right tail plot of the GARCH(1,1) residuals. The black circles show the residuals, the red circles show simulated data from the standard normal distribution, and the blue circles show simulated data from the standard Student distribution with 12 degrees of freedom.

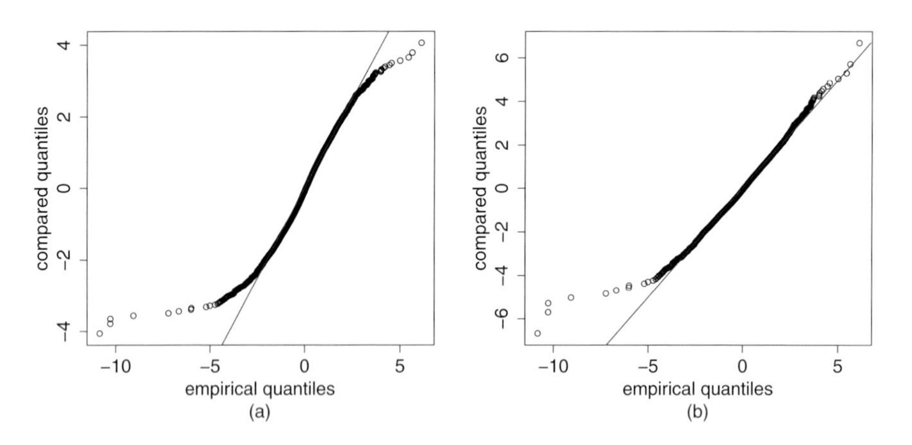

Figure 3.12 *QQ plots of the GARCH(1,1) residuals.* Panel (a) shows a QQ plot of the residuals when the comparison is to a normal distribution, and panel (b)shows a QQ plot of the residuals when the comparison is to a Student distribution with 12 degrees of freedom. The compared distributions are normalized to have variance equal to the sample variance of the residuals.

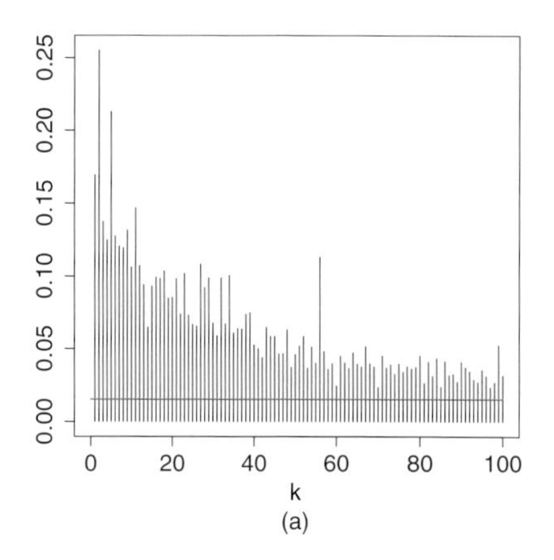

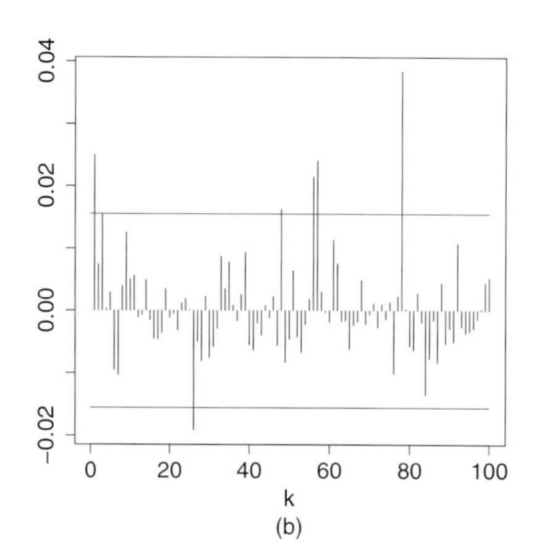

Figure 3.13 *Autocorrelations of the squared returns and the GARCH(1,1) residuals.* Panel (a) shows the sample autocorrelations of the squared S&P 500 returns for the lags $k = 1, \ldots, 100$, and panel (b) shows the sample autocorrelations of the squared residuals for the same lags. The red lines show level $\alpha = 0.05$ rejection lines for the null hypothesis $\rho_k = 0$.

well fitted with the Student distribution. The left tail of the residuals is heavier than the right tail of the residuals. In maximum likelihood estimation the normality was assumed. The normality is not plausible, so we consider the parameter estimation procedure as a quasi-maximum likelihood method.

Figure 3.13 shows the sample autocorrelations for the squared returns and for the squared residuals. The sample autocorrelation with lag k, based on data $Y_1, \ldots, Y_T$, is defined as

$$\hat{\rho}_k = \widehat{\mathrm{Cor}}(Y_t, Y_{t+k}) = \frac{\hat{\gamma}_k}{\hat{\gamma}_0}, \tag{3.67}$$

where the sample autocovariance with lag k is

$$\hat{\gamma}_k = \frac{1}{T} \sum_{t=1}^{T-k} (Y_t - \bar{Y})(Y_{t+k} - \bar{Y}),$$

and $\bar{Y} = T^{-1} \sum_{i=1}^{T} Y_t$. Panel (a) shows the sample autocorrelations of the squared returns $\widehat{\mathrm{Cor}}(Y_t^2, Y_{t+k}^2)$ and panel (b) shows the sample autocorrelations of the squared residuals $\widehat{\mathrm{Cor}}(\hat{\epsilon}_t^2, \hat{\epsilon}_{t+k}^2)$, for $k = 1, \ldots, 100$. We can see from Figure 3.13 that the squared returns have a considerable and persistent autocorrelation, whereas the squared residuals have a much smaller autocorrelation. This observation supports the assumption of the independence of innovations in the GARCH model.

The red lines in Figure 3.13 are at the heights

$$\pm z_{1-\alpha/2} T^{-1/2},$$

where z_α is the α-quantile for the standard normal distribution. We have chosen $\alpha = 0.05$, so that $z_{1-\alpha/2} \approx 1.96$. These lines can be interpreted as the rejection lines for the null hypothesis $\rho_k = 0$, because by the central limit theorem, if $Y_1, Y_2, \ldots$ are i.i.d. with mean zero, then

$$\sqrt{T}\, \hat{\rho}_k \xrightarrow{d} N(0,1),$$

as $T \to \infty$.

The Box–Ljung test can be used to test whether the autocorrelations are zero for a stationary time series $Y_1, Y_2, \ldots$. The null hypothesis is that $\rho_k = 0$ for $k = 1, \ldots, h$, where $h \geq 1$, $\rho_k = \gamma_k/\gamma_0$, and $\gamma_k = \mathrm{Cov}(Y_1, Y_{k+1})$. Let us have observed time series data $Y_1, \ldots, Y_T$. The test statistics is

$$Q(h) = T(T+2) \sum_{k=1}^{h} \frac{\hat{\rho}_k^2}{T-k},$$

where $\hat{\rho}_k$ is defined in (3.67). The test rejects the null hypothesis of zero autocorrelations if

$$Q(h) > \chi^2_{h,1-\alpha},$$

where $\chi^2_{h,1-\alpha}$ is the $1-\alpha$-quantile of the χ^2-distribution with degrees of freedom h. We can calculate the observed p-values

$$p_h = 1 - F_h(Q(h)),$$

for $h = 1, 2, \ldots$, where F_h is the distribution function of the χ^2-distribution with degrees of freedom h. Small observed p-values indicate that the observations are not compatible with the null hypothesis.

Figure 3.14 shows the results of the Box–Ljung test. Panel (a) shows as circles the values of the test statistics $Q(h)$ for $h = 1, \ldots, 100$; the red line shows the critical values $\chi^2_{h,1-\alpha}$ of the test statistics for the level $\alpha = 0.05$, and the blue line shows the critical values for the level $\alpha = 0.1$. Panel (b) shows the observed p values p_h, for $h = 1, \ldots, 100$. The red horizontal line shows the level $\alpha = 0.05$, and the blue horizontal line shows the level $p = 0.1$. We can see that the null hypothesis of zero autocorrelation is rejected for small h.

3.9.3 Moving Averages and Variance Estimation

We denote the conditional variance by

$$\sigma_t^2 = E(Y_t^2 \,|\, Y_{t-1}, Y_{t-2}, \ldots),$$

where it is assumed that $EY_t = 0$. The exponential moving average (EWMA) was defined in (3.16). The exponential moving average estimator of the conditional variance, based on observations $Y_0, \ldots, Y_{t-1}$, is

$$\hat{\sigma}_t^2 = \frac{1-\gamma}{1-\gamma^t} \sum_{k=0}^{t-1} \gamma^k Y_{t-k-1}^2, \tag{3.68}$$

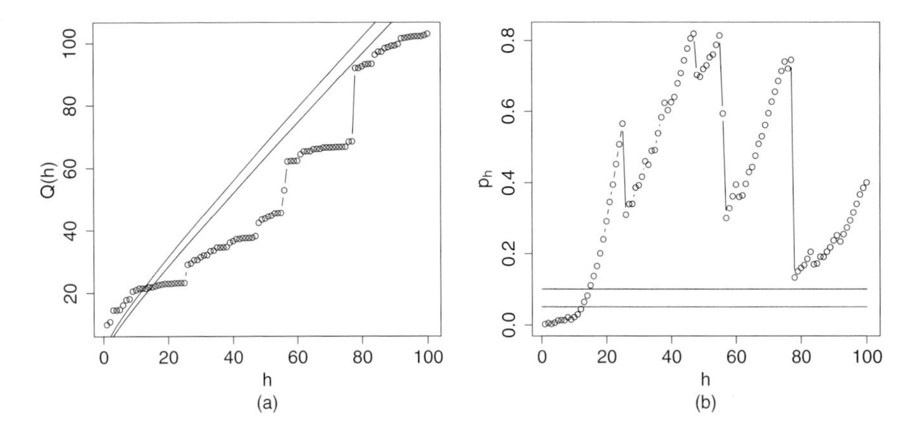

Figure 3.14 *Box-Ljung test.* We test the null hypothesis of zero autocorrelation for the GARCH(1,1) residuals. Panel (a) shows the values of the Box–Ljung test statistics $Q(h)$ for $h = 1, \ldots, 100$. The red line shows the critical values $\chi^2_{h,1-\alpha}$ for the level $\alpha = 0.05$, and the blue line for the level $\alpha = 0.1$. Panel (b) shows the observed p-values p_h for $h = 1, \ldots, 100$. The red line is at level $\alpha = 0.05$ and the blue line is at level $\alpha = 0.1$.

where $\gamma = \exp(-1/h)$, and $h > 0$ is the smoothing parameter.

We use S&P 500 data, described in Section 1.6.1, to compare the exponentially weighted moving average estimates to the GARCH(1,1) estimates. We apply later in Section 3.11.1 exponentially weighted moving averages to volatility estimation.

The weights of the exponentially weighted moving average are

$$w_k^e = \frac{1-\gamma}{1-\gamma^t}\gamma^k, \tag{3.69}$$

as can be seen from (3.68). The GARCH(1,1) weights are

$$w_k = \hat{\alpha}_1 \hat{\beta}^k, \qquad k = 1, 2, \ldots. \tag{3.70}$$

The formula in (3.70) follows because the GARCH(1,1) estimator for the conditional variance is

$$\hat{\sigma}_t^2 = \frac{\hat{\alpha}_0}{1-\hat{\beta}} + \hat{\alpha}_1 \sum_{k=0}^{t-1} \hat{\beta}^k Y_{t-k-1}^2, \tag{3.71}$$

where the parameter estimators $\hat{\alpha}_0$, $\hat{\alpha}_1$, and $\hat{\beta}$ are calculated using the data $Y_0, \ldots, Y_{t-1}$, with the maximum likelihood method as in (3.62).

Let us compare the exponential moving average weights w_k^e defined in (3.69) with the GARCH$(1,1)$ weights w_k defined in (3.70). We have estimated the GARCH(1,1) parameters for the S&P 500 returns data in (3.66). With those estimates we get

$$\frac{\hat{\alpha}_0}{1-\hat{\beta}} = 8.72 \times 10^{-6}.$$

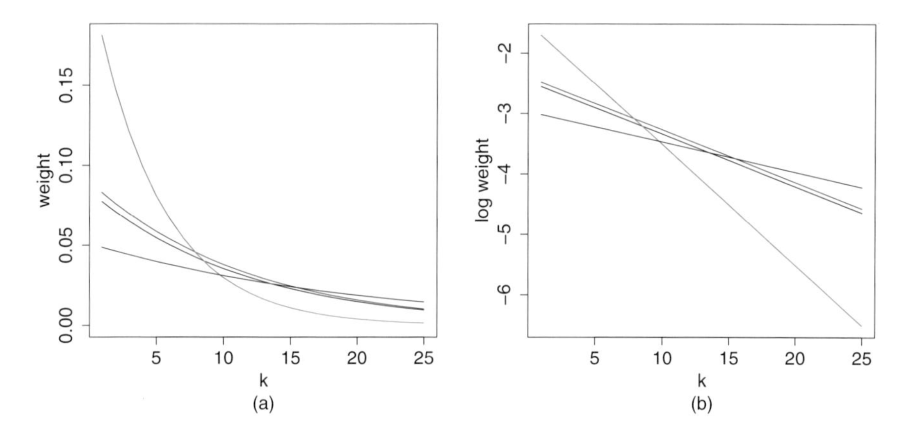

Figure 3.15 *GARCH vs. EWMA: Weights.* The weights w_k of the GARCH(1,1) estimator of volatility and the weights w_k^e of the exponentially moving average for smoothing parameters $h = 5$, $h = 11.55$, and $h = 20$ are shown. Panel (a) shows the first 25 weights, and panel (b) shows the logarithms of the first 25 weights. The black curves show the GARCH(1,1) weights. The green, red, and blue curves show the EWMA weights for $h = 5$, $h = 11.55$, and $h = 20$.

If the first term in (3.71) is considered negligible, we obtain for a large sample size t,

$$w_k^e \approx w_k,$$

if

$$\gamma \approx \hat{\beta} \Longleftrightarrow h \approx -1/\log(\hat{\beta}) = 11.55. \tag{3.72}$$

Figure 3.15 shows the first 25 weights w_k of the GARCH(1,1) estimator of the volatility and the weights w_k^e of the exponentially weighted moving average for the smoothing parameter values $h = 5$, $h = 11.55$, and $h = 20$. Panel (a) shows the first 25 weights and panel (b) shows the logarithms of the first 25 weights. The black curves show the GARCH(1,1) weights, the red curves show the EWMA weights with $h = 11.55$, the green curves show the EWMA weights with $h = 5$, and the blue curves show the EWMA weights with $h = 20$. The GARCH(1,1) weights seem close to the EWMA weights for the value $h = 11.55$, since we can hardly distinguish between the black curve and the red curve. The weights with $h = 5$ (blue curves) are decreasing fast, and the weights with $h = 20$ (green curves) are decreasing slowly.

Figure 3.16 shows the ratios $\hat{\sigma}_t^{ewma}/\hat{\sigma}_t^{garch}$, where $\hat{\sigma}_t^{ewma}$ is the moving averages estimate, and $\hat{\sigma}_t^{garch}$ is the GARCH(1,1) estimate, which is calculated in-sample, using the parameter estimates obtained from the whole sample. The smoothing parameters $h = 5$, $h = 11.55$, and $h = 20$ are considered. Panel (a) shows the case $h = 5$ with green, panel (b) shows the the case $h = 11.55$ with red, and panel (c) shows the the case $h = 20$ with blue. The ratio tends to be below one: Exponentially weighted moving average tends to give smaller volatility estimates. We have removed

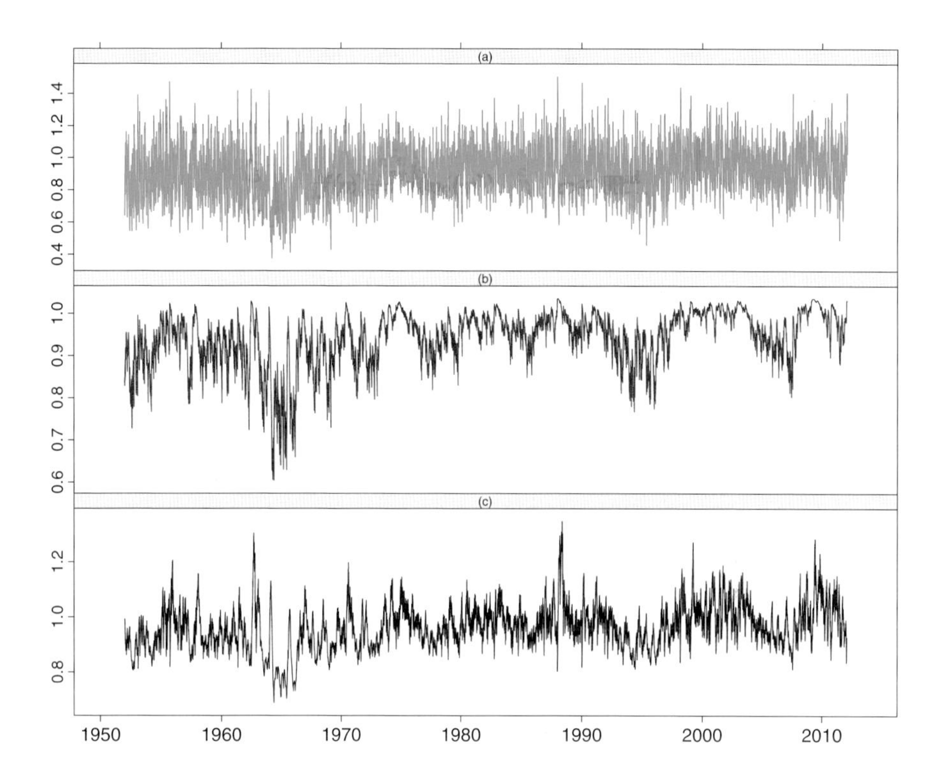

Figure 3.16 GARCH *vs.* EWMA: *Ratios of volatility estimates.* We show the ratios $\hat{\sigma}_t^{ewma}/\hat{\sigma}_t^{garch}$ of the exponentially moving average estimator of volatility to the GARCH(1,1) estimator of volatility for three smoothing parameters. Panel (a) shows the case $h = 5$ with green, panel (b) shows the the case $h = 11.55$ with red, and panel (c) shows the the case $h = 20$ with blue.

the first two years, and the time series starts at 1952-01. During the years 1950–1951 the fluctuation was larger and the ratios were at most equal to two.

Figure 3.17 shows the sample means and standard deviations of $\hat{\sigma}_t^{ewma}/\hat{\sigma}_t^{garch}$. The smoothing parameter h of $\hat{\sigma}_t^{ewma}$ takes values $1, 5, 11.55, 20, 40, 80$. Panel (a) shows the means, and panel (b) shows the standard deviations. We see that the mean is below one, and it is increasing as a function of h. We see also that the standard deviation is smallest when $h = 11.55$.

Figure 3.18 shows the left and right tail plots which compare the distributions of $\hat{\sigma}_t^{garch}$ and $\hat{\sigma}_t^{ewma}$. Tail plots are defined in Section 6.1.2. Panel (a) shows the left tail and panel (b) shows the right tail. The left tail consists of the observations smaller than the median, and the right tail consists of the observations larger than the median. The black circles show the GARCH(1,1) estimates, the red circles show the EWMA estimates with $h = 11.55$, and the blue circles show the EWMA estimates with $h = 25$. We see that the left tail of the GARCH estimates is lighter than the left

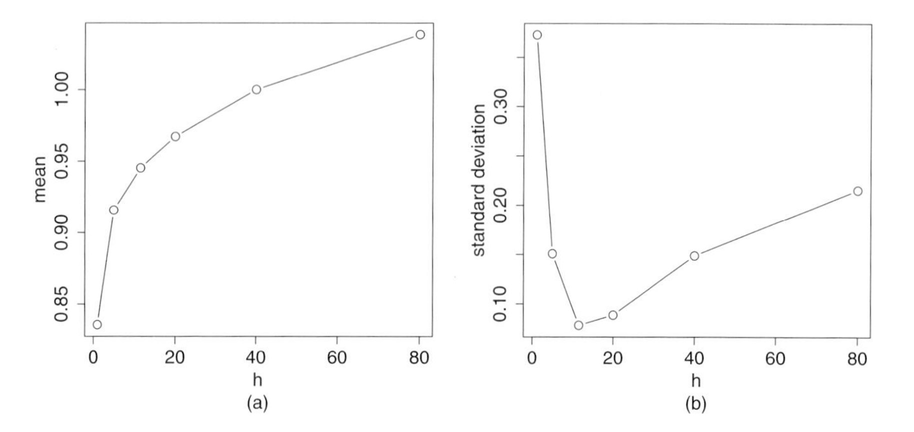

Figure 3.17 GARCH *vs.* EWMA: *Means and variances of ratios of volatility estimates.* We show the means and standard deviations of the ratios of the GARCH(1,1) estimator of volatility to the exponentially moving average estimator of volatility for smoothing parameters $h = 1, 5, 11.55, 20, 40, 80$. Panel (a) shows the means, and panel (b) shows the standard deviations.

tail of the EWMA estimates. The right tail of the GARCH estimates and the EWMA estimates with $h = 11.55$ are close to each other, but the right tail of the EWMA estimates with $h = 25$ is lighter.

Figure 3.19 shows the values of the Kolmogorov–Smirnov test statistics for comparing the distributions of $\hat{\sigma}_t^{garch}$ and $\hat{\sigma}_t^{ewma}$. The Kolmogorov–Smirnov test statistics is equal to the supremum distance between the empirical distribution functions of two samples and it is defined by

$$\text{KS} = \sup_{t \in \mathbf{R}} \left| \hat{F}(t) - \hat{G}(t) \right|,$$

where $\hat{F}$ and $\hat{G}$ are the empirical distribution functions of the two samples, as defined in (1.43).[38] The test statistics is used to test the null hypothesis of the equality of two distributions. We show the values of the test statistics for the smoothing parameters $h = 1, 5, 11.55, 20, 40, 80$ of the EWMA estimates. The test statistics indicates that the distributions are closest when $h = 40$.

3.10 CONDITIONAL COVARIANCE ESTIMATION

We define in Section 3.10.1 an estimator of the conditional covariance using state–space smoothing and local averaging. Section 3.10.2 considers GARCH estimation of

[38] We have calculated the test statistics using the R-function "ks.test" in package "stats."

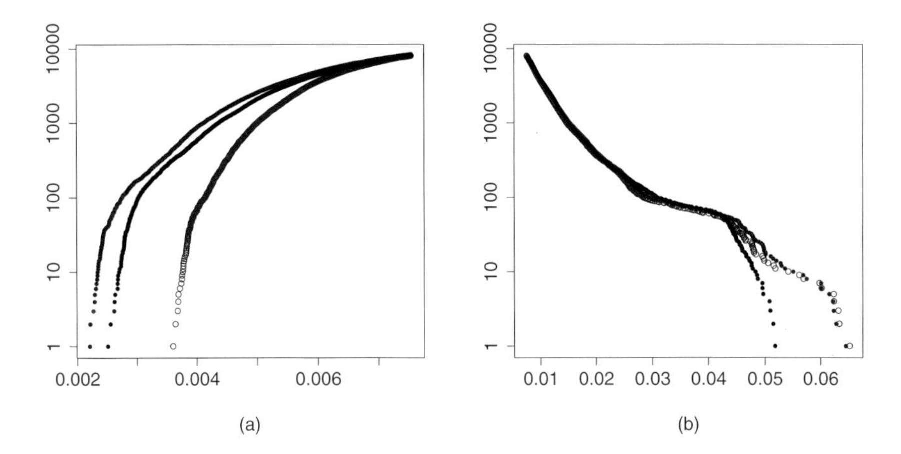

Figure 3.18 GARCH *vs.* EWMA: *Tail plots of volatility estimates.* We show the tail plots which compare the distribution of GARCH(1,1) estimates $\hat{\sigma}_t^{garch}$ and EWMA estimates $\hat{\sigma}_t^{ewma}$. Panel (a) shows the left tail plot, and panel (b) shows the right tail plot. The black circles show the GARCH(1,1) estimates, the red circles show the EWMA estimates with $h = 11.55$, and the blue circles show the EWMA estimates with $h = 25$,

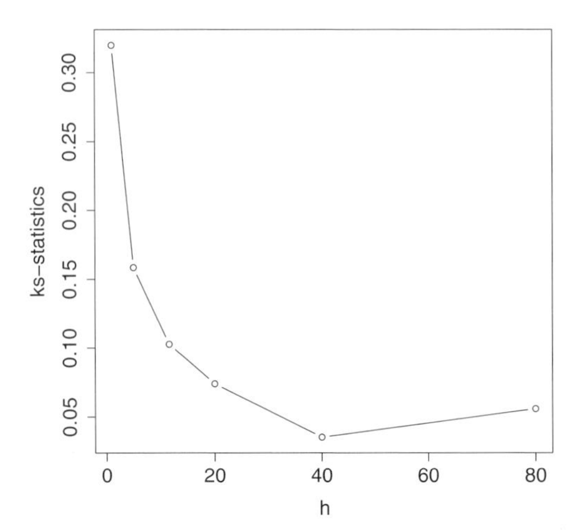

Figure 3.19 GARCH *vs.* EWMA: *Kolmogorov–Smirnov test statistics.* We show the Kolmogorov–Smirnov test statistics for comparing the distributions of $\hat{\sigma}_t^{garch}$ and $\hat{\sigma}_t^{ewma}$ as a function of smoothing parameter h of the EWMA estimates.

conditional covariance for time series data. Section 3.10.3 defines a moving average estimator of conditional covariance for time series data.

3.10.1 State–Space Smoothing and Covariance Estimation

Let $(X_1, Y_1, Z_1), \ldots, (X_n, Y_n.Z_n)$ be identically distributed data from the distribution of (X, Y, Z), where $Y, Z \in \mathbf{R}$, and $X \in \mathbf{R}^d$ is the vector of explanatory variables. Let us consider the estimation of the conditional covariance

$$f(x) = \mathrm{Cov}(Y, Z \,|\, X = x), \qquad x \in \mathbf{R}^d.$$

Covariance and correlation regression was introduced in Section 1.1.5.

We can write

$$\mathrm{Cov}(Y, Z \,|\, X = x) = E\left[(Y - f_{reg,Y}(X))\,(Z - f_{reg,Z}(X)) \,|\, X = x\right],$$

where $f_{reg,Y}(x) = E(Y \,|\, X = x)$ and $f_{reg,Z}(x) = E(Z \,|\, X = x)$ are the regression functions. Thus we can estimate the conditional covariance by

$$\hat{f}(x) = \sum_{i=1}^{n} p_i(x) \left(Y_i - \hat{f}_{reg,Y}(X_i)\right) \left(Z_i - \hat{f}_{reg,Z}(X_i)\right),$$

where $\hat{f}_{reg,Y}$ and $\hat{f}_{reg,Z}$ are estimators of the regression functions, and $p_i(x)$ are the kernel weights defined in (3.7), the regressogram weights defined in (3.5), or the nearest-neighbor weights defined in (3.30). We can also write

$$\mathrm{Cov}(Y, Z \,|\, X = x) = E\left[YZ \,|\, X = x\right] - f_{reg,Y}(x) f_{reg,Z}(x).$$

Thus we can estimate the conditional covariance by

$$\hat{f}(x) = \sum_{i=1}^{n} p_i(x)\, Y_i Z_i - \hat{f}_{reg,Y}(x) \hat{f}_{reg,Z}(x).$$

We can use the same local averaging to estimate f, $f_{reg,Y}$, and $f_{reg,Z}$. In this case

$$\begin{aligned}\hat{f}(x) &= \sum_{i=1}^{n} p_i(x) \left(X_i Y_i - \sum_{j=1}^{n} p_j(x)\, X_j \sum_{k=1}^{n} p_k(x)\, Y_k \right)^2 \\ &= \sum_{i=1}^{n} p_i(x)\, X_i Y_i - \sum_{i=1}^{n} p_i(x)\, X_i \sum_{j=1}^{n} p_j(x)\, Y_j .\end{aligned}$$

3.10.2 GARCH and Covariance Estimation

In the multivariate GARCH model the stochastic process Y_t is a vector process with d components. It is assumed that Y_t is strictly stationary and

$$Y_t = \Sigma_t^{1/2} \epsilon_t, \qquad t = 0, \pm 1, \pm 2, \ldots, \tag{3.73}$$

where $\Sigma_t^{1/2}$ is a square root of a positive definite covariance matrix Σ_t, Σ_t is measurable with respect to the sigma-algebra generated by $Y_{t-1}, Y_{t-2}, \ldots$, and ϵ_t is a d-dimensional i.i.d. process with $E\epsilon_t = 0$ and $\mathrm{Var}(\epsilon_t) = I_d$, where I_d is the $d \times d$ identity matrix (strict white noise).

The square root of Σ_t can be defined by writing the eigenvalue decomposition $\Sigma_t = Q_t \Lambda_t Q_t'$, where Λ_t is the diagonal matrix of the eigenvalues of Σ_t and Q_t is the orthogonal matrix whose columns are the eigenvectors of Σ_t. Then we define $\Sigma_t^{1/2} = Q_t \Lambda_t^{1/2} Q_t'$, where $\Lambda_t^{1/2}$ is the diagonal matrix obtained from Λ_t by taking componentwise square roots. We can define $\Sigma_t^{1/2}$ also as a Cholesky factor of Σ_t.

Multivariate GARCH processes are reviewed in McNeil et al. (2005, Section 4.6), Bauwens, Laurent & Rombouts (2006), and Silvennoinen & Teräsvirta (2009). Below we write the models only for the case $d = 2$, so that $Y_t = (Y_{t,1}, Y_{t,2})$. The multivariate GARCH models are denoted with MGARCH(p, q). We restrict ourselves to the first order models with $p = q = 1$. The multivariate GARCH models are based on (3.73) but differ in the definition of the recursive formula for Σ_t.

MGARCH Models First we define the VEC model and two restrictions of it: the diagonal VEC model and the BEKK model. Then we define the constant correlation model and the dynamic conditional correlation model.

Let us denote $\sigma_{t,1}^2 = \mathrm{Var}(Y_{t,1})$, $\sigma_{t,2}^2 = \mathrm{Var}(Y_{t,2})$, and $\sigma_{t,12} = \mathrm{Cov}(Y_{t,1}, Y_{t,2})$. The VEC model and the diagonal VEC model were introduced in Bollerslev, Engle & Wooldridge (1988). The VEC model assumes that

$$\begin{aligned}
\sigma_{t,1}^2 &= a_0 + a_1 Y_{t-1,1}^2 + a_2 Y_{t-1,2}^2 + a_3 Y_{t-1,1} Y_{t-1,2} \\
&\quad + b_1 \sigma_{t-1,1}^2 + b_2 \sigma_{t-1,2}^2 + b_3 \sigma_{t-1,12}, \\
\sigma_{t,2}^2 &= c_0 + c_1 Y_{t-1,1}^2 + c_2 Y_{t-1,2}^2 + c_3 Y_{t-1,1} Y_{t-1,2} \\
&\quad + d_1 \sigma_{t-1,1}^2 + d_2 \sigma_{t-1,2}^2 + d_3 \sigma_{t-1,12}, \\
\sigma_{t,12} &= e_0 + e_1 Y_{t-1,1}^2 + e_2 Y_{t-1,2}^2 + e_3 Y_{t-1,1} Y_{t-1,2} \\
&\quad + f_1 \sigma_{t-1,1}^2 + f_2 \sigma_{t-1,2}^2 + f_3 \sigma_{t-1,12}.
\end{aligned}$$

This model has 21 parameters $a_0, \ldots, f_3$. Since the model has a large number of parameters, it is useful to consider models with less parameters. The diagonal VEC model has only 9 parameters and assumes that

$$\sigma_{t,1}^2 = a_0 + a_1 Y_{t-1,1}^2 + b \sigma_{t-1,1}^2, \tag{3.74}$$

$$\sigma_{t,2}^2 = c_0 + c_1 Y_{t-1,2}^2 + d \sigma_{t-1,2}^2, \tag{3.75}$$

$$\sigma_{t,12} = e_0 + e_1 Y_{t-1,1} Y_{t-1,2} + f \sigma_{t-1,12}. \tag{3.76}$$

Thus, in the diagonal VEC model the components of Y_t follow univariate GARCH models. The BEKK (Baba–Engle–Kraft–Kroner) model was introduced in Engle & Kroner (1995). The model has 11 parameters and it can be written more easily with the matrix notation as

$$\Sigma_t = G_0 + G' Y_{t-1} Y_{t-1}' G + H' \Sigma_{t-1} H,$$

where G_0 is a symmetric 2×2 matrix and G and H are 2×2 matrices. The BEKK model is obtained from the VEC model by restricting the parameters. We can express the parameters $a_1, \ldots, f_3$ of the VEC model in terms of the parameters of the BEKK model as follows:

$$\begin{aligned} a_1 &= G_{11}^2, a_2 = G_{12}^2, a_3 = 2G_{11}G_{12}, b_1 = H_{11}^2, b_2 = H_{12}^2, b_3 = 2H_{11}H_{12}, \\ c_1 &= G_{22}^2, c_2 = G_{21}^2, c_3 = 2G_{22}G_{21}, d_1 = H_{22}^2, d_2 = H_{21}^2, d_3 = 2H_{22}H_{21}, \\ e_1 &= G_{11}G_{21}, e_2 = G_{22}G_{12}, e_3 = G_{11}G_{22} + G_{12}G_{21}, \\ f_1 &= H_{11}H_{21}, f_2 = H_{22}H_{12}, f_3 = H_{11}H_{22} + H_{12}H_{21}, \end{aligned}$$

where we denote the elements of G by G_{ij} and the elements of H by H_{ij}.

The recursive formula for Σ_t can be written by using the correlation matrix P_t. Let Δ_t be the diagonal matrix of the standard deviations of Σ_t. The correlation matrix P_t, corresponding to Σ_t, is such that $\Sigma_t = \Delta_t P_t \Delta_t$.

The constant correlation MGARCH model, introduced in Bollerslev (1990), is such that the components of Y_t follow univariate GARCH models, and the correlation matrix is constant. That is, $\Sigma_t = \Delta_t P \Delta_t$ and $\Delta_t = \mathrm{diag}(\sigma_{t,1}, \sigma_{t,2})$, where P is the constant correlation matrix. The constant correlation GARCH model assumes the univariate GARCH models for the components, as in (3.74) and (3.75), and

$$\rho_t = \rho.$$

The dynamic conditional correlation MGARCH model, introduced in Engle (2002), is such that the components of Y_t follow univariate GARCH models and

$$\rho_t = e_0 + e_1 \tilde{Y}_{t-1,1} \tilde{Y}_{t-1,2} + f \rho_{t-1}, \tag{3.77}$$

where $\tilde{Y}_t = \Delta_t^{-1} Y_t$, $e_0, e_1, f \geq 0$, $e_1 + f < 1$.

Covariance in MGARCH Models The recursive equation (3.76) in the stationary diagonal VEC model implies that

$$\sigma_{t,12} = \frac{e_0}{1-f} + e_1 \sum_{k=1}^{\infty} f^{k-1} Y_{t-k,1} Y_{t-k,2}.$$

This follows similarly as in the case of GARCH(1,1) model; see (3.59). The recursive equation (3.77) in the stationary dynamic conditional correlation GARCH model implies similarly that

$$\rho_t = \frac{e_0}{1-f} + e_1 \sum_{k=1}^{\infty} f^{k-1} \tilde{Y}_{t-k,1} \tilde{Y}_{t-k,2},$$

where $\tilde{Y}_t = (Y_{t,1}/\sigma_{t,1}, Y_{t,2}/\sigma_{t,2})$.

Given the observations $Y_1 = (Y_{1,1}, Y_{1,2}), \ldots, Y_T = (Y_{T,1}, Y_{T,2})$, we estimate the parameters, similarly to GARCH(p,q) estimation in (3.61), by maximizing the conditional modified likelihood,

$$\log_e \tilde{L}_r(a_0, a_1, \ldots, e_1, f) = -\frac{1}{2} \sum_{t=r+1}^{T} \log_e |\tilde{\Sigma}_t| + \sum_{t=r+1}^{T} \log f_\epsilon\left(\tilde{\Sigma}_t^{-1/2} Y_t\right),$$

where $r \geq 1$, f_ϵ is the density of the standard normal bivariate distribution $N(0, I_2)$, and $\tilde{\Sigma}_t$ is the truncated covariance, with elements $\tilde{\sigma}^2_{t,1}$, $\tilde{\sigma}^2_{t,2}$, $\tilde{\sigma}_{t,12}$, where

$$\tilde{\sigma}_{t,12} = \frac{e_0}{1-f} + e_1 \sum_{k=1}^{t} f^{k-1} Y_{t-k,1} Y_{t-k,2},$$

and $\tilde{\sigma}^2_{t,1}$, $\tilde{\sigma}^2_{t,2}$ are defined similarly.

Given the data $Y_0, \ldots, Y_{t-1}$, the MGARCH(1,1) estimator for the conditional variance is

$$\hat{\sigma}_{t,12} = \frac{\hat{e}_0}{1-\hat{f}} + \hat{e}_1 \sum_{k=0}^{t-1} \hat{f}^k Y_{t-k-1,1} Y_{t-k-1,2}, \tag{3.78}$$

where the parameter estimators $\hat{e}_0$, $\hat{e}_1$, and $\hat{f}$ are are calculated with the maximum likelihood method, analogously to the calculation in (3.62).

3.10.3 Moving Averages and Covariance Estimation

Let $Y_t = (Y_{t,1}, Y_{t,2})$ be a vector time series with mean zero components. We denote the conditional covariance by

$$\sigma_{t,12} = E(Y_{t,1} Y_{t,2} \mid Y_{t-1}, Y_{t-2}, \ldots).$$

The exponential moving average was defined in (3.16). The exponential moving average estimator of the conditional covariance, based on observations $Y_0, \ldots, Y_{t-1}$ is

$$\hat{\gamma}_t = \frac{1-\gamma}{1-\gamma^t} \sum_{k=0}^{t-1} \gamma^k Y_{t-k-1,1} Y_{t-k-1,2},$$

where $\gamma = \exp(-1/h)$, and $h > 0$ is the smoothing parameter. The moving average gives an estimator of conditional covariance which is alternative to the MGARCH estimator in (3.78).

3.11 APPLICATIONS IN RISK MANAGEMENT

We apply kernel estimation in variance, covariance, and quantile estimation. Quantile estimation has a direct application in risk management because quantile estimates give value at risk estimates and estimates of economic capital; see Section 1.5.1. We use variance estimates to construct quantile estimates. Covariance estimates can be applied to estimate the variance of the returns of a portfolio, because the variance of portfolio returns involves the covariance of the returns of the portfolio components, in addition to the variance of the individual portfolio components.

3.11.1 Volatility Estimation

The term *volatility* means the standard deviation of the returns, but sometimes the variance of the returns is called the volatility of the returns. We study the estimation of the conditional variance and the conditional standard deviation using the S&P 500 returns data, described in Section 1.6.1. We denote the net returns by $Y_t = (P_t - P_{t-1})/P_{t-1}$, where P_t is the price of the index, and the observed historical net returns are denoted by $Y_1, \ldots, Y_T$. We want to estimate

$$\sigma_t^2 = E\left(Y_t^2 \,|\, Y_{t-1}, Y_{t-2}, \ldots\right).$$

We assume that the conditional expectation of Y_t is zero, so that σ_t^2 is the conditional variance. Figure 1.3 shows the prices and the net returns of the S&P 500 index. The fitting of the GARCH(1,1) model for the S&P 500 returns was studied in Section 3.9.2, and a comparison with GARCH(1,1) and exponential moving averages was made in Section 3.9.3.

First we compare visually exponential moving average estimates and GARCH(1,1) estimates by looking at the time series of the estimates. Second we study the properties of various performance measures, with the help of GARCH(1,1) estimates. Third we use two performance measures to compare exponential moving average estimates to GARCH(1,1) estimates. Fourth we discuss smoothing parameter selection for exponentially weighted moving averages. Fifth we compare state–space kernel smoothing estimates to the GARCH(1,1) estimates and construct kernel estimators of the news impact curve.

EWMA and GARCH(1,1) Estimates of Volatility The exponential moving average was defined in (3.16), and the exponential moving average for conditional variance estimation was defined in (3.68). The formula for the exponential moving average estimator of conditional variance is

$$\hat{\sigma}_t^2 = \frac{1-\gamma}{1-\gamma^t} \sum_{i=0}^{t-1} \gamma^{t-i-1} Y_i^2, \tag{3.79}$$

where $\gamma = \exp(-1/h)$, and $h > 0$ is the smoothing parameter.

The GARCH model was defined in Section 3.9.2; and the GARCH(1,1) model, in particular, was defined in (3.65). We estimate the conditional variance so that the GARCH(1,1) model is fitted only after there are one year of observations. The first estimate of the variance is the sample variance

$$\hat{\sigma}_{t_0+1}^2 = \frac{1}{t_0} \sum_{t=1}^{t_0} Y_t^2, \tag{3.80}$$

where $t_0 = 251$. After that we use the recursion

$$\hat{\sigma}_t^2 = \hat{\alpha}_{t0} + \hat{\alpha}_{t1} Y_{t-1}^2 + \hat{\beta}_t \hat{\sigma}_{t-1}^2, \tag{3.81}$$

for $t = t_0 + 2, \ldots, T + 1$, where the parameter estimates are calculated sequentially, so that at time t only observations $Y_1, \ldots, Y_t$ are used in estimation. The prediction with the sequential estimation of parameters is called out-of-sample prediction.

Figure 3.20 shows a time series of the S&P 500 returns together with a time series of the sequentially calculated annualized standard deviations, along with three time series of exponential moving average estimates of the annualized conditional standard deviations of the S&P 500 returns. By the annualized standard deviation we mean $\sqrt{250}$ times the standard deviation. Panel (a) shows the S&P 500 returns (black line), and panel (b) shows the sequentially calculated annualized sample standard deviation (red line). Panel (c) shows the exponential moving average with the smoothing parameter $h = 1000$ (blue line), in panel (d) the smoothing parameter is $h = 25$ (brown line), and in panel (e) the smoothing parameter is $h = 0.45$ (purple line). The estimates are calculated with data starting at 1950-01-03, but we show the estimates starting at 1951-01-02. The annualized standard deviation of the complete time series of the net returns is 15.4%.

The exponential moving average can be conveniently used to make volatility estimates at different time scales, which is illustrated in Figure 3.20. We choose a large smoothing parameter to show large-scale phenomena and a small smoothing parameter to show small-scale phenomena. In Figure 3.20, panel (b) and panel (c) show the long-term rising trend of the volatility. Panel (d) and panel (e) display the short time scale volatility behavior, showing the short period volatility bursts.

Figure 3.21 shows the estimated values $\sqrt{250} \times \hat{\sigma}_t$ when we use the GARCH(1,1) formula of (3.81). We noted in (3.72) that the exponential moving averages with smoothing parameter $h = 11.55$ are close to the GARCH(1,1) estimates of the conditional variance. We can now compare the time series of the exponentially weighted moving average estimates in Figure 3.20 to the time series of the GARCH(1,1) estimates in Figure 3.21. The comparison indicates that the degree of smoothing of GARCH(1,1) is between the degree of smoothing with smoothing parameters $h = 0.45$ and $h = 25$ of the exponentially moving average.

We apply later the conditional standard deviation estimates in conditional quantile estimation and in portfolio selection. It is not obvious that the same degree of smoothing is optimal for both of these problems, and thus exponential moving average can be useful, because in the moving average estimator the degree of smoothing can be conveniently adjusted to suite the problem at hand.

Performance Measures In Section 1.9.2 we have presented several performance measures that can be used to compare volatility estimates. We want first to get information about the properties of the different performance measures. We have defined in (1.125) a class of performance measures

$$\mathrm{MDE}^{(p,q)} = \frac{1}{T - t_0 + 1} \sum_{t=t_0}^{T} \left| E|Z|^p \, \hat{\sigma}_t^p - |Y_t|^p \right|^{1/q}, \tag{3.82}$$

where $p, q > 0$ are parameters, MDE is an acronym for the mean of deviation errors, and $Z \sim N(0,1)$. See (1.126) for the closed-form expression of $E|Z|^p$.

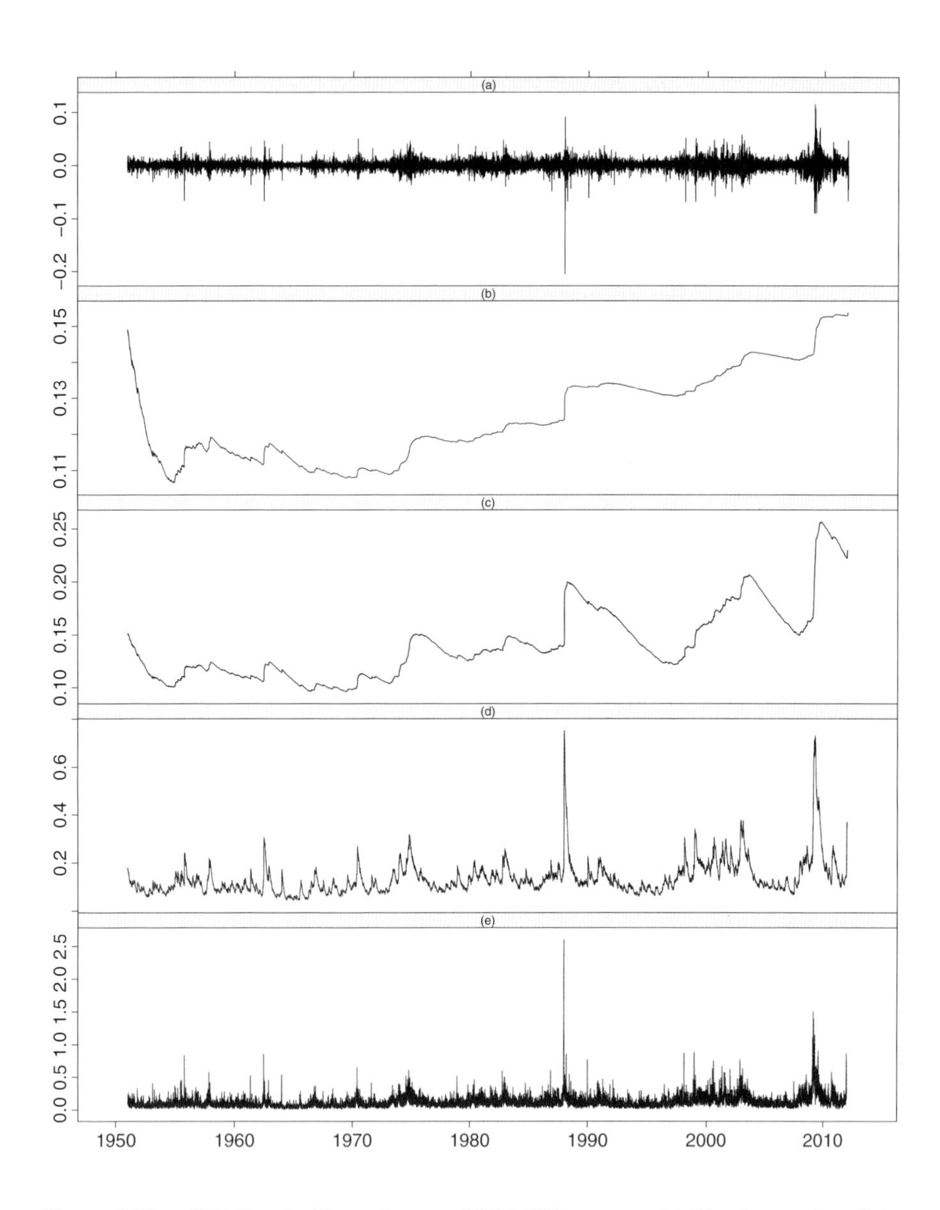

Figure 3.20 *EWMA volatility estimates of S&P 500 returns.* (a) The time series of the annualized S&P 500 returns (black line). (b) The sequential annualized sample standard deviation (red line). (c) The annualized exponential moving average estimate of the conditional standard deviation with the smoothing parameter $h = 1000$ (blue line). (d) The moving average estimate with $h = 25$ (brown line). (e) The moving average estimate with $h = 0.45$ (purple line).

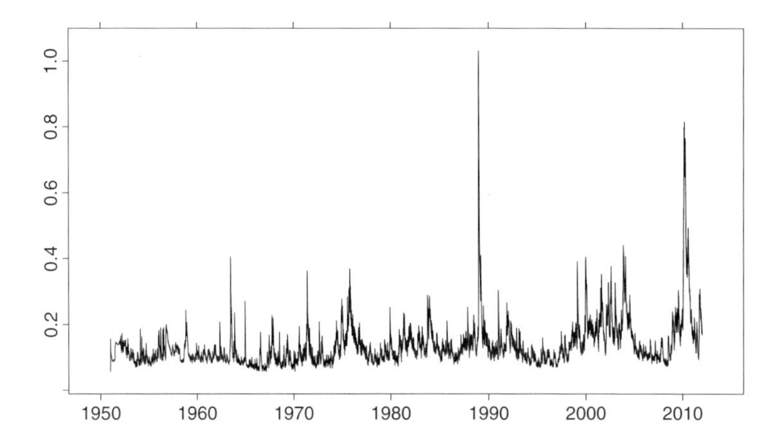

Figure 3.21 *GARCH(1,1) volatility estimates of S&P 500 returns.* Annualized conditional volatility estimates $\sqrt{250}\,\hat{\sigma}_t$, when the GARCH(1,1) model is used.

Figure 3.22 illustrates the effect of parameters p and q to the performance measure defined in (3.82). We show the time series $|E|Z|^p\hat{\sigma}_t^p - |Y_t|^p|^{1/q}$, where $\hat{\sigma}_t$ are the sequential GARCH(1,1) estimates of the conditional volatility. Panel (a) shows the case $p = 2, q = 1$, panel (b) shows the case $p = 2, q = 2$, panel (c) shows the case $p = 1, q = 1$, and panel (d) shows the case $p = 1, q = 2$. We can see that the prediction errors tend to be large when the volatility is large. Panel (a) shows that large errors dominate the performance measure when $p = 2$ and $q = 1$. Panel (b) and panel (c) show that the cases $p = 2, q = 2$ and $p = 1, q = 1$ are quite similar. Panel (d) shows that using $p = 1$ and $q = 2$ diminishes the influence of outliers in the performance measure.

We show next that when we measure the performance with the mean of deviation errors, then for some parameter values p and q the performance is better with the identically zero estimate $\hat{\sigma}_t \equiv 0$ than with the GARCH(1,1) estimates or EWMA estimates. We study this phenomenon in Figure 3.23.

Figure 3.23 shows a contour plot of the function

$$G(p,q) = \frac{\mathrm{MDE}^{(p,q)}(\mathrm{GARCH}(1,1))}{\mathrm{MDE}^{(p,q)}(\mathrm{NULL})} = \frac{\sum_{t=t_0}^{T} |E|Z|^p\hat{\sigma}_t^p - |Y_t|^p|^{1/q}}{\sum_{t=t_0}^{T} |Y_t|^{p/q}}, \tag{3.83}$$

where $\hat{\sigma}_t$ are the sequential out-of-sample GARCH(1,1) estimates, and we show the range $0.1 \le p, q \le 3$. The region with the dark gray color shows the area where the function G is larger than one. In the region where G is larger than one, the zero estimator $\hat{\sigma}_t \equiv 0$ has a smaller deviation error than the GARCH(1,1) estimator and is judged to be better than the GARCH(1,1) estimator. The region where G is larger than one is approximately equal to the region where p and q are larger than 1. Figure 3.23 suggests that we should prefer p and q values from the light gray region,

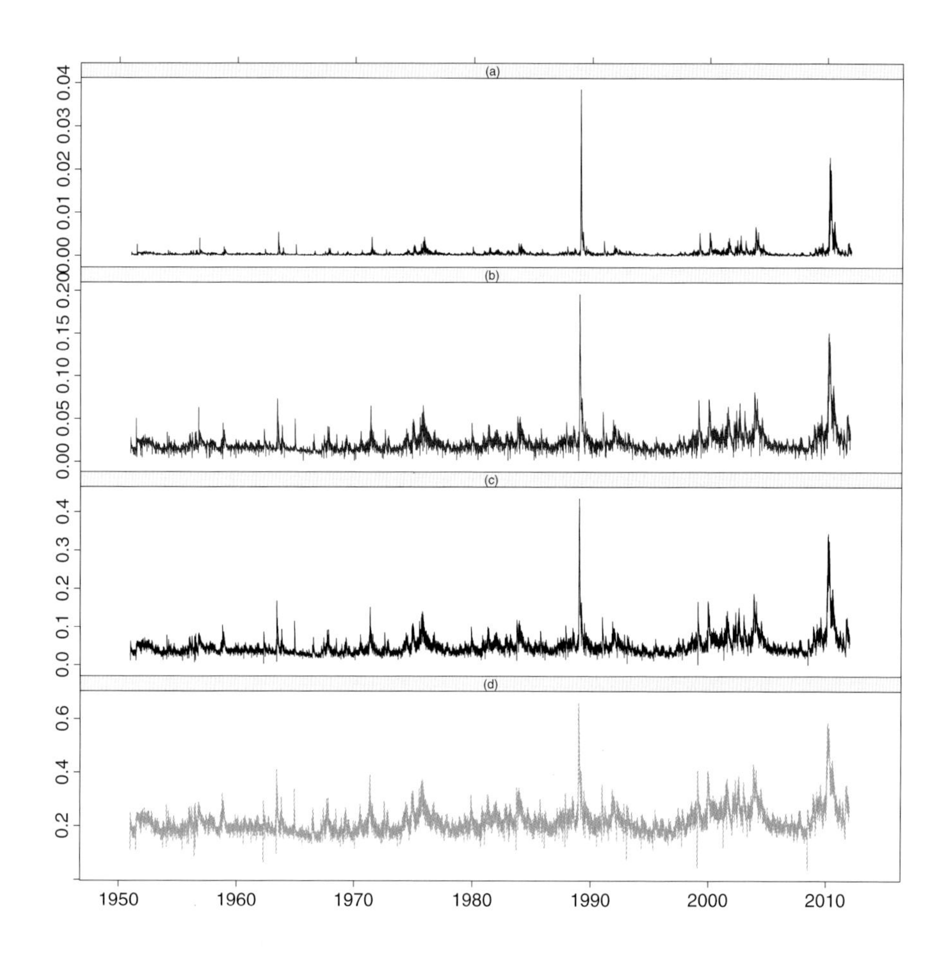

Figure 3.22 *Prediction errors of GARCH(1,1).* Shown are time series of prediction errors $|E|Z|^p\hat{\sigma}_t^p - |Y_t|^p|^{1/q}$, when $\hat{\sigma}_t$ are GARCH(1,1) estimates of the volatility. (a) $p = 2, q = 1$; (b) $p = 2, q = 2$; (c) $p = 1, q = 1$; (d) $p = 1, q = 2$.

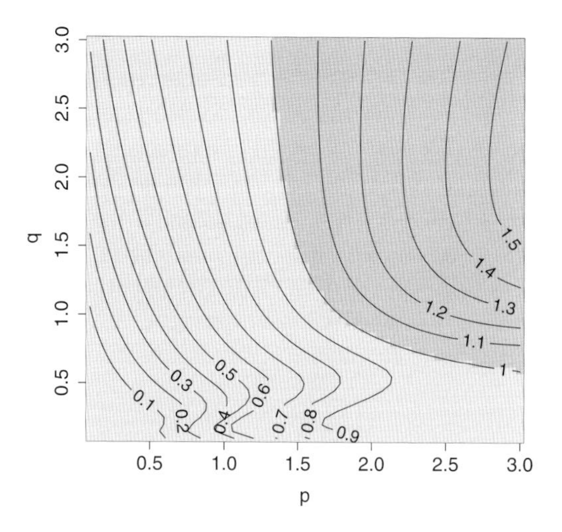

Figure 3.23 *Comparison of performance measures.* A contour plot of the function G, defined in (3.83), for the range $0.1 \leq p, q \leq 3$. The dark gray region is the area of (p, q)-values where the null estimate $\hat{\sigma}_t \equiv 0$ is judged to be better than the GARCH(1,1) estimate, by the performance measure $\mathrm{MDE}^{(p,q)}$.

where G is smaller than one. For example, values $p = 1$ and $q = 2$ would do, but not $p = q = 2$.

The performance measure

$$\mathrm{MARE}^{(p)} = \frac{1}{T - t_0 + 1} \sum_{t=t_0}^{T} \left| \frac{|Y_t|^p}{E|Z|^p \hat{\sigma}_t^p} - 1 \right| \tag{3.84}$$

was mentioned in (1.127), where $p > 0$, and $Z \sim N(0, 1)$. We shall use both $\mathrm{MARE}^{(p)}$ with $p = 2$ and $\mathrm{MDE}^{(p,q)}$ with $p = 1$ and $q = 2$ to compare the exponentially weighted moving average and GARCH(1,1).

Comparing the Performance of EWMA and GARCH(1,1) Figure 3.24 compares the performance of exponentially weighted smoothing average with the performance of GARCH(1,1). We use the performance measure $\mathrm{MDE}^{(p,q)}$ with $p = 1$ and $q = 2$, defined in (3.82), and the performance measure $\mathrm{MARE}^{(p)}$ with $p = 2$, defined in (3.84).

Figure 3.24(a) shows the ratios

$$\frac{\mathrm{MDE}^{(p,q)}(\mathrm{EWMA}(h))}{\mathrm{MDE}^{(p,q)}(\mathrm{GARCH}(1,1))},$$

where $p = 1$ and $q = 2$, the values of the smoothing parameter h are in the range $\{1, 2, \ldots, 30, 40\}$.[39] Here $\mathrm{MDE}^{(p,q)}(\mathrm{EWMA}(h))$ is the mean deviation error of the exponentially weighted moving average, with the smoothing parameter h, and $\mathrm{MDE}^{(p,q)}(\mathrm{GARCH}(1,1))$ is the mean deviation error of the GARCH(1,1) estimate.[40] We see that there is little difference between the exponentially weighted smoothing average and the GARCH(1,1) estimates. The red horizontal line is drawn at level one; the values below the red line indicate a better performance for the exponentially weighted moving average. We see additionally that the exponentially weighted moving average is robust with respect to the choice of the smoothing parameter, since the error is below the error of GARCH(1,1) at least in the range $h = 3 - 20$. The minimum error is reached when $h = 8$.

Figure 3.24(b) shows the ratios

$$\frac{\mathrm{MARE}^{(p)}(\mathrm{EWMA}(h))}{\mathrm{MARE}^{(p)}(\mathrm{GARCH}(1,1))},$$

where $p = 2$, the values of the smoothing parameter h are in the range $\{4, \ldots, 80\}$,[41] and $\mathrm{MARE}^{(p)}$ is defined in (3.84).[42] The GARCH(1,1) estimator is slightly better with the MARE criterion; the error curve is over the red line drawn at level one, for all values of h. The error of the moving average estimator increases slowly, when h increases. The minimum error is obtained when $h = 40$.

Smoothing Parameter Selection for Moving Averages We can use Figure 3.24 to choose the smoothing parameter. We conclude that $\mathrm{MDE}^{(1,2)}$ criterion would recommend about $h = 8$ and $\mathrm{MARE}^{(2)}$ criterion would recommend h in the range 20–40. JPMorgan (1996) contains the recommendation $\gamma = 0.94$ for the choice of the smoothing parameter, for daily data, which corresponds to $h = 16.16$, by the correspondence $h = -1/\log(\gamma)$. For the monthly data the recommendation of JPMorgan (1996) is $\lambda = 0.97$.

The previous comparison discusses the global choice of the smoothing parameter: We search for a smoothing parameter that would be optimal during the whole period (although the performance measurement was made sequentially, out-of-sample). It could also be that the optimal smoothing parameter changes in time. The smoothing parameter can be selected locally in time by

$$\hat{h}_t = \mathrm{arqmin}_{h>0}\, \mathrm{MDE}_t^{(p,q)}(h),$$

where

$$\mathrm{MDE}_t^{(p,q)}(h) = \frac{1}{t - t_0 + 1} \sum_{u=t_0}^{t} \left| E|Z|^p \hat{\sigma}_u^p(h) - |Y_u|^p \right|^{1/q},$$

[39] $h \in \{1, 2, 3, 4, 8, 12, 16, 20, 30, 40\}$.
[40] The value of $\mathrm{MDE}^{(p,q)}(\mathrm{GARCH}(1,1))$ is equal to 0.06025615 for $p = 1$ and $q = 2$.
[41] $h \in \{4, 8, 12, 16, 20, 30, 40, 60, 80\}$.
[42] The value of $\mathrm{MARE}^{(p)}(\mathrm{GARCH}(1,1))$ is equal to 1.08744 for $p = 2$.

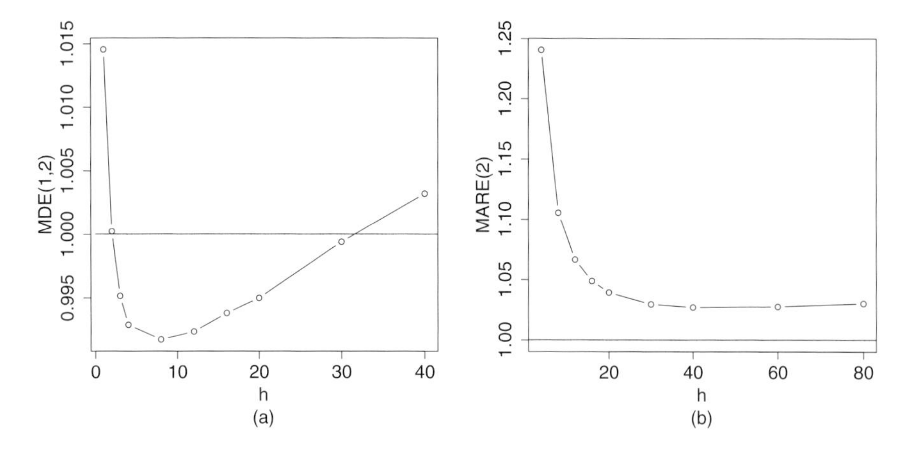

Figure 3.24 *Comparison of EWMA and GARCH(1,1).* (a) The ratio of the $\text{MDE}^{(1,2)}$ of the exponentially weighted moving average to the $\text{MDE}^{(1,2)}$ of the GARCH(1,1) estimate are shown for the smoothing parameter values $h = 1, \ldots 30, 40$. (b) The ratio of the the $\text{MARE}^{(2)}$ of the exponentially weighted moving average to the $\text{MARE}^{(2)}$ of the GARCH(1,1) estimate are shown for $h = 4, 8 \ldots, 80$. The red horizontal lines are drawn at level 1 in both panels.

and $\hat{\sigma}_u(h)$ is the volatility estimate with smoothing parameter h, calculated using data $Y_1, \ldots, Y_{u-1}$. Spokoiny (2000) discusses an other type of locally adaptive smoothing parameter selection with local constant volatility estimates.

The negative logarithmic likelihood for the GARCH model was given in (3.62). Fan & Gu (2003) propose to use the negative logarithmic pseudolikelihood

$$\text{PL}_T = \sum_{t=t_0}^{T} \left(\log \hat{\sigma}_t^2 + Y_t^2 / \hat{\sigma}_t^2 \right)$$

in choosing the smoothing parameter (the decay factor in the exponential weighted average). The smoothing parameter minimizing negative logarithmic pseudolikelihood is chosen.

State–Space Smoothing in Volatility Estimation

Previous Volatility as an Explanatory Variable In Section 3.9.1 we have defined conditional variance estimators which use local averaging. Let us define

$$Y_t = R_t, \qquad X_t = \left(\sum_{i=t-k_1}^{t-1} R_i^2, \sum_{i=t-k_1-k_2}^{t-k_1-1} R_i^2 \right), \tag{3.85}$$

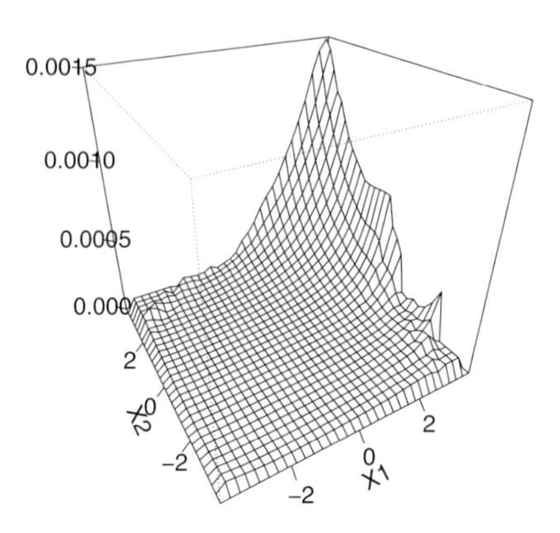

Figure 3.25 *A Kernel estimate in volatility prediction.* A perspective plot of a kernel regression estimate.

where R_t is the net S&P 500 return and $k_1, k_2 \geq 1$. We use now regression data (X_t, Y_t) to define a conditional variance estimator

$$\hat{\sigma}^2_{t+1} = \sum_{i=1}^{t} p_i(X_t)\, Y_i^2, \tag{3.86}$$

where $p_i(x)$ are the kernel weights. We choose $k_1 = k_2 = 5$ and make the copula transform of the data X_t to the standard Gaussian margins of the data X_t. In Section 2.5 we have defined the explanatory variable as $X_t = (R^2_{t-1}, \ldots, R^2_{t-p})$ and used least squares regression and ARCH modeling. The definition of the explanatory variables as in (3.85) reduces the number of explanatory variables, but still retains information from a long history.

Figure 3.25 shows a perspective plot of a kernel estimate. The smoothing parameter is $h = 1$ and the standard normal kernel is used.

Figure 3.26 shows state–space kernel smoothed annualized volatility estimates. Panel (a) shows the estimates with smoothing parameter $h = 2$, panel (b) shows estimates with $h = 0.5$ in red, and panel (c) shows estimates with $h = 0.1$ in blue. We see from panel (b) that the smoothing parameter $h = 0.5$ leads to a time series of estimates which resembles the time series of GARCH(1,1) estimates in Figure 3.21, and this time series is also similar to the time series of exponentially weighted moving average estimates with $h = 10$ (not shown). However, increasing the smoothing parameter to $h = 2$ leads to a time series of different type than the time series of exponentially weighted estimates with smoothing parameter $h = 25$, shown in Figure 3.20(d) with the brown line.

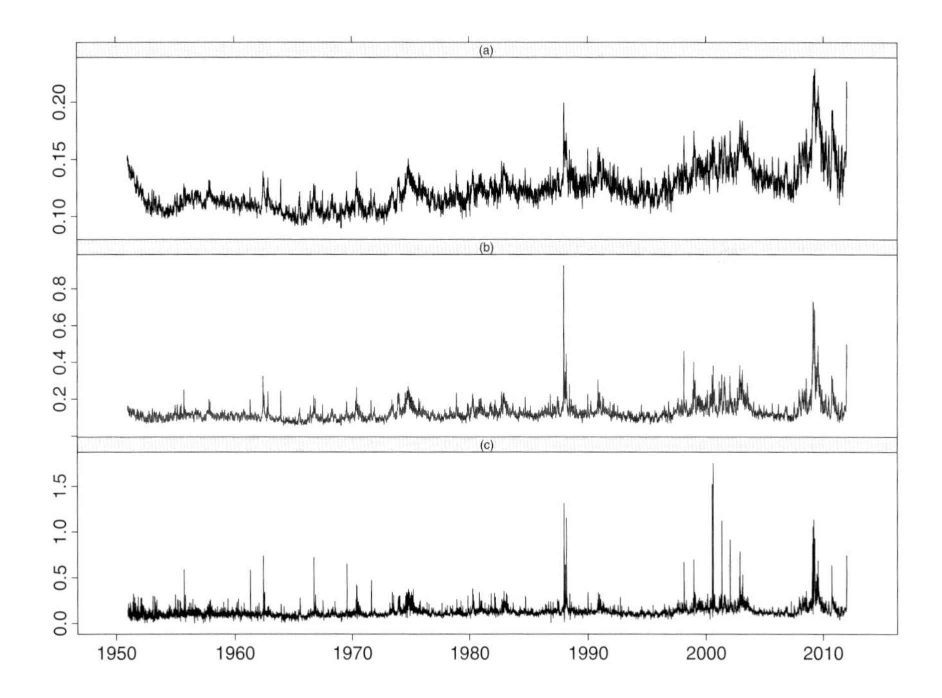

Figure 3.26 *S&P 500 volatility: State–space kernel estimates.* Annualized volatility estimates obtained with the state space kernel estimator when smoothing parameter is (a) $h = 2$ (black), (b) $h = 0.5$ (red), and (c) $h = 0.1$ (blue).

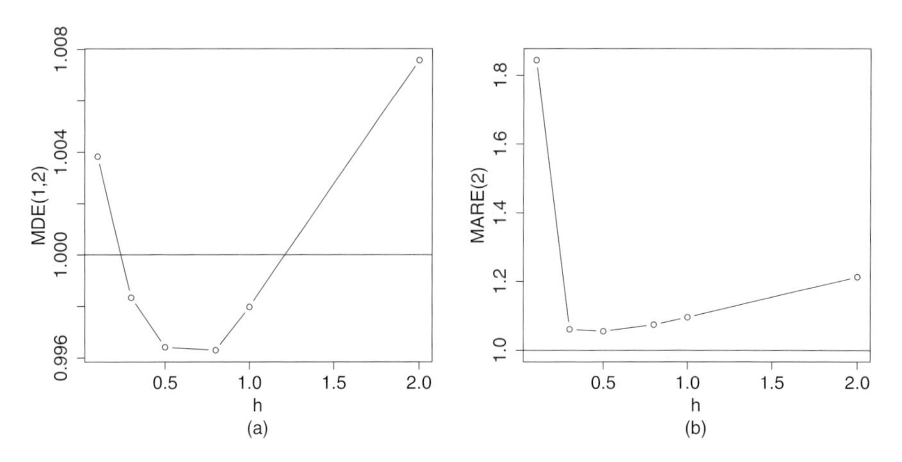

Figure 3.27 *S&P 500 volatility: The performance of state–space smoothing.* Panel (a) shows the mean absolute deviation errors ($\text{MDE}^{(1,2)}$) for different smoothing parameters h. Panel (b) shows the mean absolute ratio errors ($\text{MARE}^{(2)}$). The errors are relative to the $\text{MDE}^{(1,2)}$ and $\text{MARE}^{(2)}$ of the GARCH(1,1). The red horizontal lines are drawn at level 1 in both panels.

Figure 3.27 shows the performance of the state space smoothing estimator of the conditional variance for the smoothing parameters $h = 0.1, 0.3, 0.5, 0.8, 1, 2$. Panel (a) shows the performance measured with the mean deviation error with $p = 1$ and $q = 2$, defined in (3.82). Panel (b) shows the performance measured with the mean absolute ratio errors with exponent 2, defined in (3.84). The performance is compared to the GARCH(1,1) estimates: We have divided the MDE and the MARE values of the least squares estimator with the MDE and the MARE values of the GARCH(1,1) estimator. We see that for the MDE measure the performance is better for smoothing parameters around $h = 0.5$ than the performance of the GARCH(1,1), but for the MARE measure the performance is worse than the performance of GARCH(1,1). Comparing the performance to the performance of the exponentially weighted moving average estimators shown in Figure 3.24, we see that the state–space kernel smoothing performs slightly worse.

Previous Returns as the Explanatory Variable Let us choose

$$Y_t = R_t, \qquad X_t = R_{t-1},$$

and consider the estimation of the conditional variance with the kernel estimator, using the S&P 500 data described in Section 1.6.1. The function

$$\sigma^2(x) = \text{Var}(Y_t \mid X_t = x)$$

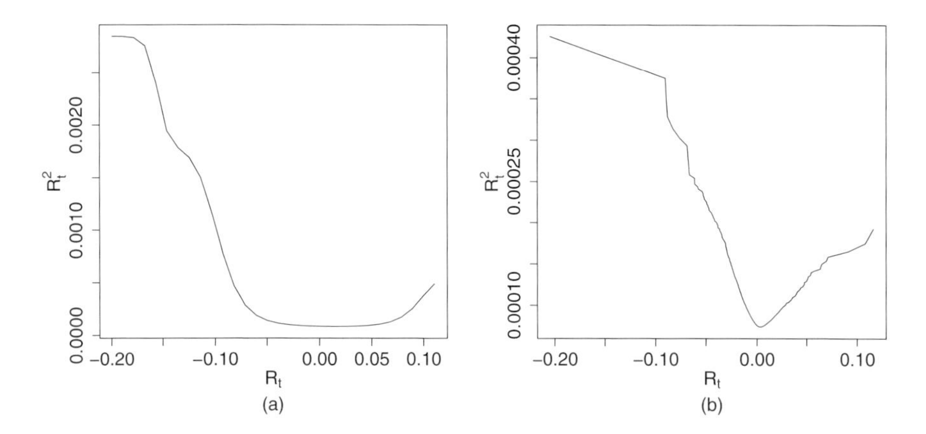

Figure 3.28 *S&P 500 volatility: News impact curves.* (a) A kernel estimate of the conditional variance. (b) A kernel estimate of the conditional variance when the values of the explanatory variable are transformed.

can be called the news impact curve, because the explanatory variable is the previous day return. We use the estimator

$$\hat{\sigma}^2(x) = \sum_{t=1}^{T} p_t(x) Y_t^2 - \hat{f}_{reg}(x)^2,$$

where $p_t(x)$ are the kernel weights, defined in (3.7), and $\hat{f}_{reg}$ is a kernel regression estimator of $E(Y_t \mid X_t = x)$.[43] An ARCH(∞) model for the estimation of the news impact curve is mentioned in (3.64). A local linear estimator of the news impact curve is shown in Figure 5.7.

Figure 3.28 shows two estimates of the news impact curve. Panel (a) shows a kernel estimator with the smoothing parameter $h = 0.025$ and the standard normal kernel. Panel (b) shows a kernel estimator where the data are first transformed to follow approximately the standard normal distribution (as in the copula transform in Section 1.7.2), then the kernel estimator with smoothing parameter $h = 1$ and the standard normal kernel is applied, and finally the x-values are transformed back to follow the original distribution.

3.11.2 Covariance and Correlation Estimation

We estimate the conditional covariance between the returns of the S&P 500 index and the returns of the Nasdaq-100 index. The S&P 500 and Nasdaq-100 index data are described in Section 1.6.2, where Figure 1.4(b) shows the scatter plot of the index

[43]The term $\hat{f}_{reg}(x)^2$ does not have any visible impact on the estimator.

returns. We can see from that figure that the index returns are highly correlated. Covariance and correlation regression was introduced in Section 1.1.5.

EWMA Estimates of Covariance and Correlation Let us denote the observed returns of the indices by $(Y_0, Z_0), \ldots, (Y_T, Z_T)$. The exponential moving average estimator of the conditional covariance is

$$\hat{\gamma}_t = \frac{1-\gamma}{1-\gamma^t} \sum_{i=0}^{t-1} \gamma^{t-i} Y_i Z_i, \tag{3.87}$$

where $\gamma = \exp(-1/h)$, and $h > 0$ is the smoothing parameter. We have assumed that the net returns have the expected value zero.

Figure 3.29 shows the time series of estimates $\hat{\gamma}_t$ of the conditional covariance between the returns of S&P 500 and Nasdaq-100. The black curve in panel (a) shows the sequentially calculated covariances. The red curve in panel (b) shows a moving average estimator with the smoothing parameter $h = 1000$. The blue curve in panel (c) has $h = 10$, and the brown curve in panel (d) has $h = 1$. We show the annualized covariance, so that the correlation estimates are multiplied by 250. We can see that the covariance estimates express a considerable time variation. However, the time variation may be due to the variation in the marginal variances. Thus we turn into the conditional correlation estimates.

There are two ways to estimate the conditional correlation, as was noted in Section 1.1.5. The first way is to first estimate the conditional covariance and then normalize the estimate with estimates of conditional standard deviation. This leads to the estimate

$$\hat{\rho}_t = \frac{\hat{\gamma}_t}{\hat{\sigma}_{t,1}\hat{\sigma}_{t,2}}, \tag{3.88}$$

where γ_t is the covariance estimate defined in (3.87) and $\hat{\sigma}_{t,1}$ and $\hat{\sigma}_{t,2}$ are estimates of the conditional standard deviation, like exponentially weighted moving averages or GARCH(1,1) estimates, studied in Section 3.11.1. The second way is to first normalize the returns with the estimated standard deviations and then calculate the condtional covariance estimate from the normalized time series. This leads to the estimate

$$\hat{\varrho}_t = \frac{1-\gamma}{1-\gamma^t} \sum_{i=0}^{t-1} \gamma^{t-i} \frac{Y_i Z_i}{\hat{\sigma}_{i,1}\hat{\sigma}_{i,2}},$$

where $\hat{\sigma}_{i,1}$ and $\hat{\sigma}_{i,2}$ are estimates of the conditional standard deviation.

Figure 3.30 shows conditional correlation estimates ρ_t, defined in (3.88). Panel (a) shows the conditional correlation estimates when the covariances and the standard deviations are calculated sequentially (black curve). Panel (b) shows the conditional correlation estimates when the covariances and the standard deviations are calculated using exponentially weighted moving averages with the smoothing parameter $h = 500$ (red curve). Panel (c) shows the case with the smoothing parameter $h = 50$ (blue curve), and panel (d) shows the case with the smoothing parameter $h = 10$ (brown curve).

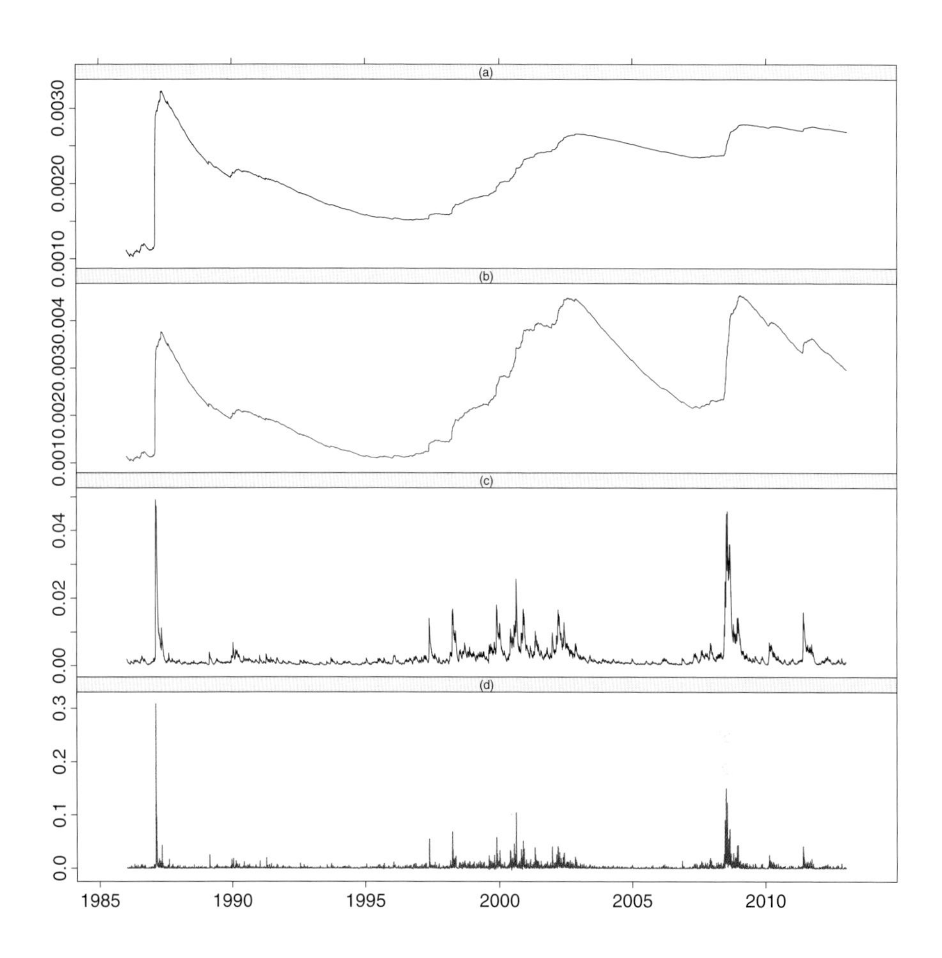

Figure 3.29 *S&P 500 and Nasdaq-100*: *Covariance estimates.* Panel (a) shows the sequentially calculated covariance estimates between the returns of S&P 500 and Nasdaq-100 (black curve). Panel (b) shows a moving average estimator with smoothing parameter $h = 1000$ (red curve). Panel (c) shows the case $h = 10$ (blue curve). Panel (d) shows the case $h = 1$ (brown curve).

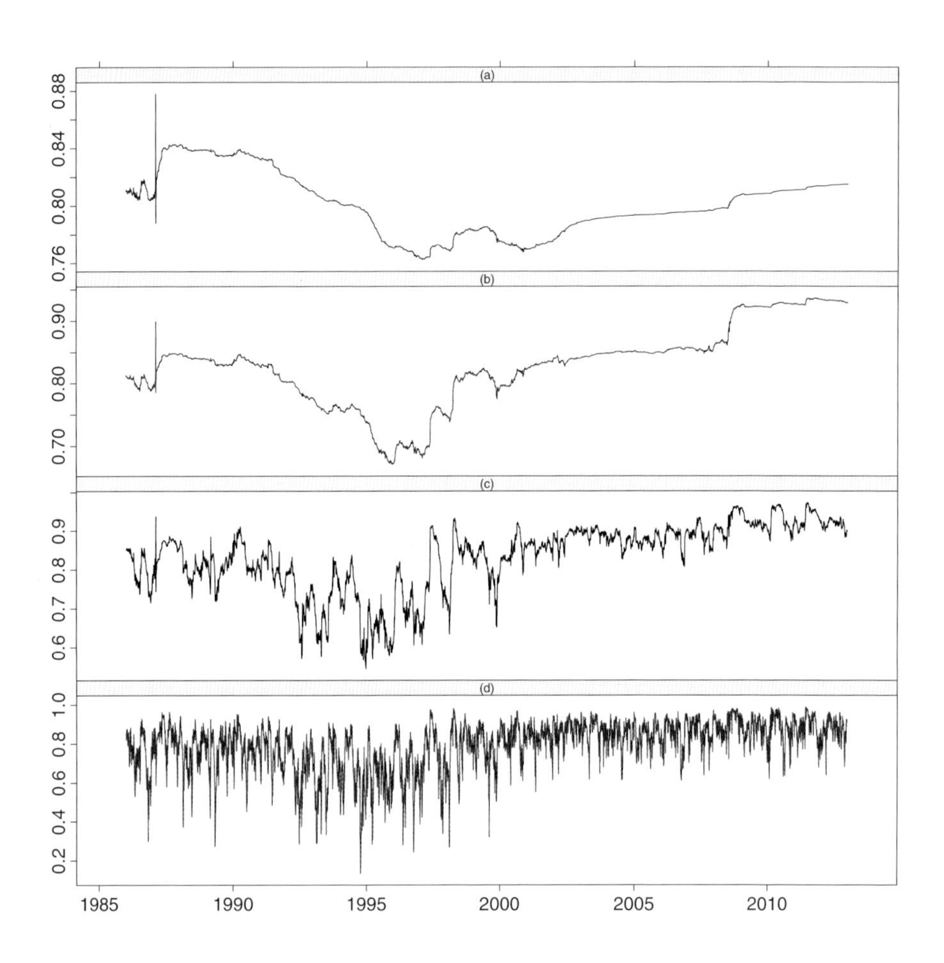

Figure 3.30 *S&P 500 and Nasdaq-100: Correlation estimates.* Panel (a) shows the time series of sequentially calculated correlation estimates between the returns of S&P 500 and Nasdaq-100 (black curve). Panel (b) shows a moving average estimator with $h = 500$ (red curve). Panel (c) shows the case $h = 50$ (blue curve). Panel (d) shows the case $h = 10$ (brown curve).

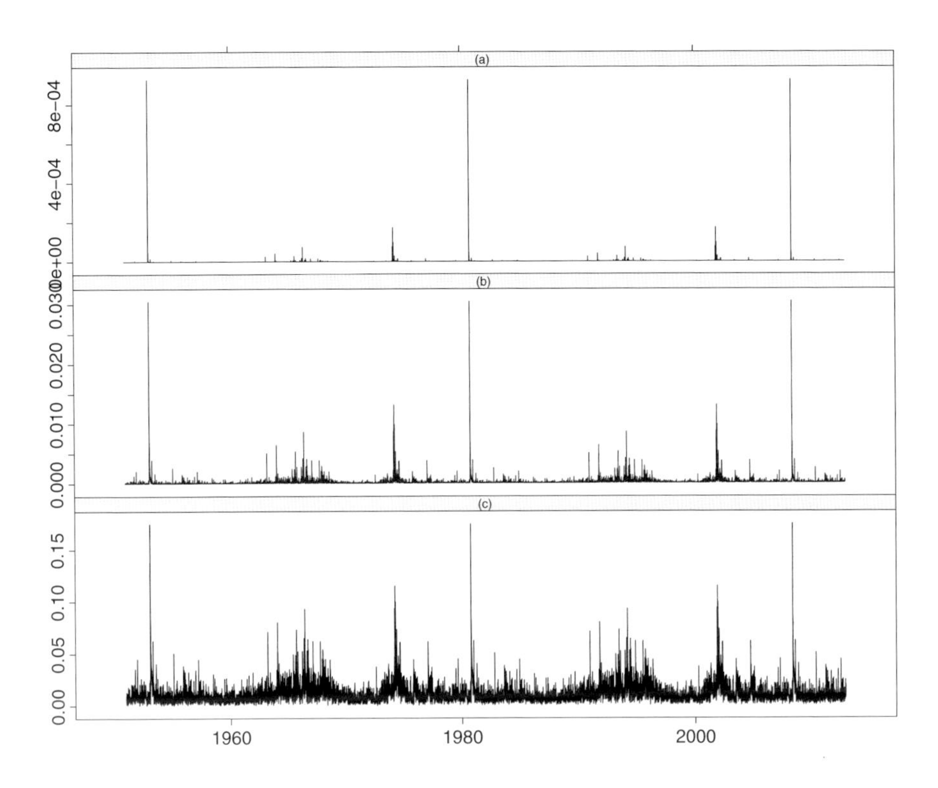

Figure 3.31 *S&P 500 and Nasdaq-100: Prediction errors.* Time series of prediction errors $|\hat{\gamma}_t - Y_t Z_t|^{1/q}$ of the covariance estimates are shown. (a) $q = 0.5$; (b) $q = 1$; (c) $q = 2$.

Performance Measures The performance of conditional covariance estimators can be measured with the mean of deviation errors $\text{MDE}^{(q)}$, defined in (1.128). The definition of this performance measure is

$$\text{MDE}^{(q)}(\hat{\gamma}) = \frac{1}{T - t_0 + 1} \sum_{t=t_0}^{T} |\hat{\gamma}_t - Y_t Z_t|^{1/q},$$

where $q > 0$. We have to choose a reasonable value for the parameter q.

Figure 3.31 shows the time series of prediction errors $|\hat{\sigma}_{t,12} - Y_t Z_t|^{1/q}$ for three values of q. Panel (a) shows the case $q = 0.5$ (black curve), panel (b) shows the case $q = 1$ (red curve), and panel (c) shows the case $q = 2$ (blue curve). We can see that the time series are inhomogeneous for the values $q = 0.5$ and $q = 1$, but the choice $q = 2$ leads to a more homogeneous time series.

Figure 3.32 shows the ratios

$$\frac{\text{MDE}^{(q)}(\hat{\gamma})}{\text{MDE}^{(q)}(0)} \tag{3.89}$$

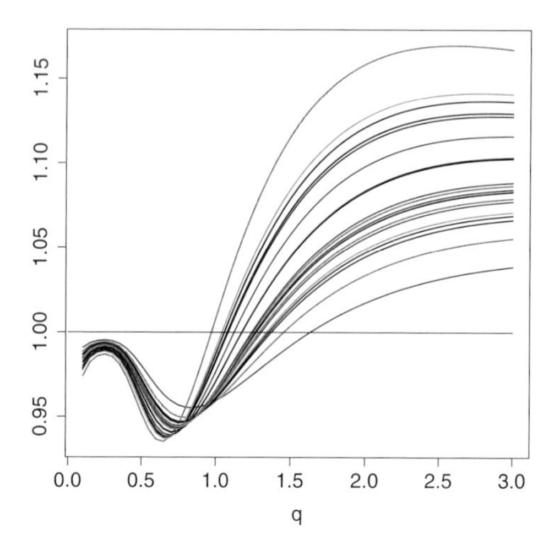

Figure 3.32 *S&P 500 and Nasdaq-100: Comparison of performance measures.* The ratio defined in (3.89) is shown as a function of q, for 21 smoothing parameters $h \in \{2, 3, \ldots, 30, 40\}$.

as a function of $q \in [0.1, 3]$. The expression $\mathrm{MDE}_T^{(q)}(0)$ means the mean of the deviation errors of the identically zero estimator. We show 21 curves, corresponding to a selection of smoothing parameters $h \in \{2, 3, \ldots, 30, 40\}$. We can see that for larger values of q the performance of a moving average is worse than the performance of the zero estimate. In particular, using $q = 2$ leads to a performance measure which prefers the identically zero estimator.

Smoothing Parameter Selection Figure 3.33 shows the normalized mean deviation errors $\mathrm{MDE}^{(q)}$ as a function of the smoothing parameter h, for the values $q = 0.5$ (black a), $q = 1$ (red b), and $q = 2$ (blue c). The mean deviation errors are normalized by dividing by the minimal value of the deviation errors. The minimal values occured at $h = 13$ for $q = 0.5$, at $h = 15$ for $q = 1$, and at $h = 5$ for $q = 2$. The h-axis is logarithmic.

3.11.3 Quantile Estimation

Quantile regression was introduced in Section 1.1.6. We study quantile estimation with the S&P 500 returns data described in Section 1.6.1. We estimate the conditional quantiles

$$Q_p(Y_t \mid Y_{t-1}, Y_{t-2}, \ldots),$$

with level $0 < p < 1$, using the observed historical returns.

We study the performance of the GARCH(1,1) volatility estimates in quantile estimation and compare their performance with the moving average estimators. The

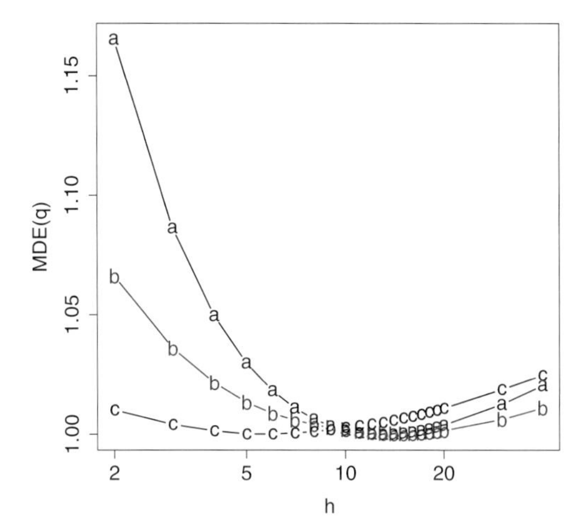

Figure 3.33 *S&P 500 and Nasdaq-100: Smoothing parameter selection.* The normalized mean deviation errors are shown as a function of the smoothing parameter h. The three curves correspond to the choices $q = 0.5$ (black curve with labels "a"), $q = 1$ (red curve with labels "b"), and $q = 2$ (blue curve with labels "c").

sequential GARCH(1,1) volatility estimates $\hat{\sigma}_t^{garch}$ were defined in (3.80) and (3.81). Exponentially weighted moving average $\hat{\sigma}_t^{ewma}$ for the estimation of conditional variance was defined in (3.68). We use also the name "EWMA(h) estimator" to refer to the exponentially weighted moving average estimator with smoothing parameter h.

The fitting of the GARCH(1,1) model for the S&P 500 data was studied in Section 3.9.2. The exponentially weighted moving average estimator was compared with GARCH(1,1) in Section 3.9.3. The volatility estimation for the S&P 500 data was studied in Section 3.11.1.

Collection of Quantile Estimators We study conditional quantile estimators which are based on conditional standard deviation estimators. This method was defined in (1.30), and it puts

$$\hat{Q}_p(Y_t \mid Y_{t-1}, Y_{t-2}, \ldots) = \hat{\sigma}_t \, \hat{F}_{\epsilon_t}^{-1}(p), \tag{3.90}$$

where $\hat{\sigma}_t$ is an estimator of the conditional standard deviation and $\hat{F}_{\epsilon_t}^{-1}(p)$ is an estimator of the p-quantile of the distribution of $\epsilon_t = Y_t/\sigma_t$. Below, the estimator $\hat{\sigma}_t$ of conditional standard deviation is either a GARCH(1,1) estimator or a EWMA(h) estimator. We study three definitions for $\hat{F}_{\epsilon_t}^{-1}(p)$. First, we choose

$$\hat{F}_{\epsilon_t}^{-1}(p) = \Phi^{-1}(p), \tag{3.91}$$

where Φ is the distribution function of the standard normal distribution. Second, we choose

$$\hat{F}_{\epsilon_t}^{-1}(p) = \sqrt{\frac{\nu - 2}{\nu}}\ t_\nu^{-1}(p), \tag{3.92}$$

where t_ν is the distribution function of the t-distribution with ν degrees of freedom, $\nu > 2$. If $X \sim t_\nu$, then $\mathrm{Var}(X) = \nu/(\nu - 2)$, so that $\sqrt{(\nu-2)/\nu}\ t_\nu^{-1}(p)$ is the p-quantile of such t-distribution, which is standardized to have unit variance. Third, we choose

$$\hat{F}_{\epsilon_t}^{-1}(p) = \hat{Q}^{res}(p), \tag{3.93}$$

where $\hat{Q}^{res}(p)$ is the empirical quantile of the residuals $Y_t/\hat{\sigma}_t$. Empirical quantiles were defined in (1.26).

The distribution of the GARCH(1,1) residuals $Y_t/\hat{\sigma}_t$ was studied in Section 3.9.2, where Figure 3.11 and Figure 3.12 show the tail plots and QQ-plots of the residuals. The maximum likelihood estimator of the GARCH(1,1) model is defined with the assumption of standard normal innovations, but we noted that the residuals are better fitted with the t-distribution and thus it makes sense to try to use quantiles from the t-distribution. We take the degrees of freedom equal to 12. The method of using empirical quantiles of the residuals was suggested in Fan & Gu (2003).

The Performance Measure The performance measurement is explained in Section 1.9.4. We measure the performance of a quantile estimator by looking at the differences $p - \hat{p}$ for p close to zero and $\hat{p} - p$ for p close to one, where $\hat{p}$ is defined in (1.129) as

$$\hat{p} = \frac{1}{T - t_0} \sum_{t=t_0+1}^{T} I_{(-\infty, \hat{q}_t]}(Y_t),$$

where $\hat{q}_t = \hat{Q}_p(Y_t \mid Y_{t-1}, Y_{t-2}, \ldots)$ and $1 \le t_0 \le T - 1$. We take $t_0 = 250$ so that the performance measurement starts after one year of observations has accumulated.

GARCH(1,1)-Based Quantile Estimators GARCH(1,1)-based quantile estimators are defined by (3.90), where $\hat{\sigma}_t$ is estimated with the GARCH(1,1) method, and the residual quantile is determined with one of the three methods in (3.91)–(3.93).

Figure 3.34 shows the time series of estimated conditional quantiles with the level $p = 0.05$ when the GARCH(1,1) method is used to estimate the conditional variance, and the method (3.91) of choosing the standard normal distribution as the distribution of the residuals is used.

Figure 3.35 compares GARCH(1,1) quantile estimators when the residual quantile is determined with the three methods in (3.91)–(3.93). Panel (a) plots the function $p \mapsto p - \hat{p}$ in the range $p \in [0.001, 0.075]$ and panel (b) plots the function $p \mapsto \hat{p} - p$ in the range $p \in [0.925, 0.999]$. Four cases are shown: the residual distribution is the standard normal distribution, the standardized t-distribution with degrees of freedom 5 and 12, and the empirical distribution. The black curves show the case of the standard normal distribution, the blue curves show the case of the standard t-distribution with degrees of freedom 12, the red curves show the case of standard

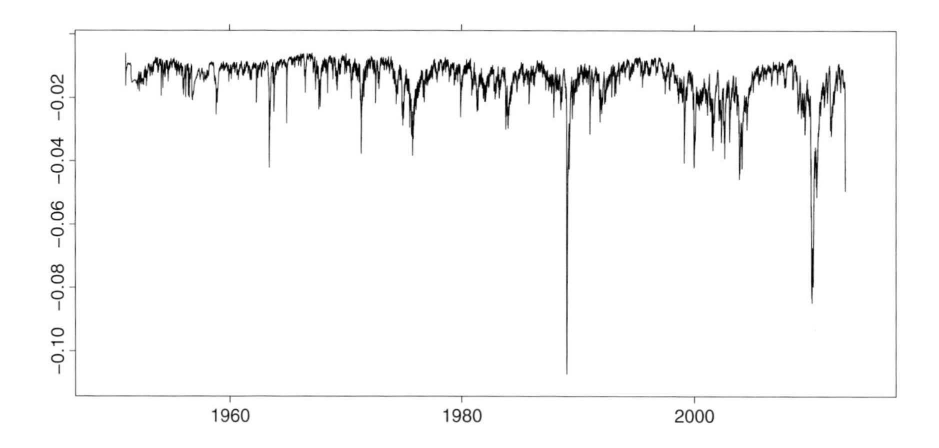

Figure 3.34 *GARCH(1,1) quantiles.* Shown is the time series of estimated quantiles with the level $p = 0.05$ for the S&P 500 returns data. The quantiles are estimated with the GARCH(1,1) method with the residual distribution being standard normal.

t-distribution with degrees of freedom 5, and the purple curves show the case of the empirical distribution. The green lines show the level $\alpha = 0.05$ fluctuation bands, defined in (1.130)–(1.131).

Figure 3.35(a) shows that the Gaussian residuals perform well for level $p = 0.05$, but for level $p = 0.01$ using a t-distribution or the empirical distribution gives better estimates. The GARCH(1,1)-based quantile estimates are estimating the left tail of the S&P 500 return distribution too light, except when the residuals are from the t-distribution with degrees of freedom 5, in which case the tail is estimated too heavy for levels $p < 0.01$. Figure 3.35(b) shows that for the right tail the quantile estimates are more accurate than for the left tail. The standard t-distribution with degrees of freedom 12 gives a good overall performance.

EWMA-Based Quantile Estimators Exponentially weighted moving average-based quantile estimators are defined by (3.90), where $\hat{\sigma}_t$ is calculated with the EWMA method, and the residual quantile is determined with one of the three methods in (3.91)–(3.93).

Figure 3.36 shows the performance of exponentially weighted moving average for four smoothing parameters: $h = 5$ (black), $h = 10$ (red), $h = 30$ (blue), and $h = 100$ (purple). Panel (a) plots the function $p \mapsto p - \hat{p}$ in the range $p \in [0.001, 0.075]$, and panel (b) plots the function $p \mapsto \hat{p} - p$ in the range $p \in [0.925, 0.999]$. The green horizontal line is drawn at level zero, and it is accompanied with the level $\alpha = 0.05$ fluctuation bands. The smoothing parameters $h = 10$ and $h = 30$ give the best results for large p. However, for small p the smoothing parameter $h = 100$ gives the best results.

Figure 3.37 shows the performance of the exponentially weighted moving average estimator with the smoothing parameter $h = 30$ for four residual distributions.

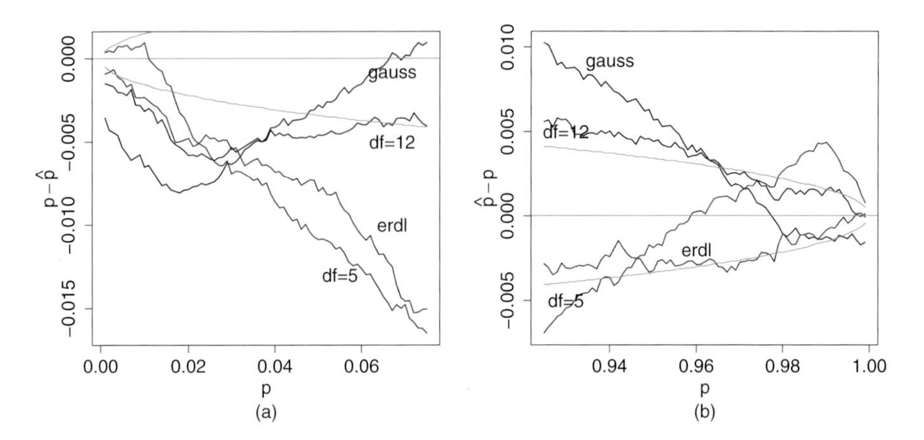

Figure 3.35 *Performance of GARCH(1,1) quantile estimators.* (a) Functions $p \mapsto p - \hat{p}$ for $p \in [0.001, 0.075]$. (b) Functions $p \mapsto \hat{p} - p$ for $p \in [0.925, 0.999]$. The black curves show the performance of the estimator with the standard normal residuals, the blue curves with the t-distribution with degrees of freedom 12, the red curves with the degrees of freedom 5, and the purple curves show the case of the empirical distribution.

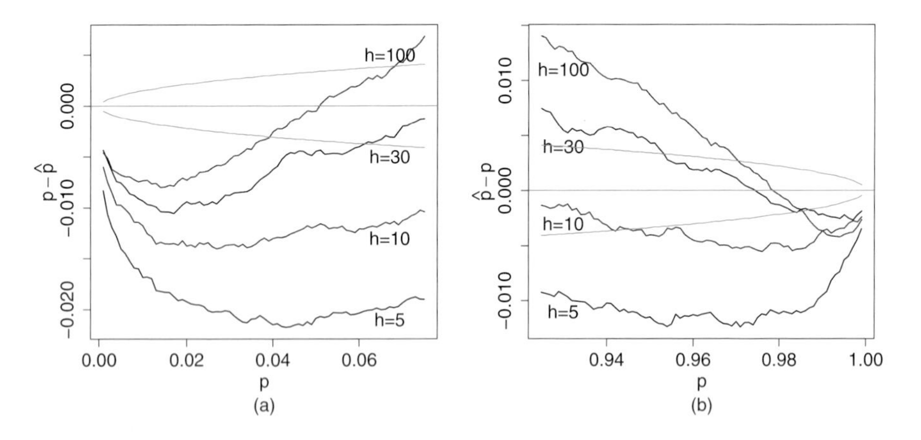

Figure 3.36 *EWMA(h) quantile estimator: Smoothing parameter selection.* Panel (a) shows the curves $p \mapsto p - \hat{p}$ for $p \in [0.001, 0.075]$, and panel (b) shows the curves $p \mapsto \hat{p} - p$ for the cases $p \in [0.925, 0.999]$. The smoothing parameters $h = 5, 10, 30, 100$ are shown with the colors black, red, blue, and purple.

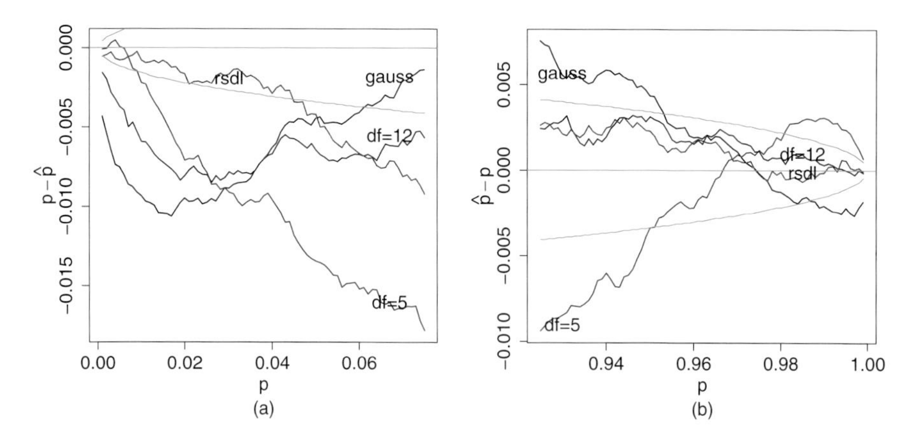

Figure 3.37 *EWMA(h) quantile estimator: Selection of residual distribution.* (a) The curves $p \mapsto p - \hat{p}$ for $p \in [0.001, 0.075]$. (b) The curves $p \mapsto \hat{p} - p$ for $p \in [0.925, 0.999]$. The residual distributions standard normal, standard t-distribution with degrees of freedom 12, degrees of freedom 5, and the empirical distribution are shown with the colors black, red, blue, and green.

Panel (a) shows the curves $p \mapsto p - \hat{p}$ for $p \in [0.001, 0.075]$, and panel (b) shows the curves $p \mapsto \hat{p} - p$ for the cases $p \in [0.925, 0.999]$. The black curve shows the standard normal residual distribution, the blue curve shows the standard t-distribution with degrees of freedom 12, the red curve shows degrees of freedom 5, and the purple curve shows the case of using empirical distribution. For the left tail the empirical residuals give the best result, except when $p \geq 0.05$, when the Gaussian residual give the best result. For the right tail the emprical residuals and the standard t-distribution with degrees of freedon 12 give the best results.

State–Space Kernel Smoothing-Based Quantile Estimators We have defined in (3.86) a state–space smoothing estimator $\hat{\sigma}_t^{state}$ for the conditional standard deviation. This estimator is now applied to quantile estimation.

Figure 3.38 shows the performance of the quantile estimators

$$\hat{Q}_p(Y_t \,|\, Y_{t-1}, \ldots) = \hat{\sigma}_t^{state} \Phi^{-1}(p).$$

We show the performance for the six smoothing parameters $h = 0.1, 0.3, 0.5, 0.8, 1, 2$. Panel (a) shows the curves $p \mapsto p - \hat{p}$ when $0.001 \leq p \leq 0.075$, and panel (b) shows the curves $p \mapsto \hat{p} - p$ for $0.925 \leq p \leq 0.999$. We can see that for the left tail the results are worse than the results of GARCH(1,1) or the results of exponentially weighted moving average. For the right tail the results are comparable when $p < 0.97$.

Figure 3.39 shows the performance of the quantile estimators

$$\hat{Q}_p(Y_t \,|\, Y_{t-1}, \ldots) = \hat{\sigma}_t^{state} \hat{Q}^{res}(p),$$

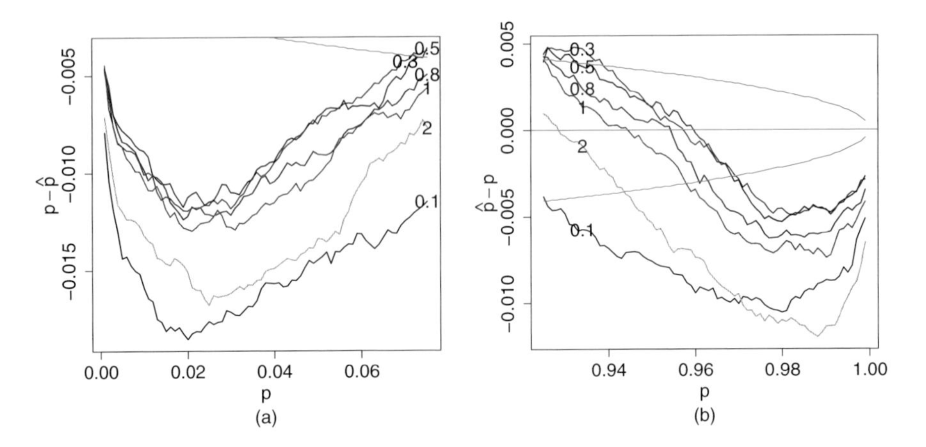

Figure 3.38 *State–space kernel quantile estimator: Gaussian residuals.* (a) The curves $p \mapsto p - \hat{p}$ for $p \in [0.001, 0.075]$. (b) The curves $p \mapsto \hat{p} - p$ for $p \in [0.925, 0.999]$. The smoothing parameters are $h = 0.1, 0.3, 0.5, 0.8, 1, 2$. The h-axis is logarithmic.

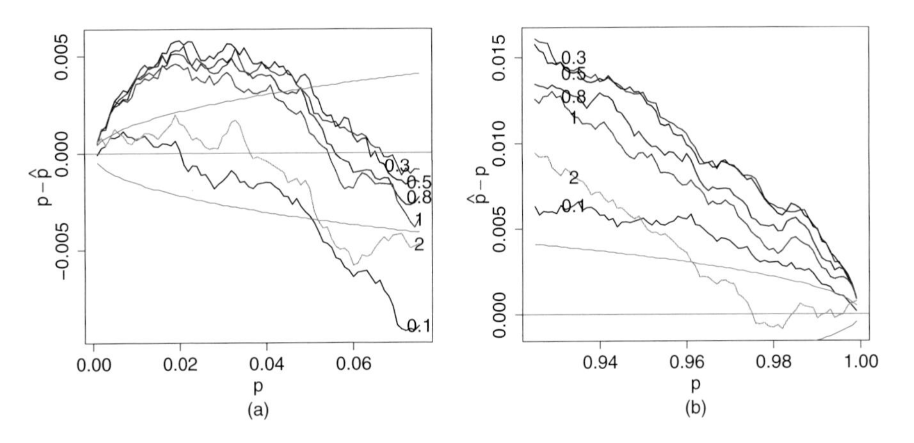

Figure 3.39 *State–space kernel quantile estimator: Empirical residuals.* (a) The curves $p \mapsto p - \hat{p}$ for $p \in [0.001, 0.075]$. (b) The curves $p \mapsto \hat{p} - p$ for $p \in [0.925, 0.999]$. The smoothing parameters are $h = 0.1, 0.3, 0.5, 0.8, 1, 2$. The h-axis is logarithmic.

where $\hat{Q}^{res}(p)$ is the empirical quantile of the residuals $Y_t/\hat{\sigma}_t$. Otherwise the setting is the same as in Figure 3.38 For the left tail the use of empirical quantiles of the residuals leads to much better performance. The performance is best for smoothing parameters $h = 1$ and $h = 2$. In these case the results are even better than the results of GARCH(1,1) and the exponentially weighted moving average. For the right tail the Gaussian distribution of the residuals gives a better performance.

3.12 APPLICATIONS IN PORTFOLIO SELECTION

Section 1.5.3 introduced the basic concepts of portfolio selection and showed how regression function estimation and classification can be used in portfolio selection. We discuss now three applications of kernel estimation in portfolio selection. Section 3.12.1 discusses the use of kernel regression estimators, Section 3.12.2 discusses the use of classification with kernel density estimation, and Section 3.12.3 discusses the use of kernel regression estimators combined with moving average estimators of volatility and correlation.

3.12.1 Portfolio Selection Using Regression Functions

We consider an application of regression function estimation to portfolio selection. We apply the rule (1.98) for the choice of the portfolio vector with the help of a regression function estimate. In this portfolio selection method the regression function estimation is used to predict the utility transformed returns of the portfolio components.

We choose as portfolio components the S&P 500 index and the Nasdaq-100 index. The S&P 500 and Nasdaq-100 data are described in Section 1.6.2.

We consider two variations of the approach of using regression function estimation: The explanatory variables are the previous returns of the portfolio components, or the explanatory variables are obtained by a transformation of the previous returns of the portfolio components.

We choose the space of portfolio vectors to be $B = \{(1,0),(0,1)\}$ in the rule (1.98). Thus, either everything is invested into the S&P 500 index or everything is invested into the Nasdaq-100 index. The investment is made to the S&P 500 index if the estimate of the utility transformed return of S&P 500 is larger than the estimate of the utility transformed return of Nasdaq-100, otherwise the investment is made to the Nasdaq-100 index.

Autoregression We use the previous day returns as the explanatory variables. Let us denote with $S_t^{(1)}$ the daily closing prices of the S&P 500 index and with $S_t^{(2)}$ the daily closing prices of the Nasdaq-100 index. Let the gross returns be denoted by

$$R_t^{(1)} = \frac{S_t^{(1)}}{S_{t-1}^{(1)}}, \qquad R_t^{(2)} = \frac{S_t^{(2)}}{S_{t-1}^{(2)}},$$

and let the utility tranformed returns be denoted by

$$Y_t^{(1)} = u_\gamma\left(R_t^{(1)}\right), \qquad Y_t^{(2)} = u_\gamma\left(R_t^{(2)}\right), \tag{3.94}$$

where the utility function $u_\gamma : (0,\infty) \to \mathbf{R}$ is a power utility function, defined in (1.95), with $\gamma \geq 1$. We need to estimate the regression functions

$$f_{sp500}(x_1, x_2) = E\left(Y_t^{(1)} \,\middle|\, R_{t-1}^{(1)} = x_1, R_{t-1}^{(2)} = x_2\right), \tag{3.95}$$

and

$$f_{ndx100}(x_1, x_2) = E\left(Y_t^{(2)} \,\middle|\, R_{t-1}^{(1)} = x_1, R_{t-1}^{(2)} = x_2\right). \tag{3.96}$$

We use the kernel regression estimator, defined in (3.6). The Gaussian kernel was used and the data were standardized to have the marginal sample standard deviations equal to one.[44]

Figure 3.40 shows the annualized Sharpe ratios of the kernel portfolio as a function of the smoothing parameter h for different risk aversion parameters γ. We calculate the annualized Sharpe ratio with the formula

$$\text{Sharpe}_\gamma(h) = \sqrt{250}\,\frac{\bar{R}_\gamma(h)}{\widehat{\text{sd}}(R_\gamma(h))}, \tag{3.97}$$

where $\bar{R}_\gamma(h)$ is the sample mean of the portfolio net returns and $\widehat{\text{sd}}(R_\gamma(h))$ is the sample standard deviation of the portfolio net returns.[45] Figure 3.40 shows the three curves $h \mapsto \text{Sharpe}_\gamma(h)$, for $\gamma = 1, 25, 50$, when $h \in [0.01, 10]$. The black curve with label "a" shows the case $\gamma = 1$, the green curve with label "b" shows the case $\gamma = 25$, and the purple curve with label "c" shows the case $\gamma = 50$. The x-axis is logarithmic. The portfolio returns were calculated sequentially, out-of-sample, so that at time t we used only the observations available at time t. The Sharpe ratio of S&P 500 is shown with a red line, and the Sharpe ratio of Nasdaq-100 is shown with a blue line.

We can see from Figure 3.40 that the Sharpe ratios of the kernel portfolios are larger than the Sharpe ratios of S&P 500 and Nasdaq-100, for a wide range of h values. For risk aversion $\gamma = 1$ the best Sharpe ratio is 0.75 with smoothing parameter $h = 0.09$. For risk aversion $\gamma = 25$ the best Sharpe ratio is 0.80 with smoothing parameter $h = 0.08$. For risk aversion $\gamma = 50$ the best Sharpe ratio is 0.76 with smoothing parameter $h = 0.06$. The parameters $h = 0.08$ and $\gamma = 25$ gave the best overall annualized Sharpe ratio of 0.80. The S&P 500 index and Nasdaq-100 index have Sharpe ratios 0.50 and 0.56.

Figure 3.41 shows the annualized mean returns and the standard deviations of the kernel portfolio as a function of the smoothing parameter h, for several risk aversion paremeters γ. Panel (a) shows the three curves $h \mapsto 250\,\bar{R}_\gamma(h)$, for risk aversion paremeters $\gamma = 1, 25, 50$, where $h \in [0.01, 10]$. Panel (b) shows the four curves $h \mapsto \sqrt{250}\,\widehat{\text{sd}}(R_\gamma(h))$. The x-axis is logarithmic. The mean return and the sample standard deviation of the S&P 500 are shown with a red line, and those of Nasdaq-100 are shown with a blue line.

We can see from Figure 3.41 that the mean return and the standard deviation of Nasdaq-100 are higher than the mean return and the standard deviation of S&P 500, although the Sharpe ratios of S&P 500 and Nasdaq-100 are close to each other. Figure 3.41(a) shows that the kernel portfolio returns are close to the returns of Nasdaq-100 and Figure 3.41(b) shows that the standard deviations of the kernel

[44] The original standard deviations are 0.012 and 0.018.

[45] We have not used the excess returns in the calculation of Sharpe ratios, as is usually done. An excess return is the return of a portfolio minus the return of a risk-free investment.

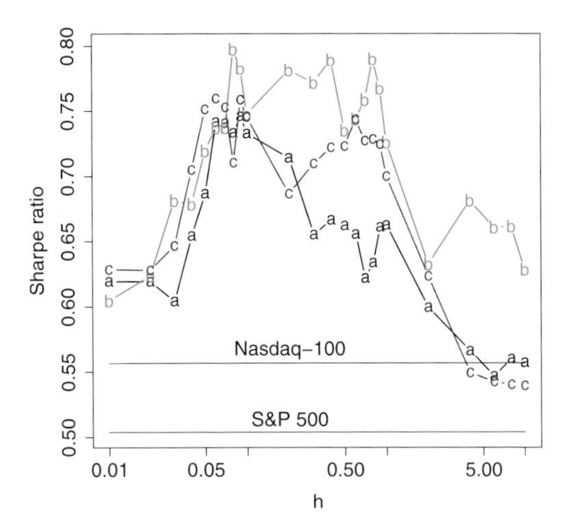

Figure 3.40 *Sharpe ratios.* Annualized Sharpe ratios of the kernel portfolios are shown as a function of smoothing parameter $h \in [0.1, 10]$, for risk aversion parameters $\gamma = 1, 25, 50$ (black "a," green "b," and purple "c"). The red line shows the Sharpe ratio of S&P 500, whereas the blue line shows that of Nasdaq-100. The x-axis is logarithmic.

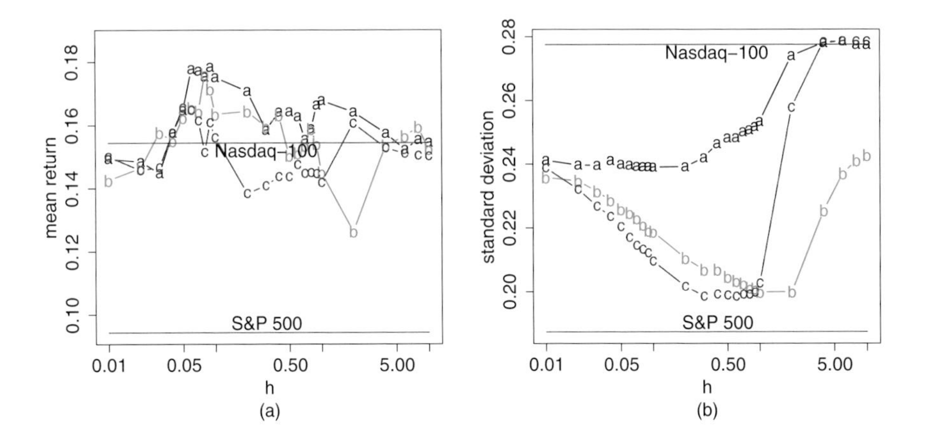

Figure 3.41 *Mean returns and standard deviations.* (a) Annualized means of the kernel portfolios are shown as a function of smoothing parameter $h \in [0.1, 10]$, for risk aversion paremeters $\gamma = 1, 25, 50$ (black "a", green "b", and purple "c"). (b) Annualized standard deviations of the kernel portfolios are shown. The red lines show the S&P 500 means, and standard deviations, whereas the blue lines show those of Nasdaq-100. The horizontal axis is logarithmic.

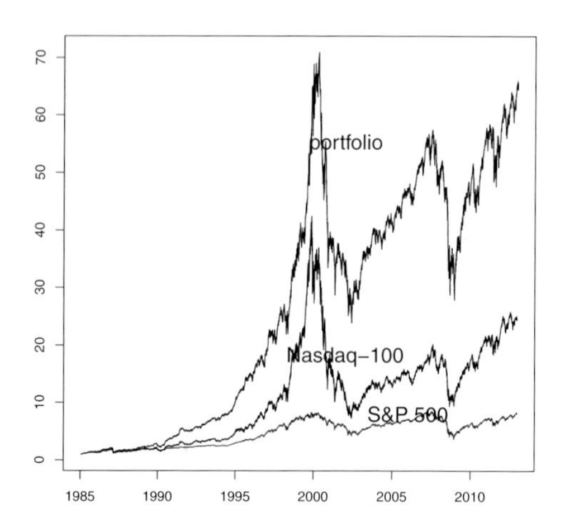

Figure 3.42 *Cumulative wealth.* The time series of the cumulative wealth of the kernel portfolio (black), Nasdaq-100 (blue), and S&P 500 (red) are shown. The kernel strategy used the smoothing parameter $h = 0.08$ and the risk aversion parameter $\gamma = 25$.

portfolio are between the standard deviations of Nasdaq-100 and S&P 500, for a wide interval of h-values. Thus, the kernel portfolio is able to obtain the same return as the Nasdaq-100, but with a lower volatility. The best overall annualized Sharpe ratio 0.80 was obtained with the annualized mean return of 17.6% and with the annualized standard deviation of 22.1%, using the parameters $h = 0.08$ and $\gamma = 25$. The annualized mean of S&P 500 returns is 9.4%, and the annualized standard deviation of S&P 500 returns is 18.7%. For Nasdaq-100 the mean is 15.4% and the standard deviation is 27.8%.

Figure 3.42 shows the cumulative wealth obtained by the kernel strategy (black) as compared to the cumulative wealths of the S&P 500 index (red) and Nasdaq-100 index (blue). The wealth is set to one at the beginning of the period. For the kernel strategy we took smoothing parameter $h = 0.08$ and risk aversion parameter $\gamma = 25$.

Figure 3.43 shows perspective plots of the kernel regression function estimates. Panel (a) shows a kernel estimate of the regression function f_{sp500}, and panel (b) shows a kernel estimate of the regression function f_{ndx100}. We have used the smoothing parameter values $h = 0.08$ and the risk aversion parameter $\gamma = 25$ for both estimates. The regression functions are drawn on the rectangle $[-17.2, 9.7] \times [-8.5, 10.6]$, which is the range of the observations. The perspective plots of the regression estimates do not reveal any interesting differences between the two estimates. The observations from the explanatory variables are concentrated on the diagonal and the modes of the regression functions estimates occuring in the corner do not have satistical or practical significance. Thus we are led to show a comparison of the regression functions estimates. The kernel regression estimates change in time

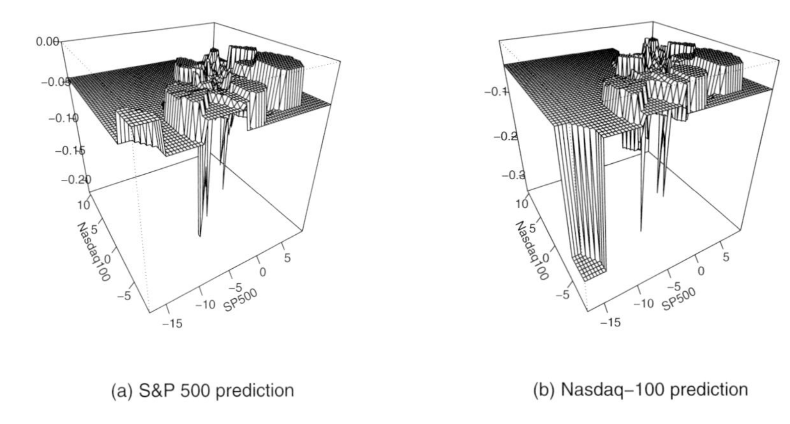

(a) S&P 500 prediction (b) Nasdaq−100 prediction

Figure 3.43 *Prediction of S&P 500 and Nasdaq-100.* (a) A kernel estimate of the regression function f_{sp500}. (b) A kernel estimate of the regression function f_{ndx100}. The regression functions are defined in (3.95). The smoothing parameter is $h = 0.08$ and the risk aversion parameter is $\gamma = 25$.

when new observations are added, but we show the estimates calculated using the complete data.

Figure 3.44 compares the values of the regression function estimates for f_{sp500} and f_{ndx100} shown in Figure 3.43. We color with red those points x where $\hat{f}_{sp500}(x) > \hat{f}_{ndx100}(x)$, that is, those points where the prediction of the S&P 500 utility transformed return is larger than the prediction of the Nasdaq-100 utility transformed return. The other points, where $\hat{f}_{sp500}(x) < \hat{f}_{ndx100}(x)$, are colored blue. Panel (a) uses a regular grid, and panel (b) uses the observed values of the explanatory variables. Panel (b) shows that there are hardly any observations at the nondiagonal corners of the rectangle. Recall that we choose the portfolio vector with the rule (1.98), when the space of portfolio vectors is $B = \{(0,1),(1,0)\}$. This means that everything is invested into the S&P 500 index if the estimate for the utility transformed return of S&P 500 is larger than the estimate of the utility transformed return of Nasdaq-100; otherwise everything is invested in Nasdaq-100 index. Thus the red points show the values of the predictors which imply investing everything in the S&P 500 index, and the blue points show the values of the predictors which imply investing everything in the Nasdaq-100 index. The decision rule shown is the final rule obtained with the complete data, but the out-of-the sample rules used in calculating the Sharpe ratios change during the observation period.

We study next whether the smoothing parameters maximizing the Sharpe ratio are close to the smoothing parameters minimizing the prediction errors. The minimization of the prediction errors with the cross-validation criterion was discussed in Section 1.9.1; see (1.117)–(1.118). We evaluated the smoothing parameters of the

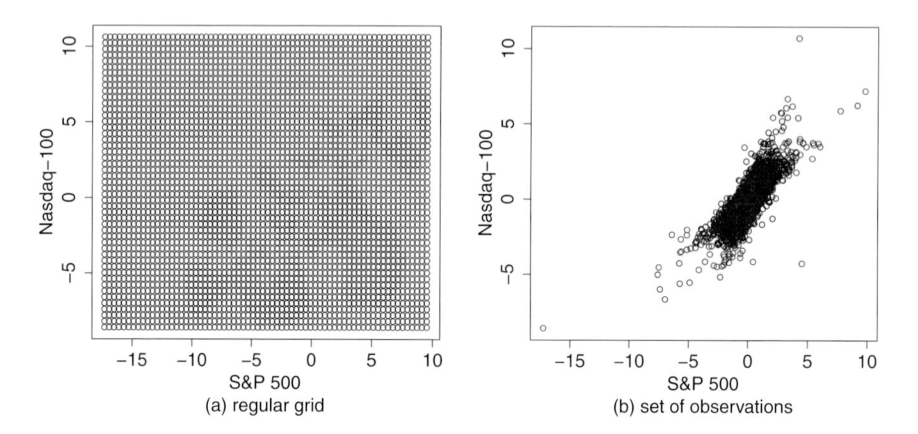

Figure 3.44 *A decision rule of the kernel strategy.* (a) Comparison of the regression function estimates on a regular grid. (b) Comparison of the regression function estimates on the empirical grid of observed x-values. The red points are such that the S&P 500 index is chosen, and the blue points are such that the Nasdaq-100 index is chosen. The comparison is based on the regression function estimates of Figure 3.43.

kernel estimators in Figure 3.40 by calculating the Sharpe ratios for kernel strategies using different smoothing parameters.

Figure 3.45 shows the means of the absolute prediction errors as a function of the smoothing parameter h, for risk aversions $\gamma = 1, 25, 50$. The mean absolute prediction error is defined in (1.118) as

$$\mathrm{MAPE}(h) = \frac{1}{T - t_0} \sum_{t=t_0}^{T-1} \left| Y_{t+1} - \hat{f}_{t,h}(X_t) \right|. \tag{3.98}$$

Panel (a) concerns the prediction of the utility transformed returns of S&P 500, and panel (b) concerns the prediction of the utility transformed returns of Nasdaq-100. Thus $Y_t = Y_t^{(1)}$ in panel (a) and $Y_t = Y_t^{(2)}$ in panel (b), where the utility transformed returns are defined in (3.94). The regression function estimate is $\hat{f}_{t,h} = \hat{f}_{sp500,t,h}$ in panel (a) and $\hat{f}_{t,h} = \hat{f}_{ndx100,t,h}$ in panel (b), where the regression functions are defined in (3.95)–(3.96) and h is the smoothing parameter of the kernel estimator. We show the functions $h \mapsto \mathrm{MAPE}(h)/\min_h \mathrm{MAPE}(h)$ for $h \in [0.01, 10]$. The risk aversions are $\gamma = 1$ (black curve with label "a"), $\gamma = 25$ (green curve with label "b"), and $\gamma = 50$ (purple curve with label "c").

For the prediction of the S&P 500 utility transformed returns the MAPE-optimal smoothing parameters are $h = 4$, $h = 4$, and $h = 2$ for the risk aversions $\gamma = 1$, $\gamma = 25$, and $\gamma = 50$. For the prediction of the Nasdaq-100 the MAPE-optimal smoothing parameters are $h = 10$, $h = 2$, and $h = 2$ for the same risk aversions. We conclude that the smoothing parameters recommended by the cross validation

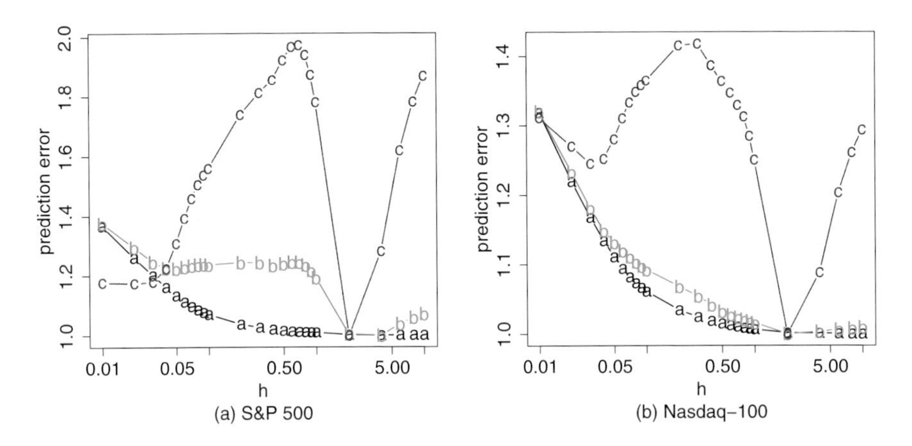

Figure 3.45 *Cross validation.* Shown are (a) the mean absolute prediction errors as a function of the smoothing parameter h, for the prediction of the utility transformed S&P 500 returns and (b) the mean absolute prediction errors for the prediction of the utility transformed Nasdaq-100 returns. The risk aversion parameters $\gamma = 1$ (black "a"), $\gamma = 25$ (green "b"), and $\gamma = 50$ (purple "c") are used.

are different from the smoothing parameters recommended by the Sharpe ratios in Figure 3.40. The optimal smoothing parameters in terms of the Sharpe ratio were $h = 0.09$, $h = 0.08$, and $h = 0.06$.

Autoregression with Transformed Explanatory Variables We illustrate how a suitable transformation of explanatory variables can help to interpret and improve the results. We use the copula transform, as defined in (1.105).

Figure 3.46 shows the annualized Sharpe ratios of the kernel portfolio as a function of the smoothing parameter h for different risk aversion parameters γ. We calculate the Sharpe ratio with the formula (3.97) and denote by $\text{Sharpe}_\gamma(h)$ the Sharpe ratio of the portfolio returns, by $\bar{R}_\gamma(h)$ the sample mean of the portfolio net returns, and by $\widehat{\text{sd}}(R_\gamma(h))$ the sample standard deviation of the portfolio net returns, when the smoothing parameter is h and the risk aversion parameter is γ. Figure 3.46 shows the three curves $h \mapsto \sqrt{250}\,\text{Sharpe}_\gamma(h)$, for $\gamma = 1, 25, 50$, when $h \in [0.01, 10]$. The curve with label "a" shows the case $\gamma = 1$, the curve with label "b" shows the case $\gamma = 25$, and the curve with label "c" shows the case $\gamma = 50$. The x-axis is logarithmic. The portfolio returns are calculated sequentially. The Sharpe ratio of S&P 500 is shown with a red line, and the Sharpe ratio of Nasdaq-100 is shown with a blue line.

We can see from Figure 3.46 that the Sharpe ratios behave in a quite similar way as in the kernel portfolio without the copula transform, shown in Figure 3.40. Now the parameters $h = 0.5$ and $\gamma = 25$ give the best annualized Sharpe ratio of 0.84, with the annualized mean return of 17.6% and with the annualized standard deviation of

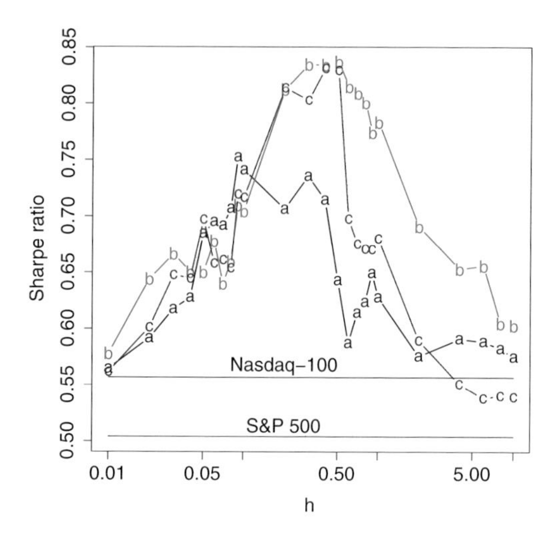

Figure 3.46 *Sharpe ratios with copula transform.* Annualized Sharpe ratios of the kernel portfolios are shown as a function of smoothing parameter $h \in [0.1, 10]$, for risk aversion parameters $\gamma = 1, 25, 50$ (black "a," green "b," and purple "c"). The red line shows the Sharpe ratio of S&P 500, whereas the blue line shows the Sharpe ratio of Nasdaq-100. The x-axis is logarithmic.

21.0%. The Sharpe ratio is now slightly better than the Sharpe ratio of 0.80, obtained without the copula transform. For $\gamma = 1$ the best Sharpe ratio is 0.75 obtained with $h = 0.09$ and for $\gamma = 50$ the best Sharpe ratio is 0.83 obtained with $h = 0.4$.

Figure 3.47 shows the mean returns and the standard deviations of the kernel portfolio as a function of the smoothing parameter h, for different risk aversion parameters γ. Panel (a) shows the three curves $h \mapsto 250\, \bar{R}_\gamma(h)$, for risk aversion parameters $\gamma = 1, 25, 50$, where $h \in [0.01, 10]$, and panel (b) shows the three curves $h \mapsto \sqrt{250}\, \widehat{\mathrm{sd}}(R_\gamma(h))$. The x-axis is logarithmic. The mean return and the sample standard deviation of the S&P 500 are shown with a red line, and those of Nasdaq-100 are shown with a blue line. Figure 3.47 looks quite the same as Figure 3.41, where the performance of the kernel portfolio without the copula transform was shown. The mean return of the best strategy is the same as without the copula transform, but the standard deviation was decreased to 21.0% from the previous 22.1%.

Figure 3.48 shows the cumulative wealth reached when using the kernel regression portfolio selection with the copula transform (black curve). The wealth is compared to the wealth reached by S&P 500 (red) and by Nasdaq-100 (blue). Smoothing parameter $h = 0.5$ and the risk aversion $\gamma = 25$ were used.

Figure 3.49 shows kernel regression function estimates obtained using the copula transform. Panel (a) shows an estimate of the regression function f_{sp500}, and panel (b) shows an estimate of the regression function f_{ndx100}. We have used the smoothing parameter value $h = 0.5$ and risk aversion $\gamma = 25$ for both estimates. The regression

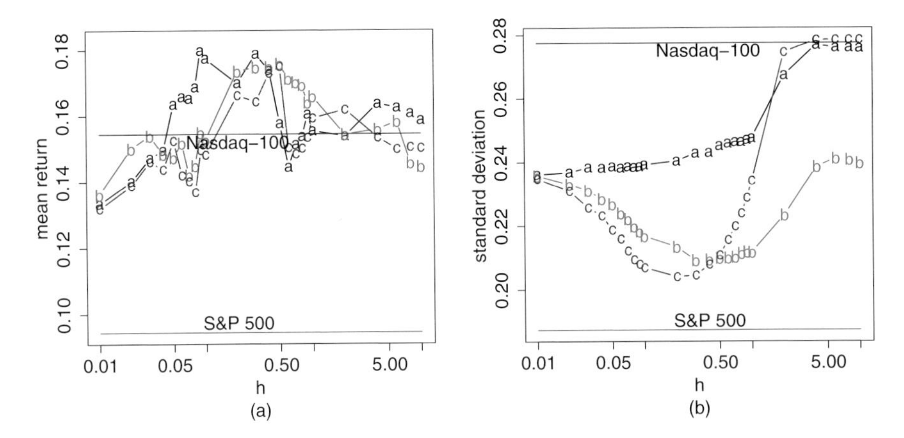

Figure 3.47 *Mean returns and standard deviations with copula transform.* (a) Annualized means of the kernel portfolios as a function of smoothing parameter $h \in [0.1, 10]$, for risk aversion parameters $\gamma = 1, 25, 50$ (black "a," green "b," and purple "c"). (b) Annualized standard deviations of the kernel portfolios. The red lines show the S&P 500 Sharpe means and standard deviations, whereas the blue lines show those of Nasdaq-100. The x-axis is logarithmic.

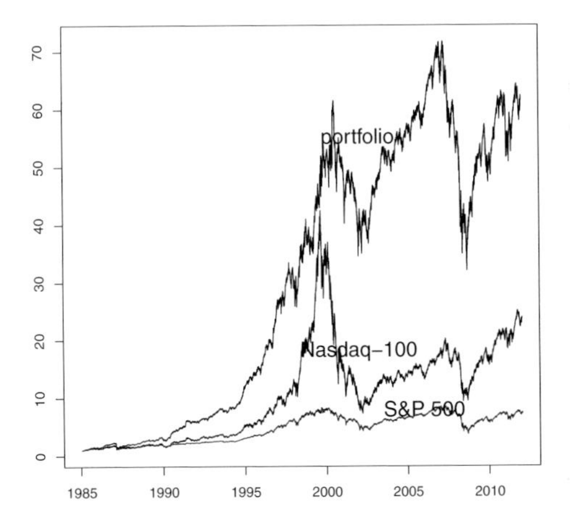

Figure 3.48 *Cumulative wealth with copula transform.* The time series of the cumulative wealth is shown for the kernel portfolio (black), Nasdaq-100 (blue), and S&P 500 (red). The kernel strategy used the smoothing parameter $h = 0.5$ and the risk aversion parameter $\gamma = 25$.

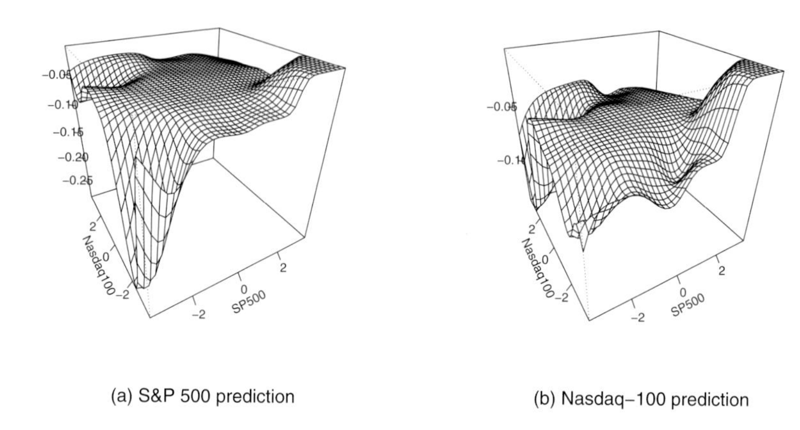

(a) S&P 500 prediction

(b) Nasdaq–100 prediction

Figure 3.49 *Prediction of S&P 500 and Nasdaq-100 with copula transform.* (a) A kernel estimate of the regression function f_{sp500}. (b) A kernel estimate of the regression function f_{ndx100}. The regression functions are defined in (3.95). The smoothing parameter is $h = 0.5$ and the risk aversion parameter is $\gamma = 25$.

functions have the global maximum at the lower right corner, where the previous day return of Nasdaq-100 was low and the previous day return of S&P 500 was high. However, the area where the regression function takes the highest values is such that there are hardly any observations of the explanatory variables. The regression function estimates change in time, when new observations are added, but we show the final estimates.

Figure 3.50 compares the values of the regression function estimates for f_{sp500} and f_{ndx100}, shown in Figure 3.43. We color with red those points x where $\hat{f}_{sp500}(x) > \hat{f}_{ndx100}(x)$, that is, those points where the prediction of the S&P 500 utility transfomed return is larger than the prediction of the Nasdaq-100 utility transformed return. The other points, where $\hat{f}_{sp500}(x) < \hat{f}_{ndx100}(x)$, are colored blue. Panel (a) shows a regular grid, and panel (b) shows only the observed values of the explanatory variables, transformed with the copula transform.

Figure 3.51 shows the means of the absolute prediction errors as a function of smoothing parameter h, for risk aversion parameters $\gamma = 1, 25, 50$, when the copula transform was used. The mean absolute prediction error $\mathrm{MAPE}(h)$ is defined in (3.98). The setting is the same as in Figure 3.45: panel (a) concerns the prediction of the utility transformed returns of S&P 500, and panel (b) concerns the prediction of the utility transformed returns of Nasdaq-100. We show the functions $h \mapsto \mathrm{MAPE}(h)/\min_h \mathrm{MAPE}(h)$ for $h \in [0.01, 10]$. The risk aversions are $\gamma = 1$ (black curve with label "a"), $\gamma = 25$ (green curve with label "b"), and $\gamma = 50$ (purple curve with label "c").

For the prediction of S&P 500 the MAPE-optimal smoothing parameters are $h = 10$, $h = 2$, and $h = 0.01$ for the risk aversions $\gamma = 1$, $\gamma = 25$, and $\gamma = 50$. For

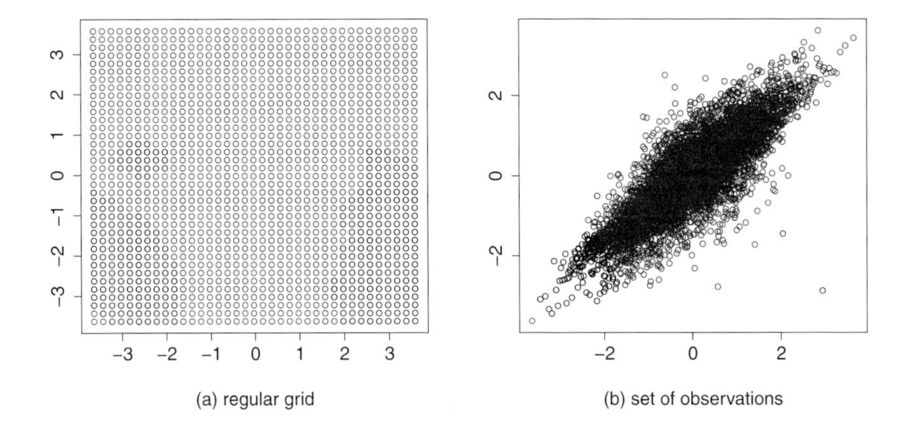

Figure 3.50 *A Decision rule of the kernel strategy wiht copula transform.* (a) Comparison of the regression function estimates on a regular grid. (b) Comparison of the regression function estimates on the empirical grid of observed x-values. The red points are such that the S&P 500 index is chosen, and the blue points are such that the Nasdaq-100 index is chosen. The comparison is based on the regression function estimates of Figure 3.49.

the prediction of Nasdaq-100 the MAPE-optimal smoothing parameters are $h = 10$, $h = 2$, and 2 for the same risk aversions. The Sharpe ratio optimal smoothing parameters were $h = 0.09$, $h = 0.5$, and $h = 0.4$.

3.12.2 Portfolio Selection Using Classification

Section 1.5.3 contains a description how classification can be used in portfolio selection. In particular, (1.100) shows how to define the class labels when classification techniques will be used in portfolio selection. We apply now classification in portfolio selection, when the portfolio components are the S&P 500 index and the Nasdaq-100 index. The S&P 500 and Nasdaq-100 data are described in Section 1.6.2.

The collection of classification data $(X_t, Y_t), t = 1, \ldots, n$, is obtained by choosing $Y_t = 0$ when the return at time $t+1$ is higher for the S&P 500 index than for the Nasdaq-100 index. Otherwise we take $Y_t = 1$. The explanatory variables X_t are the previous returns of S&P 500 and Nasdaq-100:

$$X_t = \left(R_t^{(1)}, R_t^{(2)}\right), \qquad R_t^{(i)} = \frac{S_t^{(i)}}{S_{t-1}^{(i)}}, \quad i = 1, 2,$$

where $S_t^{(1)}$ is the price of the S&P 500 index and $S_t^{(2)}$ is the price of the Nasdaq-100 index. Note that with the classification approach we are not able to introduce a risk aversion parameter, as in the case of the regression approach, where a utility transformed return was predicted. The portfolios obtained by classification correspond to using the risk aversion parameter $\gamma = 1$.

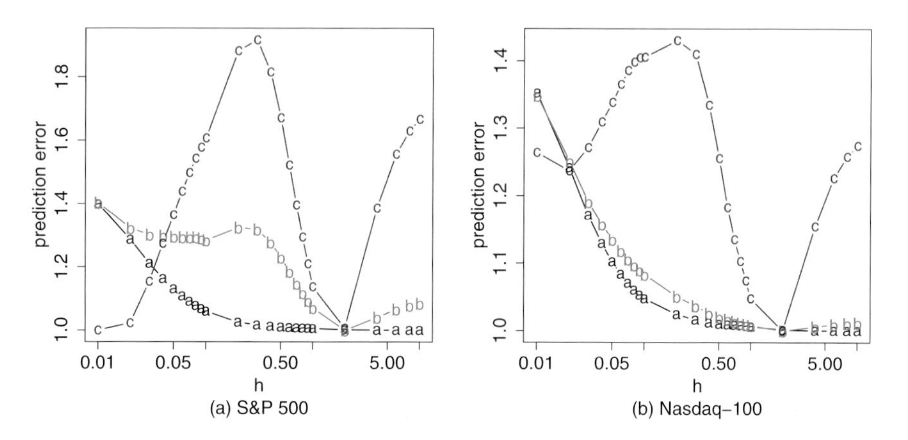

Figure 3.51 *Cross validation with copula transform.* Shown are (a) the mean absolute prediction errors as a function of the smoothing parameter h, for the prediction of the utility transformed S&P 500 returns. (b) The mean absolute prediction errors for the prediction of the utility transformed Nasdaq-100 returns. The risk aversion parameters $\gamma = 1$ (black "a"), $\gamma = 25$ (green "b"), and $\gamma = 50$ (purple "c") are used. The horizontal axis is logarithmic.

We choose the space of portfolio weights to be $B = \{(1,0),(0,1)\}$ so that either everything is invested into the S&P 500 index or everything is invested into the Nasdaq-100 index. We make the copula transformation to the explanatory variables. The transformation was defined in (1.105). We use the observations $Z_t = (Z_t^1, Z_t^2)$, obtained from $X_t = (X_t^1, X_t^2)$ by the copula transform.

Figure 3.52 shows a scatter plot of vectors Z_t. The class labels are indicated with colors, so that the red points show the observations made at a day, when the next day return of S&P 500 was higher than the return of Nasdaq-100, and the blue points show the observations made at a day, when the next day return of S&P 500 was smaller than the return of Nasdaq-100. We can see that the classes are not well separated, but the red and blue points are completely mixed. There are 3319 red points and 3604 blue points.

Section 1.4 contains an introduction to classification and Section 3.4.1 contains definitions of classifiers based on kernel density estimation and kernel regression estimation, which lead to the same classification functions. The density rule in (3.33) specializes in the two-class case to the classification rule

$$\hat{g}(x) = \begin{cases} 1, & \text{if } n_1 \hat{f}_{X|Y=1}(x) \geq n_0 \hat{f}_{X|Y=0}(x), \\ 0, & \text{otherwise}, \end{cases} \tag{3.99}$$

where $\hat{f}_{X|Y=0}$ and $\hat{f}_{X|Y=1}$ are kernel estimators of the class density functions, and n_0 and n_1 are the class 0 and class 1 frequencies. We have defined the class kernel

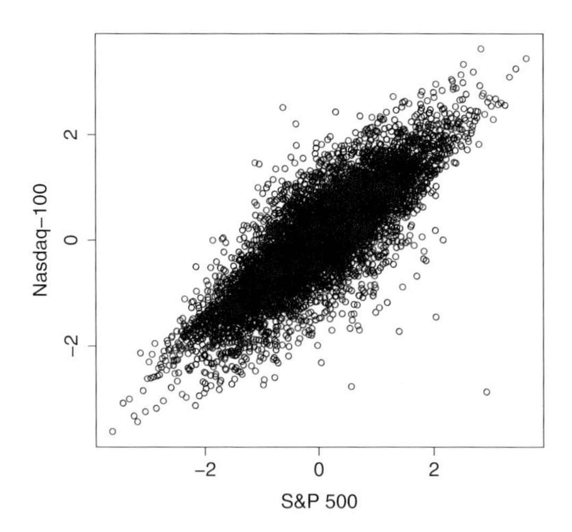

Figure 3.52 *Classification data.* A scatter plot of the copula transformed returns of S&P 500 and Nasdaq-100. The red points indicate the days when the next day closing return of S&P 500 was higher and the blue points indicate the days when the next day return of Nasdaq-100 was higher.

density estimators in (3.35). The regression rule in (3.36) puts in the two-class case

$$\hat{g}(x) = \begin{cases} 1, & \text{if } \hat{p}_1(x) \geq \hat{p}_0(x), \\ 0, & \text{otherwise}, \end{cases} \tag{3.100}$$

where $\hat{p}_0$ and $\hat{p}_1$ are the kernel regression estimates of $P(Y = 0 \mid X = x)$ and $P(Y = 1 \mid X = x)$. The kernel regression function estimate was defined in (3.6). As noted in in Section 3.4.1, classification rules (3.99) and (3.100) are identical.

Figure 3.53 shows contour plots of kernel density estimates of the class densities. Panel (a) shows a density estimate of the S&P 500 class, and panel (b) shows a density estimate of the Nasdaq-100 class. The smoothing parameter $h = 0.2$ and the Gaussian kernel were used. The differences between the estimates seem to be very small.

Figure 3.54 shows a kernel regression function estimate of the regression function $f(x) = P(Y = 1 \mid X = x)$. Thus, $f(x)$ is the conditional probability that the next day return is higher for Nasdaq-100. Panel (a) shows a perspective plot, and panel (b) shows a contour plot. The smoothing parameter $h = 0.2$ and the Gaussian kernel were used.

Figure 3.55 shows the estimated final classification sets. Panel (a) shows the classification sets calculated with a regular grid, and panel (b) shows how the empirical classification rule classifies the observed values of the predictive variables. The red points are such that the classifier chose S&P 500, and the blue points are such that the classifier chose Nasdaq-100. The classifier is changing in time as new observations

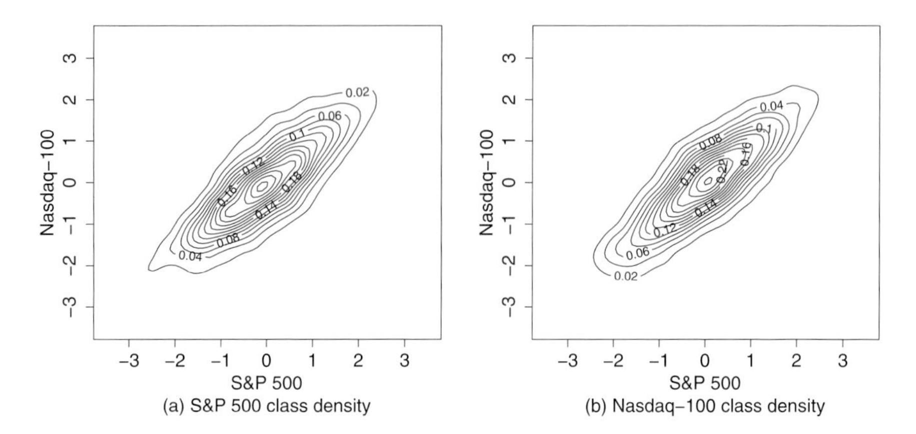

Figure 3.53 *Class density function estimates.* Contour plots of kernel density estimates of class densities. (a) A density estimate of the S&P 500 class. (b) A density estimate of the Nasdaq-100 class.

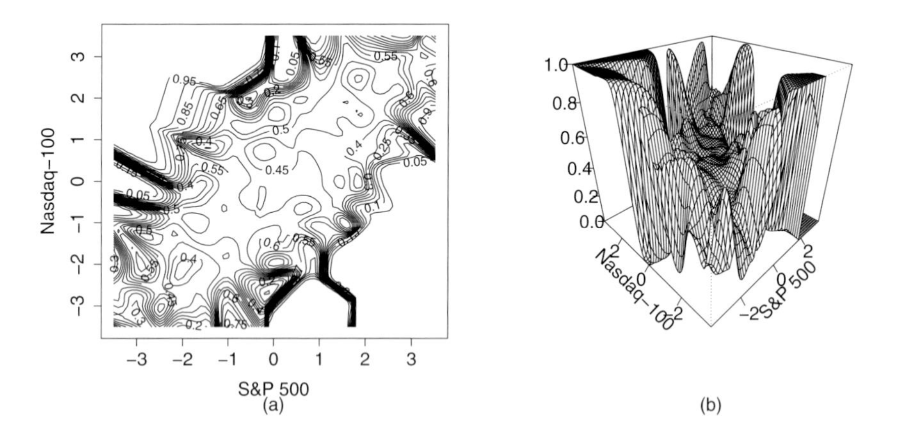

Figure 3.54 *Nasdaq-100 class probability estimate.* A kernel estimate of the conditional class probability $P(Y = 1 \mid X = x)$. (a) A perspective plot. (b) A contour plot.

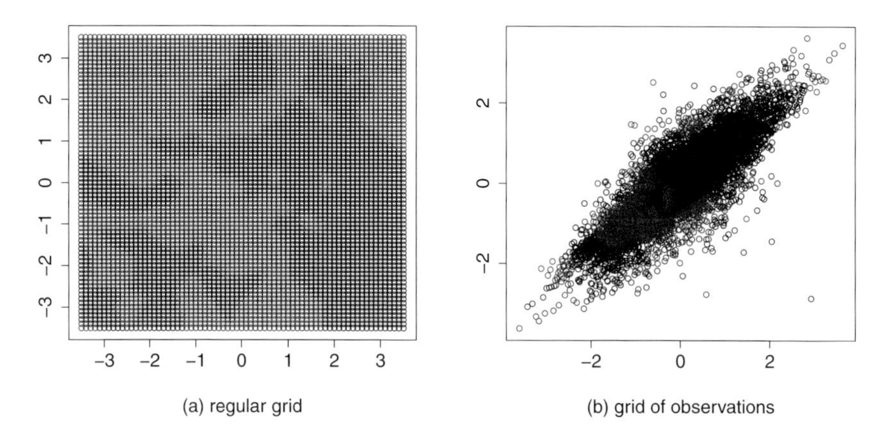

Figure 3.55 *Classification rule.* The red points are classified to S&P 500, and the blue points are classified to Nasdaq-100. (a) A regular grid. (b) A grid of observations.

are included, and we show the final classification rule estimated with the complete data.

Figure 3.56 shows the annualized Sharpe ratios of the kernel classification portfolio as a function of the smoothing parameter h, and it makes a comparison to the annualized Sharpe ratios of the kernel regression portfolios. In regression portfolios, risk aversion $\gamma = 1$ (logarithmic utility function) was used, and the copula transform was made. The black curve with symbol "a" corresponds to the classification portfolios, and the brown curve with symbol "b" corresponds to the regression portfolios. Figure 3.56 shows the curve $h \mapsto \text{Sharpe}(h)$, when $h \in [0.01, 10]$ and the Sharpe ratio is calculated with the formula (3.97). The x-axis is logarithmic. The portfolio returns are calculated sequentially, out-of-sample. The Sharpe ratio of the S&P 500 is shown with a red line, and the Sharpe ratio of Nasdaq-100 is shown with a blue line. The smoothing parameter $h = 0.2$ gives the best annualized Sharpe ratio of 0.75 for the classification portfolio, with the annualized mean return of 17.5% and with the annualized standard deviation of 23.3%.

We can see from Figure 3.56 that the Sharpe ratios are robust with respect to the choice of the smoothing parameter. We see also that the Sharpe ratios of kernel classification portfolios and kernel regression portfolios behave much the same way and have the same maximums, but the kernel classification seems to be more robust with respect to the choice of the smoothing parameter. Note that Figure 3.46 already showed the Sharpe ratios of the kernel regression portfolios, and the figure shows also Sharpe ratios of kernel regression portfolio with several risk aversion parameters. The Sharpe ratios of kernel regression portfolios without the copula transform are shown in Figure 3.40.

Figure 3.57 shows the mean returns and the standard deviations of the kernel classification portfolio as a function of the smoothing parameter h, and it makes a

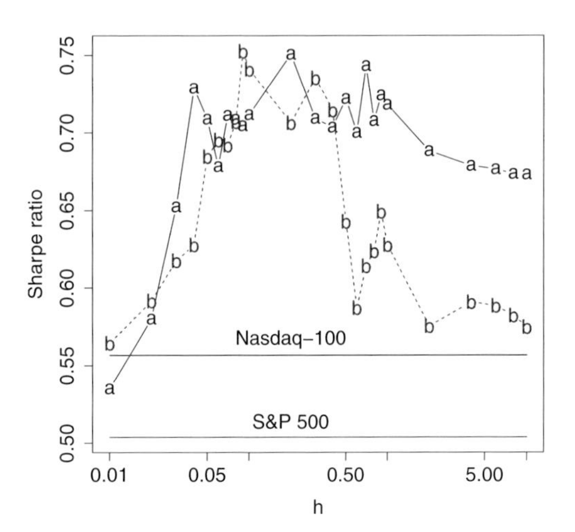

Figure 3.56 *Sharpe Ratios of kernel classification.* Annualized Sharpe ratios of the kernel classification portfolios (black "a") and kernel regression portfolios (brown "b") are shown as a function of smoothing parameter $h \in [0.01, 10]$. The red line shows the Sharpe ratio of S&P 500, and the blue line shows the Sharpe ratio of Nasdaq-100. The x-axis is logarithmic.

comparison to the mean returns and standard deviations of the regression portfolios. Panel (a) shows the curves $h \mapsto 250\,\bar{R}(h)$, where $h \in [0.01, 10]$, and $\bar{R}(h)$ is the average net return. Panel (b) shows the curves $h \mapsto \sqrt{250}\,\widehat{\mathrm{sd}}(R(h))$, where $\widehat{\mathrm{sd}}(R(h))$ is the sample standard deviation of the portfolio returns. The x-axis is logarithmic. The mean return and the sample standard deviation of the S&P 500 are shown with a red line, and those of Nasdaq-100 are shown with a blue line. Note that Figure 3.47 already showed the mean returns and standard deviations of regression portfolios, and Figure 3.41 showed the case of regression portfolios without the copula transform. We can see that the mean returns behave similarily for the classification and regression portfolios, but the standard deviations are smaller for the classification portfolios, at least for the large smoothing parameters.

Figure 3.58 shows the cumulative wealth of the kernel classification portfolio (black) as compared to the cumulative wealth obtained by the S&P 500 index (red) and the Nasdaq-100 index (blue). Now the Sharpe ratio is 0.73 with the annualized mean of 16.8% and with the annualized standard deviation of 23.1%.

Figure 3.59 shows the classification errors as a function of smoothing parameter h. We have discussed the evaluation of classifiers in Section 1.9.6; see (1.133) and (1.134) for the time series formulas. We use these formulas to see whether the performance measurement with the Sharpe ratios, as in Figure 3.56, gives different results than performance measurement with classification error. Panel (a) shows the

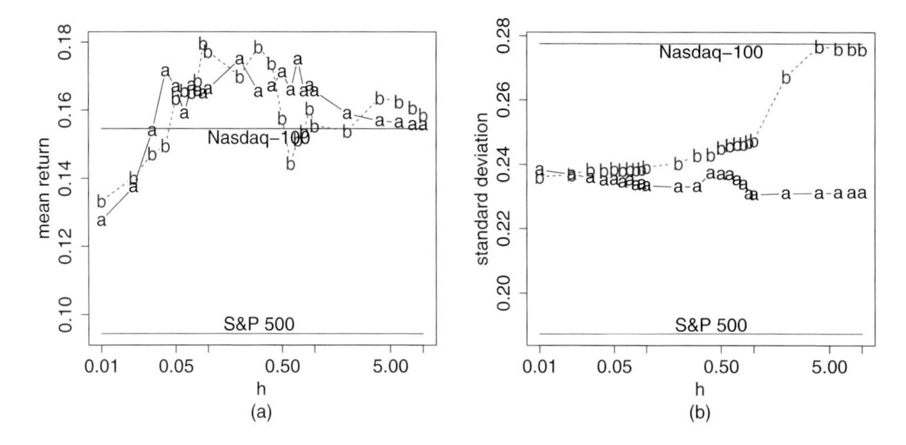

Figure 3.57 *Mean returns and standard deviations of kernel classification.* (a) Annualized means of the classification and regression portfolios as a function of smoothing parameter $h \in [0.01, 10]$. (b) Annualized standard deviations of the classification and regression portolios. The black curves with "a" correspond to classification portfolios, and the brown curves with "b" correspond to regression portfolios. The red lines show the S&P 500 means and standard deviations, whereas the blue lines show those of Nasdaq-100.

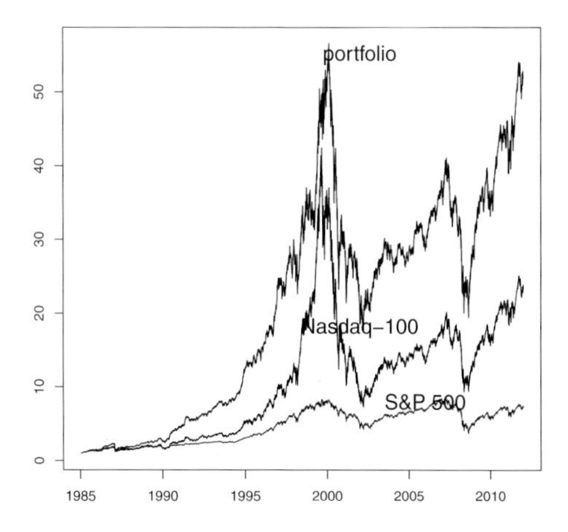

Figure 3.58 *Wealth of classification portfolio.* The red curve shows the cumulative wealth of the portfolio chosen with kernel classification. The black curve shows the wealth when invested in the S&P 500 index, and the blue curve shows the wealth when invested in the Nasdaq-100 index.

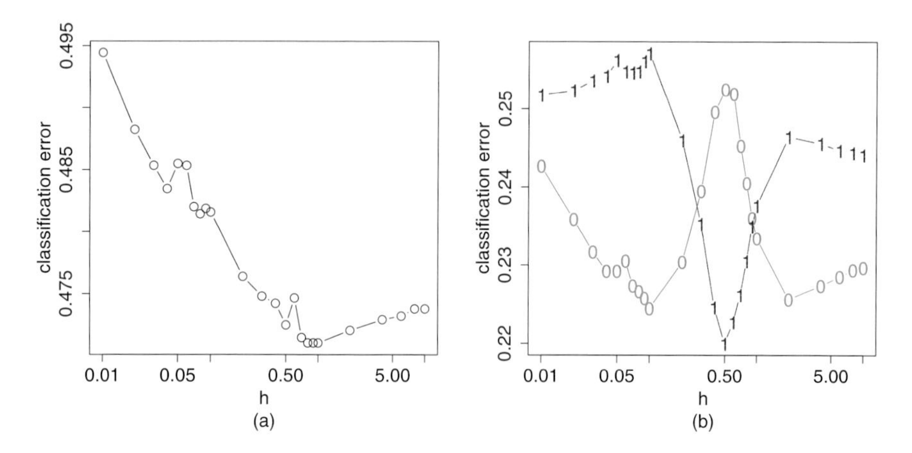

Figure 3.59 *Kernel classification errors.* (a) The classification error of kernel classification as a function of smoothing parameter $h \in [0.01, 10]$. (b) Type one and type two classification errors of kernel classification. The green curve with label "0" shows the classification error when the true class is S&P 500. The purple curve with label "1" shows the classification error when the true class is Nasdaq-100. The x-axis is logarithmic.

function $h \mapsto P_{error}(h)$, where

$$P_{error} = \frac{1}{T - t_0} \sum_{t=t_0}^{T-1} I_{\{\hat{g}_t^*(X_{t+1})\}^c}(Y_{t+1}),$$

where $\hat{g}_t^*$ is a classifier constructed using the data $(X_1, Y_1), \ldots, (X_t, Y_t)$, and $t_0 = 10$. That is, we estimate the error probability $P(\hat{g}(X) \neq Y)$ for different choices of h. Panel (b) shows the functions $h \mapsto P_{error}^{(0)}(h)$ and $h \mapsto P_{error}^{(1)}(h)$, where

$$P_{error}^{(k)}(h) = \frac{1}{T - t_0} \sum_{t=t_0}^{T-1} I_{\{\hat{g}_t^*(X_{t+1})\}^c}(k)\, I_{\{k\}}(Y_{t+1}),$$

which estimates $P(\hat{g}(X) \neq Y \,|\, Y = k)$ for $k = 0, 1$. Now $P_{error}^{(0)}(h)$ estimates the probability that the observation was classified to Nasdaq-100 class although the next-day return was better for S&P 500, and $P_{error}^{(1)}(h)$ estimates the probability that the observation was classified to S&P 500 class although the next-day return was better for Nasdaq-100.

We can see from Figure 3.59(a) that the error probabilities are only slightly under 0.5. Furthermore, the optimal smoothing parameter in terms of classification error is $h = 0.9$, which is much larger than the smoothing parameter $h = 0.2$, which is optimal in terms of the Sharpe ratio. Figure 3.59(b) shows that the error components do not behave monotonically as a function of h.

3.12.3 Portfolio Selection Using Markowitz Criterion

We continue to study portfolio selection when the portfolio components are the S&P 500 index and the Nasdaq-100 index. The S&P 500 and Nasdaq-100 data are described in Section 1.6.2.

We have introduced mean-variance preferences in (1.101), and derived the optimal portfolio vector in (1.102), for the case of two risky assets. The formula for the optimal portfolo vector in the conditional setting is

$$w(x) = \frac{1}{\gamma} \frac{\mu_2(x) - \mu_1(x) - \gamma(\sigma_{12}(x) - \sigma_1^2(x))}{\sigma_1^2(x) + \sigma_2^2(x) - 2\sigma_{12}(x)},$$

where $\gamma > 0$ is the risk aversion parameter. The portfolio weight $w(x)$ can take negative values and be larger than one, but we restrict the weight to the interval $[0, 1]$ and use the weight $\min\{\max\{w(x), 0\}, 1\}$. The conditional expectations are defined by

$$\mu_1(x) = f_{sp500}(x_1, x_2) = E\left(R_t^{(1)} \,\middle|\, R_{t-1}^{(1)} = x_1, R_{t-1}^{(2)} = x_2\right),$$

and

$$\mu_2(x) = f_{ndx100}(x_1, x_2) = E\left(R_t^{(2)} \,\middle|\, R_{t-1}^{(1)} = x_1, R_{t-1}^{(2)} = x_2\right),$$

where $R_t^{(1)}$ is the gross return of S&P 500, and $R_t^{(2)}$ is the gross return of Nasdaq-100. The conditional variances are defined by

$$\sigma_k^2(x) = \text{Var}\left(R_t^{(k)} \,\middle|\, R_{t-1}^{(k)} = x_{t-1}, R_{t-2}^{(k)} = x_{t-2}, \ldots\right), \qquad k = 1, 2,$$

and the conditional covariance is defined by

$$\sigma_{12}(x) = \text{Cov}\left(R_t^{(1)}, R_t^{(2)} \,\middle|\, (R_{t-1}^{(1)}, R_{t-1}^{(2)}) = (x_{t-1}, y_{t-1}), \ldots\right).$$

We estimate the conditional means $\mu_1(x), \mu_2(x)$ using the state–space smoothing with the kernel estimator, defined in (3.6). The conditional variances $\sigma_1^2(x), \sigma_2^2(x)$, and the conditional covariance $\sigma_{12}(x)$ are estimated by the time–space smoothing using the kernel weighted moving averages, defined in (3.79) and (3.87). Since we are estimating five functions, we have to choose five smoothing parameters. We reduce the problem to the choice of two smoothing parameters: We choose the smoothing parameter h_μ for the estimation of the two means, and we choose the smoothing parameter h_σ for the estimation of the two variances and the covariance. In the state–space smoothing the kernel is the standard Gaussian kernel, and in the time–space smoothing the exponential moving average is used.

Figure 3.60 shows the annualized Sharpe ratios of the Markowitz portfolio as a function of the smoothing parameters h_μ and h_σ. Risk aversion parameter $\gamma = 10$ was used, and the copula transform was made. Panel (a) shows the functions $h_\mu \mapsto \text{Sharpe}(h_\mu, h_\sigma)$, for two choices of h_σ: $h_\sigma = 25$ and $h_\sigma = \infty$, when $h_\mu \in [0.01, 10]$. The curve with label "1" corresponds to the case $h_\sigma = 25$, and the curve with label "2" corresponds to the case $h_\sigma = \infty$. Smoothing parameter $h_\sigma = \infty$ means that the

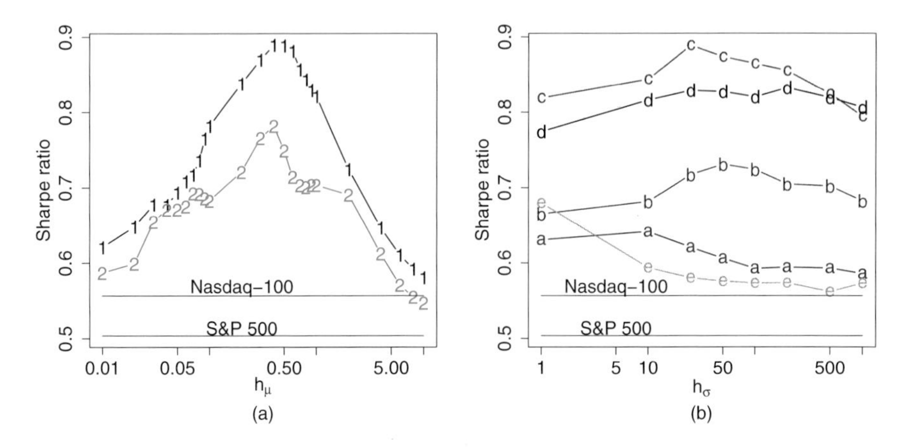

Figure 3.60 *Sharpe ratios for Markowitz portfolios.* Annualized Sharpe ratios of the Markowitz portfolios. (a) Functions $h_\mu \mapsto \text{Sharpe}(h_\mu, h_\sigma)$; the black curve with "1" shows the case $h_\sigma = 25$ and the green curve with "2" shows the case $h_\sigma = \infty$. (b) Functions $h_\sigma \mapsto \text{Sharpe}(h_\mu, h_\sigma)$; the curves with labels "a"–"e" show the cases $h_\mu = 0.01, \ldots, 10$. The red line shows the Sharpe ratio of S&P 500, and the blue line shows the Sharpe ratio of Nasdaq-100. The x-axis is logarithmic.

variances are estimated with sequential sample variances and the covariance with the sequential sample covariance. Panel (b) shows the functions $h_\sigma \mapsto \text{Sharpe}(h_\mu, h_\sigma)$, for five choices of h_μ: $h_\mu = 0.01, 0.07, 0.4, 0.9, 10$, when $h_\sigma \in [1, 1000]$.[46] The five choices of h_μ are shown with the five curves labeled with "a"–"e" in the increasing order of h_μ. The Sharpe ratio is calculated with the formula (3.97). The x-axis are logarithmic. The portfolio returns are calculated sequentially, out-of-sample. The Sharpe ratio of the S&P 500 is shown with a red line, and the Sharpe ratio of Nasdaq-100 is shown with a blue line.

We see from Figure 3.60(a) that it is possible to improve significantly the Sharpe ratio by using a moving average estimates of variances and the covariance. Namely, the curve labeled with "1" shows higher Sharpe ratios, and corresponds to the case $h_\sigma = 25$, than the curve labeled with "2," which corresponds to the use of sequentially calculated variances and covariance. Furthermore, the smoothing parameters $h_\mu = 0.4$ and $h_\sigma = 25$ give the annualized Sharpe ratio of 0.89, with the annualized mean return of 18.3% and with the annualized standard deviation of 20.6%. We see from Figure 3.60(b) that the Sharpe ratio is robust with respect to the choice of smoothing parameter h_σ, which controls the smoothing of moving averages.

Figure 3.61 shows the annualized mean returns and standard deviations of the Markowitz portfolio as a function of the smoothing parameter h_μ. The same parameters as in Figure 3.60 were used. Panel (a) shows the functions $h_\mu \mapsto \bar{R}(h_\mu, h_\sigma)$, for

[46]More precisely, $h_\sigma \in \{1, 10, 25, 50, 100, 200, 500, 1000\}$.

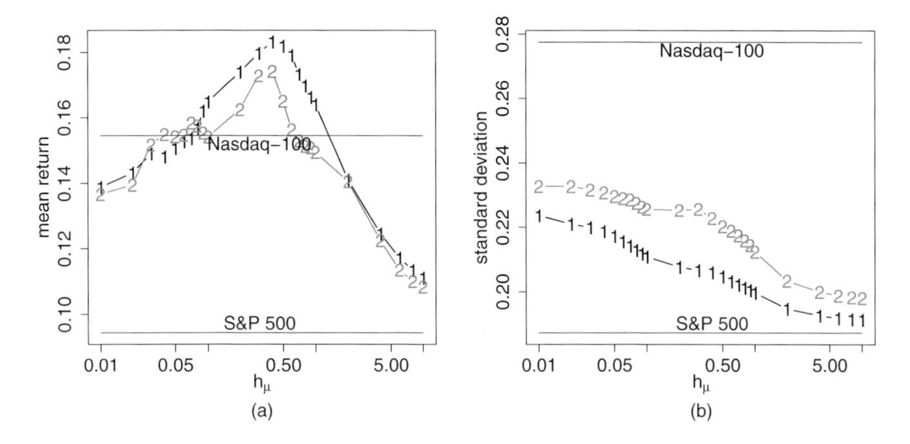

Figure 3.61 *Mean returns and standard deviations for Markowitz portfolios.* (a) Functions $h_\mu \mapsto 250\,\bar{R}(h_\mu, h_\sigma)$ are shown, where $\bar{R}$ is the mean return. (b) Functions $h_\mu \mapsto \sqrt{250}\,\widehat{\mathrm{sd}}(R(h_\mu, h_\sigma))$ are shown, where $\widehat{\mathrm{sd}}(R)$ is the sample standard deviation of returns. The smoothing parameter is $h_\sigma = 25$ (black "1") and $h_\sigma = \infty$ (green "2"). The red lines shows the mean and standard deviation of S&P 500, and the blue lines the mean and standard deviation of Nasdaq-100. The x-axis is logarithmic.

two choices of h_σ: $h_\sigma = 25$ and $h_\sigma = \infty$, when $h_\mu \in [0.01, 10]$. Panel (b) shows the functions $h_\mu \mapsto \widehat{\mathrm{sd}}(R(h_\mu, h_\sigma))$. The black curves with "1" show the case with $h_\sigma = 25$, and the green curves with "2" show the case with $h_\sigma = \infty$. The mean return and the standard deviation of the S&P 500 are shown with red lines, and those of Nasdaq-100 are shown with blue lines.

Figure 3.62 shows the annualized mean returns and standard deviations of the Markowitz portfolio as a function of the smoothing parameter h_σ. The same parameters as in Figure 3.60 were used. Panel (a) shows the functions $h_\sigma \mapsto \bar{R}(h_\mu, h_\sigma)$, for five choices of h_μ: $h_\mu = 0.01, 0.07, 0.4, 0.9, 10$, when $h_\sigma \in [1, 1000]$. Panel (b) shows the functions $h_\sigma \mapsto \widehat{\mathrm{sd}}(R(h_\mu, h_\sigma))$. The curves with labels "a"–"e" show the cases $h_\mu = 0.01, \ldots, 10$. The mean return and the standard deviation of the S&P 500 are shown with red lines, and those of Nasdaq-100 are shown with blue lines.

Figure 3.63 shows the cumulative wealths obtained with Markowitz strategies. The wealth is one at the starting date 1985-10-01. The black curve shows the cumulative wealth of the conditional Markowitz strategy when $h_\mu = 0.4$ and $h_\sigma = 25$. The purple curve shows the case of $h_\mu = 0.4$ and $h_\sigma = \infty$, meaning that variances and the covariance were estimated sequntially and not with moving averages. The red curve shows the cumulative wealth for the S&P 500 index, and the blue curve shows the cumulative wealth for the Nasdaq-100 index. The Markowitz portfolio with $h_\mu = 0.4$ and $h_\sigma = 25$ has the Sharpe ratio 0.90 with the annualized mean return of 18.3% and the annualized standard deviation of 20.6%. The Markowitz

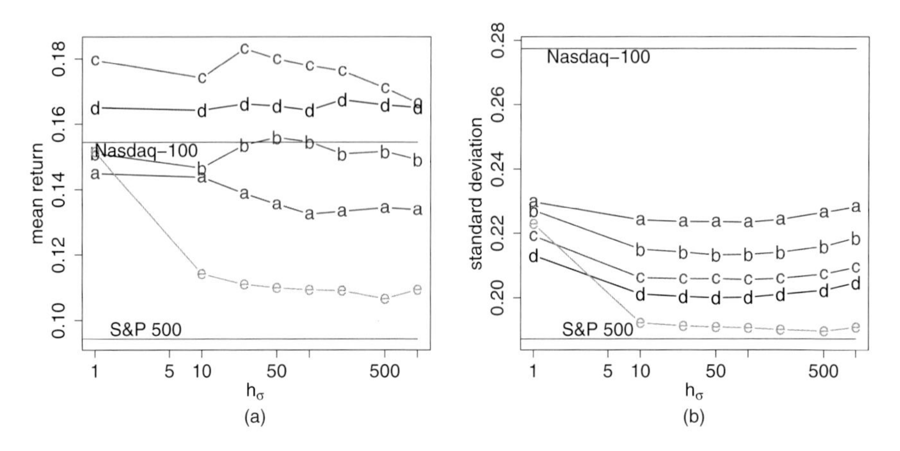

Figure 3.62 *Mean returns and standard deviations for Markowitz portfolios.* (a) Functions $h_\sigma \mapsto 250\,\bar{R}(h_\mu, h_\sigma)$ are shown, where $\bar{R}$ is the mean return. (b) Functions $h_\sigma \mapsto \sqrt{250}\ \widehat{\mathrm{sd}}(R(h_\mu, h_\sigma))$ are shown, where $\widehat{\mathrm{sd}}(R)$ is the sample standard deviation of returns. The curves with labels "a"–"e" show the cases $h_\mu = 0.01, \ldots, 10$. The red lines shows the mean and standard deviation of S&P 500, and the blue lines the mean and standard deviation of Nasdaq-100. The x-axis is logarithmic.

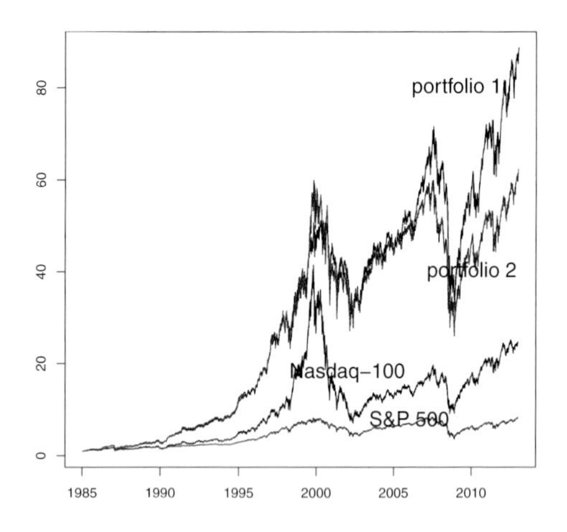

Figure 3.63 *Wealth of Markowitz portfolios.* The cumulative wealth of the Markowitz portfolio with $h_\mu = 0.4$ and $h_\sigma = 25$ (black) and $h_\mu = 0.4$ and $h_\sigma = \infty$ (brown). The blue curve shows the cumulative wealth of the Nasdaq-100 index, and the red curve shows that of the S&P 500 index.

portfolio with $h_\mu = 0.4$ and $h_\sigma = \infty$ has the Sharpe ratio 0.78 with the annualized mean return of 17.4% and the annualized standard deviation of 22.3%.

CHAPTER 4

SEMIPARAMETRIC AND STRUCTURAL MODELS

Parametric models impose restrictive assumptions on the data-generative mechanism. These assumption can be relaxed by increasing the number of parameters, but the estimation of a large number of parameters is difficult. Completely nonparametric models can suffer from the "curse of dimensionality" if the number of explanatory variables is large. However, it is possible to create models that are between parametric models and completely nonparametric models. Semiparametric models contain a parametric part and a nonparametric part, and structural models make qualitative restriction on the underlying data-generative mechanism. We do not present any precise definition of the concepts semiparametric model and structural model. Note that the ordinary linear model can also be considered as a semiparametric model if there are no parametric assumptions on the distribution of the error term.

The single-index model is considered in Section 4.1. In the single-index model the regression function $f(x) = E(Y \mid X = x)$ is a composition of a linear function and a univariate link function. The linear function makes the parametric part, and the link function makes the nonparametric part. We define the minimization estimator, quasi-maximum likelihood estimator, an iterative algorithm estimator, and the derivative, average derivative, and weighted average derivative estimators.

Multivariate Nonparametric Regression and Visualization. By Jussi Klemelä

The additive model is considered in Section 4.2. In the additive model the regression function is a sum of univariate component functions. The additive model can be called a structural model, because it does not contain a finite-dimensional parameter, but it makes a qualitative restriction on the form of the regression function. We define the backfitting, smooth backfitting, and marginal integration estimators.

Section 4.3 contains descriptions of the partially linear model and several extensions and combinations of the single-index, additive, and partially linear models.

We can view semiparametric estimation as a visualization method: Even when the true regression function does not satisfy the model assumptions, fitting a semiparametric model gives some information about the underlying regression function. In the case of the marginal integration estimation in the additive model, we have a clear interpretation of the estimators of the additive components: These are the marginal effects, or partial dependence functions, up to an additive constant.

Even when the single-index model and the additive models are in some sense simpler than the completely nonparametric model, the estimators can be computationally more demanding than for example a kernel estimator. The estimators in semiparametric models require typically iterative optimization, which can lead to computational complexity.

4.1 SINGLE-INDEX MODEL

4.1.1 Definition of the Single-Index Model

In the single-index model it is assumed that the regression function $f(x) = E(Y \mid X = x)$ satisfies

$$f(x) = g(x'\theta), \qquad x \in \mathbf{R}^d, \tag{4.1}$$

where $g : \mathbf{R} \to \mathbf{R}$ is the unknown link function and $\theta \in \mathbf{R}^d$ is the unknown index vector. Note that, unlike in the generalized linear model (2.58), the link function g is now unknown and needs to be estimated. The index $x'\theta$ aggregates the influence of the observed values $x = (x_1, \ldots, x_d)$ of the explanatory variables into one number. Examples of economic indexes include a stock index, inflation index, cost-of-living index, and price index.

The vector θ is not uniquely defined: The use of the vector $c\theta$ and the link function $g_c(u) = g(u/c)$, with some $c > 0$, leads to the same regression function f. To guarantee uniqueness, we shall assume that $\|\theta\| = 1$. (We could also assume that the first component of θ is equal to one, for example.) Also, the sign of the coefficient vector θ is not unique, because the use of the vector $-\theta$ and the link function $g_-(u) = g(-u)$ leads to the same regression function. Single-index models and their identification is discussed extensively in Horowitz (2009, Chapter 2).

4.1.2 Estimators in the Single-Index Model

For a given $\theta \in \mathbf{R}^d$, the link function g can be estimated by applying univariate nonparametric regression function estimation. When we observe regression data

$(X_1, Y_1), \ldots, (X_n, Y_n)$, where $X_i \in \mathbf{R}^d$ and $Y_i \in \mathbf{R}$, then the link function can be estimated with the univariate regression data $(X_1'\theta, Y_1), \ldots, (X_n'\theta, Y_n)$. Thus, we can estimate the regression function in the single-index model by first estimating the parameter vector θ and then estimating the link function g. We consider both the minimization estimation approach and the average derivative approach.

M-Estimation Approach In the minimization-estimation (M-estimation) approach, one finds for each fixed θ a nonparametric estimator $\hat{g}_\theta$ of regression function $g_\theta(t) = E(Y \mid X'\theta = t)$ and then the estimator of θ can be defined as a minimizer of the sum of squared errors:

$$\hat{\theta} = \text{argmin}_\theta \sum_{i=1}^{n} (Y_i - \hat{g}_\theta(X_i'\theta))^2. \tag{4.2}$$

The minimization can be done without restriction $\|\theta\| = 1$; and after finding a solution, we can normalize the direction vector to have length one. We can also use some other contrast function than the squared error contrast and define

$$\hat{\theta} = \text{argmin}_\theta \sum_{i=1}^{n} \psi\left(Y_i, \hat{g}_\theta(X_i'\theta)\right),$$

where ψ is a contrast function. The least squares contrast function is $\psi(y, z) = |y - z|^2$. Other examples of contrast functions are given in Section 5.1.1.

The estimator $\hat{g}_\theta$ can be taken to be a kernel estimator. Then,

$$\hat{g}_\theta(t) = \sum_{i=1}^{n} p_i(t)\, Y_i,$$

where

$$p_i(t) = \frac{K_h(t - X_i'\theta)}{\sum_{i=1}^{n} K_h(t - X_i'\theta)}, \qquad i = 1, \ldots, n, \tag{4.3}$$

$K : \mathbf{R} \to \mathbf{R}$ is the kernel function, $K_h(x) = K(x/h)/h$, and $h > 0$ is the smoothing parameter. Ichimura (1993) studied the properties of a semiparametric least squares estimator, and Delecroix & Hristache (1999) studied a more general M-estimator.

Quasi-Maximum Likelihood Under the single-index model the conditional density of Y given $\theta' X = t$ is given by

$$f_{Y \mid \theta' X = t}(y) = f_\epsilon(y - t),$$

where f_ϵ is the density of $\epsilon = Y - X'\theta$. Thus we can choose the contrast function as

$$\psi(y, z) = -f_\epsilon(y - z),$$

or

$$\psi(y, z) = -\log f_\epsilon(y - z),$$

which leads to the semiparametric maximum likelihood estimator. In general, f_ϵ and thus the conditional density are unknown, but we can estimate the conditional density by the conditional kernel estimator given in (3.42), defining

$$\hat{f}_{Y \mid \theta' X = t}(y) = \sum_{i=1}^{n} p_i(t)\, L_g(y - Y_i), \qquad y \in \mathbf{R},$$

where $L : \mathbf{R} \to \mathbf{R}$ is the kernel function, $L_g(y) = L(y/g)/g$, $g > 0$ is the smoothing parameter, and the weights $p_i(x)$ are defined similarly as in (4.3). Then the contrast function can be defined by

$$\psi(y, z) = -\log \hat{f}_{Y \mid \theta' X = z}(y). \tag{4.4}$$

Weisberg & Welsh (1994) studied the quasi-maximum likelihood version of the minimization estimation. Klein & Spady (1993) studied semiparametric maximum likelihood estimator in a binary response model. Ai (1997) uses quasi-maximum likelihood approach and replaces the unknown probability density function f_ϵ with a nonparametric estimator. Delecroix, Härdle & Hristache (2003) used semiparametric maximum likelihood with the estimated contrast function (4.4).

Iterative Method As before, we assume that for each index θ we have an estimator $\hat{g}_\theta$ of g. We can solve the minimization problem of the semiparametric least squares estimator iteratively. That is, we find $\hat{\theta}$ iteratively, when $\hat{\theta}$ is defined by (4.2). Given a current value θ_0 of θ, we can make an expansion

$$\hat{g}_{\theta_0}(\theta' X_i) \approx \hat{g}_{\theta_0}(\theta_0' X_i) + \hat{g}_{\theta_0}'(\theta_0' X_i)(\theta - \theta_0)' X_i, \tag{4.5}$$

which gives

$$\sum_{i=1}^{n} \left(Y_i - \hat{g}_{\theta_0}(\theta' X_i)\right)^2 \approx \sum_{i=1}^{n} \left(Y_i - \hat{g}_{\theta_0}(\theta_0' X_i) - \hat{g}_{\theta_0}'(\theta_0' X_i)(\theta - \theta_0)' X_i\right)^2,$$

The iteration proceeds as follows.

1. Choose an initial value θ_0.

2. For $m = 0, \ldots, M - 1$:

$$\theta_{m+1} = \operatorname{argmin}_\theta \sum_{i=1}^{n} W_i^2 \left(Z_i - \theta' X_i\right)^2, \tag{4.6}$$

where

$$W_i = \hat{g}_{\theta_m}'(\theta_m' X_i)$$

and

$$Z_i = \theta_m' X_i + \frac{Y_i - \hat{g}_{\theta_m}(\theta_m' X_i)}{\hat{g}_{\theta_m}'(\theta_m' X_i)},$$

with $\hat{g}_{\theta_m}$ being an estimator of the regression function, and $\hat{g}'_{\theta_m}$ being an estimator of the derivative of the regression function, both calculated using data $(\theta'_m X_i, Y_i)$, $i = 1, \ldots, n$.

The minimizer in (4.6) is the weighted least squares regression estimator

$$\theta_{m+1} = (\mathbf{X}'\mathbf{W}\mathbf{X})^{-1}\mathbf{X}'\mathbf{W}\mathbf{z}, \tag{4.7}$$

where $\mathbf{X} = (X_1, \ldots, X_n)'$ is the $n \times d$ design matrix whose ith row is X'_i, $\mathbf{z} = (Z_1, \ldots, Z_n)'$ is the $n \times 1$ column vector, and $\mathbf{W}$ is the $n \times n$ diagonal matrix with the diagonal elements W_i^2, $i = 1, \ldots, n$. The solution (4.7) is derived similarly as the solution of varying coefficients linear regression in (2.53). This iteration method was proposed in Hastie et al. (2001, p. 349).

The Derivative Method and the Average Derivative Method Under the single-index model (4.1) the gradient of the regression function is

$$Df(x) = \theta g'(x'\theta),$$

where we denote the gradient by $Df(x) = ((\partial(\partial x_1)f(x), \ldots, (\partial/\partial x_d)f(x))'$ and g' is the derivative of g. Thus the gradient has the same or the opposite direction as the vector θ. If we have an estimator $\widehat{Df}(x)$ of the gradient at any point x, then we can estimate θ by normalizing $\widehat{Df}(x)$ to have unit length, or we can obtain the normalization $\theta_1 = 1$ by dividing each component by the first component. We use the normalization of $\hat{\theta}$ to have unit length and define

$$\hat{\theta} = \hat{\beta}/\|\hat{\beta}\|, \qquad \hat{\beta} = \widehat{Df}(x), \tag{4.8}$$

for a point $x \in \mathbf{R}^d$. This estimator is called the derivative estimator.

We have that

$$EDf(X) = \theta E[g'(X'\theta)],$$

so that $EDf(X)$ is vector with the same or the opposite direction as the vector θ. The vector $EDf(X)$ is called the average derivative. This leads to an estimator, where we first construct an estimator $\widehat{Df}(X_i)$ at the observations X_i and then define

$$\hat{\theta} = \hat{\beta}/\|\hat{\beta}\|, \qquad \hat{\beta} = \frac{1}{n}\sum_{i=1}^{n} \widehat{Df}(X_i). \tag{4.9}$$

This estimator is called the average derivative estimator.

The average derivative estimator can be extended to a weighted average derivative estimator. Let $W : \mathbf{R}^d \to \mathbf{R}$ be a weight function and define the weighted average estimator as

$$\hat{\theta} = \hat{\beta}/\|\hat{\beta}\|, \qquad \hat{\beta} = \frac{1}{n}\sum_{i=1}^{n} W(X_i)\widehat{Df}(X_i).$$

The estimator can justified because

$$E[W(X)Df(X)] = \theta E[W(X)g'(X'\theta)].$$

Powell, Stock & Stoker (1989) define a density weighted average derivative estimator

$$\hat{\theta} = \hat{\beta}/\|\hat{\beta}\|, \qquad \hat{\beta} = -\frac{2}{n}\sum_{i=1}^{n} Y_i D\hat{f}_X(X_i),$$

where $\hat{f}_X$ is a density estimator of density f_X. In this estimator the weight is $W(x) = f_X(x)$ and the estimator can be justified by first noting that

$$E[f_X(X)Df(X)] = \int_{\mathbf{R}^d} Df(x) f_X^2(x)\, dx.$$

If f_X is zero on the boundary of the support of f_X, then integration by parts gives

$$E[f_X(X)Df(X)] = -2\int_{\mathbf{R}^d} f(x) Df_X(x) f_X(x)\, dx = -2E[Y Df_X(X)].$$

Samarov (1991), Samarov (1993), and Härdle & Tsybakov (1993) consider the average derivative method and show that θ can be estimated with rate $n^{-1/2}$ if the design density is very smooth. Hristache, Juditsky & Spokoiny (2001) consider the estimation of the index vector in a single-index model, using an iterative method, which avoids the pilot estimation of Df. Their estimator is based on the observation that $g(x'\theta)$ does not vary when x varies in a direction that is perpendicular to θ. Thus, only the directional derivative of $E(Y \mid X = x)$ in the direction of θ is needed to estimate. We have assumed that X has a continuous distribution. Horowitz (2009, Section 2.6.3, p. 37) discusses the case of X with a discrete distribution.

4.2 ADDITIVE MODEL

4.2.1 Definition of the Additive Model

In an additive model it is assumed that the regression function $f(x) = E(Y \mid X = x)$ has the form

$$f(x) = c + \sum_{j=1}^{d} g_j(x_j), \tag{4.10}$$

where $c \in \mathbf{R}$ is an intercept and $g_j : \mathbf{R} \to \mathbf{R}$, $j = 1, \ldots, d$, are univariate functions. The intercept and the functions g_j are estimated with identically distributed regression data $(X_1, Y_1), \ldots, (X_n, Y_n)$. For identifiability we assume that

$$Eg_j(X_j) = 0, \qquad j = 1, \ldots, d, \tag{4.11}$$

where X_j is the component of $X = (X_1, \ldots, X_d)$. Then we can estimate the constant c by

$$\hat{c} = \bar{Y} = \frac{1}{n}\sum_{i=1}^{n} Y_i.$$

Without the assumption (4.11) we cannot estimate the components g_j, because the model with the regression function

$$f(x) = (c-a) + (g_1 + a)(x_1) + \sum_{j=2}^{d} g_j(x_j)$$

leads to observations identically distributed as the observations from the model (4.10)

It can be shown that the difficulty of estimation in the additive model is equal to the difficulty of estimation in a univariate regression model, when the difficulty of estimation is measured with the minimax rates of convergence; see Stone (1985). An overview of additive models and their history can be found in Hastie & Tibshirani (1990) and in Härdle et al. (2004, Section 8).

4.2.2 Estimators in the Additive Model

We define the backfitting, smooth backfitting, and marginal integration estimators. Section 5.4.2 describes an algorithm for stagewise fitting of additive models.

Backfitting In the additive model we have

$$g_1(X_1) = E\left[Y - c - \sum_{l=2}^{d} g_l(X_l) \,\middle|\, X_1 \right]. \tag{4.12}$$

The backfitting algorithm is an iterative algorithm which is based on the idea that if we have estimates $\hat{g}_2, \ldots, \hat{g}_d$ for $g_2, \ldots, g_d$, and an estimate $\hat{c}$ for c, then we can apply a univariate nonparametric estimator to estimate g_1. Let

$$\tilde{Y}_i = Y_i - \hat{c} - \hat{g}_2(X_{i2}) - \cdots - \hat{g}_d(X_{id}), \qquad i = 1, \ldots, n.$$

Now we can use the data $(X_{i1}, \tilde{Y}_i)$, $i = 1, \ldots, n$, to estimate g_1. We describe below the backfitting algorithm for calculating the estimates $\hat{g}_j(x_j)$, $j = 1, \ldots, d$.

1. Initialize the $n \times d$ matrix $\hat{G}$ to have zero elements.
2. We iterate the following steps M times.

   Go through all coordinates $j = 1, \ldots, d$.

   (a) Let

   $$\tilde{Y}_i = Y_i - \bar{Y} - \sum_{l=1, l \neq j}^{d} [\hat{G}]_{il}, \qquad i = 1, \ldots, n.$$

   (b) Let $\hat{g}_j$ be a one-dimensional regression function estimate, based on data $(X_{ij}, \tilde{Y}_i)$, $i = 1, \ldots, n$.

   Evaluate $\hat{g}_j$ at the points X_{ij}, to obtain $\hat{g}_j(X_{ij})$, $i = 1, \ldots, n$.

   Put $[\hat{G}]_{ij} = \hat{g}_j(X_{ij})$.

3. Let $\hat{g}_j(x)$, $j = 1, \ldots, d$, be a regression function estimate based on data $(X_i^j, \tilde{Y}_i)$, $i = 1, \ldots, n$, where $\tilde{Y}_i = Y_i - \bar{Y} - \sum_{l=1, l \neq j}^{d} [\hat{G}]_{il}$.

Steps 1 and 2 calculate the evaluation matrix $\hat{G} = [\hat{g}_j(X_{i,j})]_{i=1,\ldots,n, j=1,\ldots,d}$, and step 3 is used to calculate $\hat{g}_j(x)$, $j = 1 \ldots, d$, at any point $x \in \mathbf{R}$. Properties of backfitting were studied in Buja, Hastie & Tibshirani (1989).

Smooth Backfitting The Nadaraya–Watson smooth backfitting estimate can be defined as a minimizer of

$$\sum_{i=1}^{n} \int \left(Y_i - c - \sum_{j=1}^{d} g_j(x_j) \right)^2 K((X_{i1} - x_1)/h, \ldots, (X_{id} - x_d)/h) \, dx,$$

where the minimization is done over $c \in \mathbf{R}$ and over functions $g_j : \mathbf{R} \to \mathbf{R}$ satisfying the constraints

$$\int g_j(x_j) \hat{f}_{X_j}(x_j) \, dx_j = 0, \qquad j = 1, \ldots, d,$$

where $\hat{f}_{X_j}$ is a kernel density estimate of the marginal density f_{X_j}. Furthermore, $K : \mathbf{R}^d \to \mathbf{R}$ is the kernel function, and $h > 0$ is the smoothing parameter. We get $\hat{c} = \bar{Y} = n^{-1} \sum_{i=1}^{n} Y_i$ and the minimizer $\hat{g}_j$ is found as

$$\hat{g}_j(x_j) = \tilde{g}_j(x_j) - \sum_{l \neq j} \int \hat{g}_l(x_l) \hat{f}_{X_l | X_j = x_j}(x_k) \, dx_l - \bar{Y}, \qquad j = 1, \ldots, d, \quad (4.13)$$

where $\tilde{g}_j$ is a univariate Nadaraya-Watson kernel regression function estimate and $\hat{f}_{X_l | X_j = x_j}(x_l)$ is an estimator of the conditional density. The estimator of the conditional density is defined by $\hat{f}_{X_j, X_l}(x_j, x_l) / \hat{f}_{X_j}(x_j)$, where $\hat{f}_{X_j, X_l}$ is a kernel density estimate of the density f_{X_j, X_l} of (X_j, X_l) and $\hat{f}_{X_j}$ is a kernel density estimate of the density f_{X_j} of X_j. Note that (4.12) implies that

$$g_j(X_j) = E[Y | X_j] - \sum_{l=1, l \neq j}^{d} E[g_l(X_l) | X_j] - c, \quad (4.14)$$

and (4.13) can be seen as a sample version of (4.14). An iterative algorithm is needed to find the estimates satisfying (4.13). At step $k + 1$ the estimate is

$$\hat{g}_j^{(k+1)}(x_j) = \tilde{g}_j(x_j) - \sum_{l \neq j} \int \hat{g}_l^{(k)}(x_l) \hat{f}_{X_l | X_j = x_j}(x_k) \, dx_l - \bar{Y}, \qquad j = 1, \ldots, d,$$

for $k = 0, 1, \ldots$. The smooth backfitting estimator was introduced in Mammen, Linton & Nielsen (1999). The practical implementation of smooth backfitting is studied in Nielsen & Sperlich (2005).

Marginal Integration Estimator The marginal integration estimator of the first component g_1 is defined in two steps. First we define a preliminary multivariate regression function estimator $\hat{f} : \mathbf{R}^d \to \mathbf{R}$ of the regression function $f(x) = E(Y \mid X = x)$, using the regression data $(X_1, Y_1), \ldots, (X_n, Y_n)$. The preliminary regression function estimator can be a kernel regression function estimator, for example. Second we define the estimator of g_1 as

$$\hat{g}_1(x_1) = \frac{1}{n} \sum_{i=1}^{n} \hat{f}(x_1, X_{i,2}, \ldots, X_{i,d}) - \hat{c}, \tag{4.15}$$

where

$$\hat{c} = \frac{1}{n} \sum_{i=1}^{n} Y_i.$$

Note that the marginal integration can be used also in the cases where the additive model does not hold, and in these cases it is almost an estimator of the partial dependence function, defined in (7.1). The estimator of the partial dependence function, as in (7.2), does not involve the term $\hat{c}$.

The marginal integration estimator can be motivated by the fact that from the identifiability condition (4.11) it follows that

$$Ef(x_1, X_2, \ldots, X_d) = E\left(c + g_1(x_1) + g_2(X_2) + \cdots + g_d(X_d)\right) = c + g_1(x_1).$$

Thus, the estimator in (4.15) is obtained by replacing the expectation by the sample mean and subtracting an estimator of intercept c. We can choose also

$$\hat{c} = \frac{1}{n} \sum_{i=1}^{n} \frac{1}{n} \sum_{j=1}^{n} \hat{f}(X_{i,1}, X_{j,2}, \ldots, X_{j,d}),$$

because $n^{-1} \sum_{j=1}^{n} \hat{f}(X_1, X_{j,2}, \ldots, X_{j,d})$ estimates $g_1(X_1) + c$, and $Eg_1(X_1) = 0$. The marginal integration estimator was introduced in Tjøstheim & Auestadt (1994) and Linton & Nielsen (1995).

4.3 OTHER SEMIPARAMETRIC MODELS

We present first the partially linear model in Section 4.3.1 and then we list in Section 4.3.2 some models related to the single-index model, additive model, and partially linear model

4.3.1 Partially Linear Model

Let (X_i, Z_i, Y_i), $i = 1, \ldots, n$, be identically distributed regression data with the distribution of (X, Z, Y). The response variable is $Y \in \mathbf{R}$ and $(X, Z) \in \mathbf{R}^p \times \mathbf{R}^q$ is the vector of the response variables. In a partially linear model the regression function $f(x, z) = E(Y \mid X = x, Z = z)$ is modeled as

$$f(x, z) = x'\beta + g(z), \qquad (x, z) \in \mathbf{R}^p \times \mathbf{R}^q,$$

where $\beta \in \mathbf{R}^p$ is an unknown vector and $g : \mathbf{R}^q \to \mathbf{R}$ is an unknown function. Note that the linear part does not contain an intercept, because it could not be identified separately from the unknown function g.

An Estimator of the Parametric Component An estimator of β is defined in two steps. First we estimate the regression functions $f_1(z) = E(X_1 \mid Z = z), \ldots, f_p(z) = E(X_p \mid Z = z)$ and $f_0(z) = E(Y \mid Z = z)$. Define $\hat{X}_i = (\hat{f}_1(Z_i), \ldots, \hat{f}_p(Z_i))'$, $i = 1, \ldots, n$, where $\hat{f}_1, \ldots \hat{f}_p$ are the regression function estimators. Similarly, let $\hat{Y}_i = \hat{f}_0(Z_i)$, $i = 1, \ldots, n$, where $\hat{f}_0$ is a regression function estimator. We can use kernel regression function estimators, defined in (3.6). Then we define

$$\hat{\beta} = \left[\sum_{i=1}^{n} \tilde{X}_i \tilde{X}_i' \right]^{-1} \sum_{i=1}^{n} \tilde{X}_i \tilde{Y}_i, \tag{4.16}$$

where

$$\tilde{X}_i = X_i - \hat{X}_i, \qquad \tilde{Y}_i = Y_i - \hat{Y}_i.$$

The estimator can be motivated by the facts that if we take conditional expectations of

$$Y = X'\beta + g(Z) + \epsilon, \tag{4.17}$$

then we get

$$E(Y \mid Z) = E(X \mid Z)'\beta + g(Z).$$

Subtracting, we get

$$Y - E(Y \mid Z) = (X - E(X \mid Z))'\beta + \epsilon.$$

This linear regression model can be solved to get an estimator of β, but the unknown conditional expectations $E(Y \mid Z)$ and $E(X \mid Z)$ has to be estimated.

An Estimator of the Nonparametric Component From (4.17) we get

$$g(Z) = E(Y - X'\beta \mid Z).$$

Inserting the estimator β from (4.16), we can define an estimator for g as a kernel estimator

$$\hat{g}(z) = \sum_{i=1}^{n} p_i(z)(Y_i - X_i'\hat{\beta}),$$

where $p_i(z)$ are the kernel weights, defined in (3.7).

4.3.2 Related Models

We list in the following some models that are extensions or combinations of the single-index model, additive model, or the partially linear model.

Models Related to the Single-Index Model We can generalize the single-index model to a nonlinear single-index model In the nonlinear single-index model the linear index $x'\theta$ is replaced with a nonlinear index $v_\theta(x)$, which depends on parameter θ, and it is assumed that

$$f(x) = g\left(v_\theta(x)\right), \qquad x \in \mathbf{R}^d,$$

where $g : \mathbf{R} \to \mathbf{R}$.

In the multiple-index model it is assumed that

$$f(x) = x_0'\beta_0 + G(x_1'\beta_1, \ldots, x_M'\beta_M), \qquad x \in \mathbf{R}^d,$$

where $G : \mathbf{R}^M \to \mathbf{R}$ is an unknown function, $M \geq 1$ is a known integer, x_k, $k = 0, \ldots, M$ are subvectors of $x = (x_1, \ldots, x_d)$, and β_k, $k = 0, \ldots, M$, are unknown vectors of the same lengths as x_k. Ichimura & Lee (1991) and Hristache, Juditsky, Polzehl & Spokoiny (2001) have studied multiple-index models. Li & Duan (1989), Duan & Li (1991), and Li (1991) consider sliced inverse regression estimation of θ, assuming that the distribution of the explanatory variables is elliptically symmetric.

Models Related to the Additive Model In a generalized additive model (GAM) it is assumed that the regression function has the form

$$f(x) = G\left(c + \sum_{i=1}^{d} g_i(x_i)\right),$$

where G is a known link function, $c \in \mathbf{R}$ is an unknown intercept, and $g_i : \mathbf{R} \to \mathbf{R}$, $i = 1, \ldots, d$, are unknown univariate functions. In an additive partially linear model the regression function is

$$f(x, z) = x'\beta + \sum_{i=1}^{q} g_i(z_i), \qquad (x, z) \in \mathbf{R}^p \times \mathbf{R}^q,$$

where $\beta \in \mathbf{R}^p$ is an unknown vector and $g_i : \mathbf{R} \to \mathbf{R}$, $i = 1, \ldots, q$, are an unknown univariate functions. In a generalized additive partial linear model the regression function is

$$f(x, z) = G\left(x'\beta + \sum_{i=1}^{q} g_i(z_i)\right), \qquad (x, z) \in \mathbf{R}^p \times \mathbf{R}^q,$$

where $G : \mathbf{R} \to \mathbf{R}$ is a known link function.

Models Related to the Partial Linear Model A generalized partial linear model (GPLM) assumes that the regression function has the form

$$f(x, z) = G(x'\beta + g(z)), \qquad (x, z) \in \mathbf{R}^p \times \mathbf{R}^q,$$

where $G : \mathbf{R} \to \mathbf{R}$ is a known link function. Carroll, Fan, Gijbels & Wand (1997) discuss the generalized partially linear single-index model, were the regression is modeled as

$$f(x, z) = x'\beta + g(z'\alpha),$$

where $\beta \in \mathbf{R}^p$ and $\alpha \in \mathbf{R}^q$ are unknown parameters and $g : \mathbf{R} \to \mathbf{R}$ is the unknown function. Zhang, Lee & Song (2002) study the semivarying coefficient model where

$$f(x, z) = x'\beta(u) + z'\alpha,$$

where $u \in \mathbf{R}$ is some component of x or z. Wong, Ip & Zhang (2008) study the model where

$$f(x, z) = x'\beta(u) + g(z'\alpha).$$

Varying coefficient model assumes that

$$f(x, u) = x'\beta(u).$$

Now we observe (X_i, U_i, Y_i), where $U_i \in \mathbf{R}$. This model was introduced in Hastie & Tibshirani (1993). We study varying coefficient linear model in Section 2.2.

CHAPTER 5

EMPIRICAL RISK MINIMIZATION

Linear least squares regression analysis is an example of empirical risk minimization. The coefficients of the least squares estimate of the regression function are chosen as the minimizers of the sum of squared errors. The least squares estimate is

$$\hat{f} = \text{argmin}_{f \in \mathcal{F}} \sum_{i=1}^{n} (Y_i - f(X_i))^2 ,$$

where $\mathcal{F}$ is the class of linear functions:

$$\mathcal{F} = \{f(x) = \alpha + \beta' x : \alpha \in \mathbf{R}, \beta \in \mathbf{R}^d\}. \tag{5.1}$$

The linear least squares regression can be generalized by choosing $\mathcal{F}$ to be some other collection of functions $f : \mathbf{R}^d \to \mathbf{R}$. In choosing class $\mathcal{F}$ we have to take into account the balancing between the bias and the variance: Choosing a large class $\mathcal{F}$ will lead to an estimator with a large variance, and choosing a small class $\mathcal{F}$ will lead to an estimator with a large bias. In addition, the choice of class $\mathcal{F}$ has an influence on the computational complexity of the empirical minimization.

Section 5.1 starts with discussing other criterions than the least squares criterion. Changing the loss function leads to estimators with different properties than the estimator obtained with the squared loss function. For example, the absolute deviation loss function leads to an estimator that is less sensitive to outliers. On the other hand, when we want to estimate functionals of the distribution other than the conditional mean, then we have to change the loss function to some other than the squared loss function. For example, if the distribution is such that the conditional mean and the conditional median are not the same, then we have to use the squared loss for mean estimation and use the absolute deviation loss for median estimation. The absolute deviation loss can be generalized to a loss suitable for the estimation of other quantiles than the median. To estimate the conditional density, we need again to define a suitable loss function.

Section 5.2 introduces local polynomial and local likelihood estimators. Local polynomial and local likelihood estimators can be seen as local averaging methods, covered in Section 3. However, since they are defined as solutions to a minimization problem involving an empirical risk, we discuss them together with other empirical risk minimizers. The local constant estimator is equal to the kernel regression estimator, if the weights of the local empirical risk are chosen to be the the kernel weights. The local linear estimator and the higher-order local polynomial estimators can be better than the kernel regression estimator in boundary regions and near jump points. Local likelihood estimators can be used in binary choice models, in Poisson count models, and in other cases where the response variable follows a parametric distribution.

Section 5.3 defines support vector machines, support vector machines can be defined as minimizers of a penalized empirical risk. The introduction of the penalization can be useful in high-dimensional cases, when there are many explanatory variables. Support vector machines were initially defined for classification, but analogous methods can be defined for regression.

Section 5.4 discusses stagewise methods for the minimization of an empirical risk. We can define an estimator either by defining it as a solution to a minimization problem or by defining an algorithm for its calculation. We define estimators in Section 5.4 by defining algorithms for their calculation. First we define forward stagewise modeling algorithms, which can also be called boosting algorithms. Second we define an algorithm for stagewise fitting of additive models, which is an alternative to backfitting and other methods presented in Section 4.2. Third we describe a projection pursuit regression algorithm.

Section 5.5 describes adaptive regressograms. Adaptive regressograms are such regressograms whose partition is chosen by minimizing an empirical risk. We define greedy regressograms, classification and regression trees, and dyadic regressograms. A method to select the partition of a regressogram can be seen as a method for variable selection. The partition should be made more granular in the direction of the more important variables, and the regressogram should be a constant function in the direction of the variables that are less important.

5.1 EMPIRICAL RISK

We define empirical risks suitable for the estimation of conditional expectations, conditional quantiles, and conditional density functions.

5.1.1 Conditional Expectation

Let $(X_1, Y_1), \ldots, (X_n, Y_n)$ be the identically distributed regression data, where $X_i \in \mathbf{R}^d$ and $Y_i \in \mathbf{R}$. We want to estimate the regression function

$$f(x) = E[Y \mid X = x], \qquad x \in \mathbf{R}^d,$$

where (X, Y) is distributed as (X_i, Y_i), $i = 1, \ldots, n$.

Let $\mathcal{F}$ be a class of functions $\mathbf{R}^d \to \mathbf{R}$ and let $\epsilon > 0$. We define the empirical risk minimizer $\hat{f} : \mathbf{R}^d \to \mathbf{R}$ to be such that

$$\gamma_n(\hat{f}) \le \inf_{g \in \mathcal{F}} \gamma_n(g) + \epsilon,$$

where $\gamma_n(g)$ is the sum of squared errors:

$$\gamma_n(g) = \sum_{i=1}^{n} (Y_i - g(X_i))^2, \qquad g : \mathbf{R}^d \to \mathbf{R}. \tag{5.2}$$

The linear least squares estimator is obtained by choosing $\epsilon = 0$, along with $\mathcal{F}$ as in (5.1). We can write

$$\sum_{i=1}^{n} (Y_i - g(X_i))^2 = \sum_{i=1}^{n} Y_i^2 - 2 \sum_{i=1}^{n} Y_i g(X_i) + \sum_{i=1}^{n} g(X_i)^2.$$

The term $\sum_{i=1}^{n} Y_i^2$ does not depend on the function g, and thus it is enough to minimize the simplified empirical risk

$$\tilde{\gamma}_n(g) = -\frac{2}{n} \sum_{i=1}^{n} Y_i \, g(X_i) + \frac{1}{n} \sum_{i=1}^{n} g(X_i)^2.$$

More generally, we can define

$$\gamma_n(g) = \sum_{i=1}^{n} \gamma(Y_i, g(X_i)),$$

where the contrast function γ can be chosen in the following ways, for example.

1. The contrast function can be a power function:

$$\gamma(y, z) = |y - z|^p,$$

for $p \ge 1$.

2. The contrast function can be an ε-sensitive loss function

$$\begin{aligned} \gamma(y,z) &= (|y-z)| - \varepsilon)\, I_{[\epsilon,\infty)}(|y-z|) \\ &= \begin{cases} 0, & \text{when } |y-z| < \varepsilon, \\ |y-z| - \varepsilon, & \text{when } |y-z| \geq \epsilon, \end{cases} \end{aligned}$$

for $\varepsilon > 0$.

3. The contrast function can be a robust loss function

$$\gamma(y,z) = \begin{cases} \frac{1}{2}(y-z)^2, & \text{when } |y-z| \leq c, \\ c|y-z| - c^2/2, & \text{when } |y-z| > c, \end{cases}$$

where $c > 0$. This loss function was defined in Huber (1964).

Figure 5.1 shows the squared error contrast function $x \mapsto x^2$ with black, the absolute error contrast function $x \mapsto |x|$ with blue, the ε-sensitive contrast function $x \mapsto (|x| - \varepsilon)I_{[\varepsilon,\infty)}(|x|)$ with red, and the robust loss function $x \mapsto (|x|^2/2)I_{[0,c]}(|x|) + (c|x| - c^2/2)I_{(c,\infty)}(|x|)$ with green. We have chosen $\varepsilon = 0.2$ and $c = 0.8$. The ε-sensitive contrast function is close to the squared error contrast function for small errors, but it behaves similarly to the absolute error contrast function for large errors. Thus the ε-sensitive contrast function penalizes less from small errors and more from large errors, and it is an intermediate contrast function between the squared error and the absolute error contrast functions. Penalizing less from large errors makes the estimates more robust against large deviations, in the sense that the estimator is not driven solely by outliers.

We noted in Section 1.1.7 that for the estimation of the conditional median $\text{med}(Y \mid X = x)$ it is natural to use the absolute deviation contrast function $\gamma(y,z) = |y-z|$.

5.1.2 Conditional Quantile

Quantiles are defined in (1.23). In (1.41) we have noted that a quantile can be characterized as

$$Q_p(Y) = \text{argmin}_{\theta \in \mathbf{R}} E\rho_p(Y - \theta),$$

where

$$\rho_p(t) = t\,[p - I_{(-\infty,0)}(t))], \qquad t \in \mathbf{R}, \tag{5.3}$$

for $0 < p < 1$.

Let $(X_1, Y_1), \ldots, (X_n, Y_n)$ be regression data and define the contrast function

$$\gamma(y,z) = \rho_p(y - z)$$

and $\gamma_n(g) = \sum_{i=1}^n \gamma(Y_i, g(X_i))$. We can define an estimator of the conditional quantile $f(x) = Q_p(Y \mid X = x)$ as

$$\hat{f} = \text{argmin}_{g \in \mathcal{G}} \gamma_n(g) = \text{argmin}_{g \in \mathcal{G}} \sum_{i=1}^n \rho_p(Y_i - g(X_i)),$$

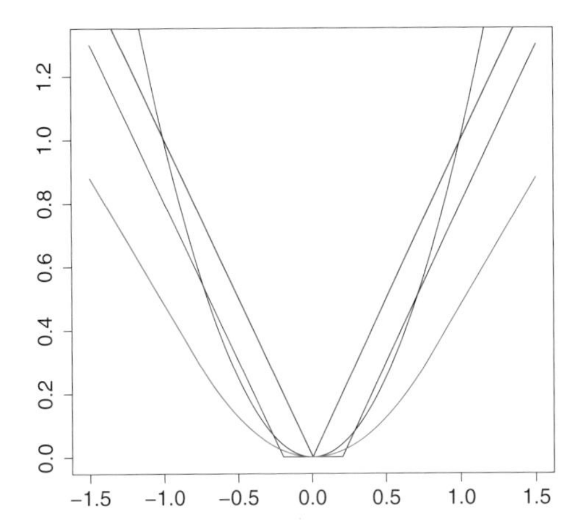

Figure 5.1 *Contrast functions.* Shown are the squared error contrast function with black, the absolute error contrast function with blue, the ε-sensitive contrast function with red, and the robust loss function with green.

where $\mathcal{G}$ is the class of all measurable functions.

Figure 5.2 shows linear conditional quantile estimates for the levels $p = 0.1, 0.2, \ldots, 0.9$. The data are the same as in Figure 1.1: the data consist of the daily S&P 500 returns $R_t = (S_t - S_{t-1})/S_{t-1}$, where S_t is the price of the index. The explanatory and the response variables as

$$X_t = \log_e \sqrt{\frac{1}{k}\sum_{i=1}^{k} R_{t-i}^2}, \qquad Y_t = \log_e |R_t|.$$

The S&P 500 index data are described more precisely in Section 1.6.1. We show also a contour plot of a kernel estimate of the density of (X_t, Y_t).

5.1.3 Conditional Density

Conditional density function was introduced in Section 1.1.8. Let $f_{Y|X=x}(y)$ be the conditional density of Y given $X = x$, where $y \in \mathbf{R}$ and $x \in \mathbf{R}^d$. The conditional density will be estimated using the identically distributed regression observations $(X_1, Y_1), \ldots, (X_n, Y_n)$. The estimator is

$$\hat{f}_{Y|X} = \text{argmin}_{g \in \mathcal{G}} \sum_{i=1}^{n} \gamma_n \left(\hat{f}_{Y|X=X_i}(Y_i) \right),$$

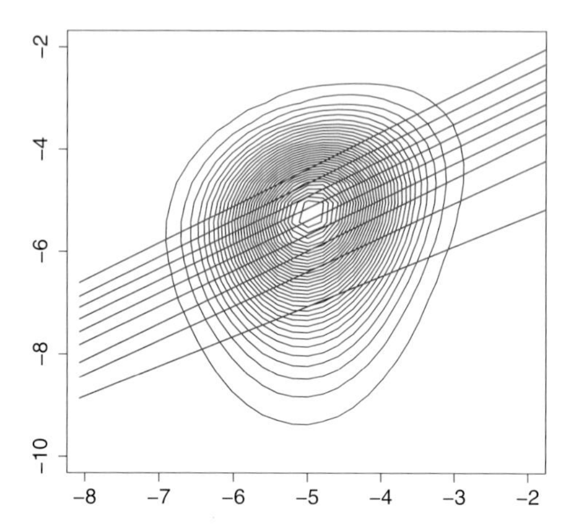

Figure 5.2 *Linear conditional quantile estimates.* Red lines show the quantile estimates for levels $p = 0, 1, \ldots, 0.9$. A contour plot of a kernel estimate of the density of (X_t, Y_t) is also shown.

where $\hat{f}_{Y|X} : \mathbf{R} \times \mathbf{R}^d \to \mathbf{R}$, $(y, x) \mapsto \hat{f}_{Y|X=x}(y)$, $\mathcal{G}$ is a class of measurable functions $\mathbf{R} \times \mathbf{R}^d \to \mathbf{R}$, and the empirical risk is

$$\gamma_n\left(\hat{f}_{Y|X}\right) = \frac{1}{n}\sum_{i=1}^{n}\left[\int_{-\infty}^{\infty} \hat{f}^2_{Y|X=X_i}(y)\,dy - 2\hat{f}_{Y|X=X_i}(Y_i)\right].$$

The empirical risk can be justified in two ways.

First, the minimization of the empirical risk is approximately equal to the minimization of the squared L_2 error

$$\left\|\hat{f}_{Y|X} - f_{Y|X}\right\|^2_{dy,dP_X(x)} = \int \left(\hat{f}_{Y|X=x}(y) - f_{Y|x=x}(y)\right)^2 f_X(x)\,dx\,dy.$$

The L_2 error is minimized with respect to $\hat{f}_{Y|X}$, when

$$\begin{aligned}\int \hat{f}^2_{Y|X=x}(y)\, f_X(x)\,dx\,dy - 2\int \hat{f}_{Y|X=x}(y) f_{Y|X=x}(y) f_X(x)\,dx\,dy \\ = E\left[\int \hat{f}^2_{Y|X=X}(y)\,dy - 2\hat{f}_{Y|X=X}(Y)\right]\end{aligned}$$

is minimized with respect to $\hat{f}_{Y|X}$, where the expectation is with respect to random variables (X, Y). The expectation can be estimated with an average, which leads to $\gamma_n(\hat{f}_{Y|X})$.

Second, let us take the expectation with respect to (X, Y), which is independent of the sample $(X_1, Y_1), \ldots, (X_n, Y_n)$ that is used to calculate the estimate:

$$
\begin{aligned}
& E\left[\gamma_n(\hat{f}_{Y|X}) - \gamma_n(f_{Y|X})\right] \\
= \; & E\left[\int \left(\hat{f}^2_{Y|X=X}(y) - f^2_{Y|X=X}(y)\right) dy - 2\left(\hat{f}_{Y|X=X}(Y) - f_{Y|X=X}(Y)\right)\right] \\
= \; & \int \left(\hat{f}^2_{Y|X=x}(y) - f^2_{Y|X=x}(y)\right) f_X(x)\, dx\, dy \\
& -2 \int \left(\hat{f}_{Y|X=x}(y) - f_{Y|X=x}(y)\right) f_{Y|X=x}(y) f_X(x)\, dx\, dy \\
= \; & \int \left(\hat{f}^2_{Y|X=x}(y) - 2\hat{f}_{Y|X=x}(y) f_{Y|X=x}(y) + f^2_{Y|X=x}(y)\right) f_X(x)\, dx\, dy \\
= \; & \int \left(\hat{f}_{Y|X=x}(y) - f_{Y|X=x}(y)\right)^2 f_X(x)\, dx\, dy
\end{aligned}
$$

and this is minimized by choosing $\hat{f}_{Y|X} = f_{Y|X}$.

5.2 LOCAL EMPIRICAL RISK

We discuss two types of local empirical risk: the risk that leads to local polynomial estimators and the local likelihood risk.

5.2.1 Local Polynomial Estimators

Local constant estimators are equal to local averaging estimators, as defined in Section 3. A local averaging estimator can be a kernel estimator, nearest-neighbor estimator, or regressogram, depending on the choice of the weights. Local linear estimators are a new class of estimators. Local linear estimators can be better in estimating approximately linear regression functions than kernel estimators. Furthermore, they can perform better on the jump points of the regression function. Local linear estimators can be generalized to local polynomial estimators. Local polynomial estimators can be used both in state–space smoothing and in time–space smoothing.

Local Constant Estimator Let us consider estimation of $E(Y \mid X = x)$ and let $(X_1, Y_1), \ldots, (X_n, Y_n)$ be regression data. The weighted least squares criterion is

$$
\gamma_n(\theta, x) = \sum_{i=1}^{n} w_i(x)\,(Y_i - \theta)^2, \qquad \theta \in \mathbf{R}, \;\; x \in \mathbf{R}^d,
$$

where the weights $w_i(x)$ depend on $X_1, \ldots, X_n$. The minimizer of the weighted least squares criterion is

$$
\mathrm{argmin}_{\theta \in \mathbf{R}} \gamma_n(\theta, x) = \sum_{i=1}^{n} p_i(x)\, Y_i,
$$

where

$$p_i(x) = \frac{w_i(x)}{\sum_{i=1}^n w_i(x)}.$$

Indeed, derivating the weighted empirical risk with respect to θ and setting this derivative to zero gives the equation $\sum_{i=1}^n w_i(x)(Y_i - \theta) = 0$, which leads to the solution.

If the weights are such that $w_i(x)$ is large when $\|X_i - x\|$ is small, and $w_i(x)$ is small when $\|X_i - x\|$ is large, then we can call $\gamma_n(\theta, x)$ a local empirical risk and the estimator

$$\hat{f}(x) = \sum_{i=1}^n p_i(x)\, Y_i, \qquad x \in \mathbf{R}^d \tag{5.4}$$

is a local constant estimator.

The estimator $\hat{f}(x)$ is a kernel estimator defined in (3.6), if the weights $w_i(x)$ are chosen as the kernel weights

$$w_i(x) = K_h(x - X_i), \tag{5.5}$$

where $K : \mathbf{R}^d \to \mathbf{R}$ is a kernel function, $K_h(x) = K(x/h)/h^d$, and $h > 0$ is the smoothing parameter.

The local constant estimator can be defined also for the conditional quantile. We can define

$$\hat{Q}_p(Y \mid X = x) = \text{argmin}_{\theta \in \mathbf{R}} \sum_{i=1}^n w_i(x)\, \rho_p(Y_i - \theta),$$

where ρ_p is defined in (5.3) and the weights $w_i(x)$ can be taken as in (5.5).

Local Linear Estimator Let $f(x) = E(Y \mid X = x)$ and $(X_1, Y_1), \ldots, (X_n, Y_n)$ be a sample from the distribution of (X, Y). The local linear estimators for $f(x)$ and $Df(x)$ are constructed by finding minimizers $\hat{\alpha}$ and $\hat{\beta}$ of

$$\sum_{i=1}^n w_i(x) \left[Y_i - \alpha - \beta'(X_i - x)\right]^2,$$

over $\alpha \in \mathbf{R}$ and $\beta \in \mathbf{R}^d$, where the weights can be defined as in (5.5). The estimator of the conditional expectation is

$$\hat{f}(x) = \hat{\alpha}, \qquad x \in \mathbf{R}^d, \tag{5.6}$$

and the estimator of its gradient is

$$\widehat{Df}(x) = \hat{\beta}, \qquad x \in \mathbf{R}^d.$$

The heuristics of the procedure comes from the Taylor approximation

$$Y_i \approx f(X_i) \approx f(x) + Df(x)'(X_i - x),$$

where $Df(x) = (\partial f(x)/\partial x_1, \ldots, \partial f(x)/\partial x_d)'$ is the gradient of $f(x)$.

We can find an explicit expression for $\hat{\alpha}(x)$ and $\hat{\beta}(x)$. The derivation is similar to the case of linear regression; see (2.10). Let us denote by $\mathbf{X}$ the $n \times (d+1)$-matrix whose ith row is $(1, (X_i - x)')$, where $X_i - x$ is a column vector of length d:

$$\mathbf{X} = \begin{bmatrix} 1 & (X_1 - x)' \\ \vdots & \vdots \\ 1 & (X_n - x)' \end{bmatrix}.$$

Let $\mathbf{y} = (Y_1, \ldots, Y_n)'$ be the column vector of length n. Let $W(x)$ be the $n \times n$ diagonal matrix with diagonal elements $w_i(x)$. Then,

$$(\hat{\alpha}, \hat{\beta}')' = [\mathbf{X}'W(x)\mathbf{X}]^{-1} \mathbf{X}'W(x)\mathbf{y}.$$

We can write

$$\hat{f}(x) = \hat{\alpha} = e_1' [\mathbf{X}'W(x)\mathbf{X}]^{-1} \mathbf{X}'W(x)\mathbf{y}, \qquad (5.7)$$

where $e_1 = (1, 0, \ldots, 0)'$ is a $(d+1) \times 1$ vector. Now we can write

$$\hat{f}(x) = \sum_{i=1}^{n} q_i(x)\, Y_i,$$

where

$$q_i(x) = e_1' [\mathbf{X}'W(x)\mathbf{X}]^{-1} \begin{bmatrix} 1 \\ X_i - x \end{bmatrix} w_i(x),$$

where $X_i - x$ is a $d \times 1$ vector.

It is to be noted that we can define the local linear estimator as a minimizer of

$$\sum_{i=1}^{n} p_i(x) [Y_i - \alpha - \beta' X_i]^2, \quad \alpha \in \mathbf{R}, \quad \beta \in \mathbf{R}^d, \quad x \in \mathbf{R}^d.$$

Now the local linear estimator is a linear function whose the coefficients depend on x:

$$\hat{f}(x) = \hat{\alpha}(x) + \hat{\beta}(x)'x, \qquad x \in \mathbf{R}^d,$$

where $\hat{\alpha}(x)$ and $\hat{\beta}(x)$ minimize the empirical risk.

One-Dimensional Local Linear Estimator In the one-dimensional case, when $d = 1$, we get, modifying (2.6),

$$\hat{\beta} = \frac{\sum_{i=1}^{n} p_i(x)(X_i - \bar{X})(Y_i - \bar{Y})}{\sum_{i=1}^{n} p_i(x)(X_i - \bar{X})^2}, \qquad \hat{\alpha} = \bar{Y} - \hat{\beta}(\bar{X} - x),$$

where

$$\bar{X} = \sum_{i=1}^{n} p_i(x)\, X_i, \qquad \bar{Y} = \sum_{i=1}^{n} p_i(x)\, Y_i.$$

We can write

$$\hat{\beta} = \frac{\sum_{i=1}^n p_i(x)(X_i - \bar{X})Y_i}{\sum_{i=1}^n p_i(x)(X_i - \bar{X})^2}. \tag{5.8}$$

Thus

$$\hat{\alpha} = \sum_{i=1}^n q_i(x)\, Y_i,$$

where

$$q_i(x) = p_i(x)\left(1 - \frac{(X_i - \bar{X})(\bar{X} - x)}{\sum_{i=1}^n p_i(x)(X_i - \bar{X})^2}\right). \tag{5.9}$$

We can also write

$$q_i(x) = \frac{w_i(x)[s_2(x) - (X_i - x)s_1(x)]}{\sum_{i=1}^n w_i(x)[s_2(x) - (X_i - x)s_1(x)]}, \tag{5.10}$$

where

$$s_1(x) = \sum_{i=1}^n w_i(x)(X_i - x), \qquad s_2(x) = \sum_{i=1}^n w_i(x)(X_i - x)^2.$$

Note that (5.10) shows that $\sum_{i=1}^n q_i(x) = 1$.[47] From (5.8) we get that

$$\hat{\beta} = \sum_{i=1}^n r_i(x)\, Y_i,$$

where

$$r_i(x) = \frac{p_i(x)(X_i - \bar{X})}{s_2(x) - s_1^2(x)}.$$

Note that $\sum_{i=1}^n r_i(x) = 0$.

Illustrations of Local Linear Estimators Figure 5.3 illustrates the estimation of a one-dimensional linear regression function, and it shows the effect of the smoothing parameter. Panel (a) shows the true regression function and the data. The true regression function is the linear function $f(x) = 2x$. The data $(X_1, Y_1), \ldots, (X_n, Y_n)$ is simulated with $n = 200$, $Y_i = f(X_i) + \epsilon_i$, where $\epsilon_i \sim N(0,1)$ and $X_i \sim N(0,1)$, where X_i and ϵ_i are uncorrelated. Panel (b) shows local linear estimates of f with smoothing parameters $h = 0.08, \ldots, 5$, when the kernel is the standard Gaussian density function. Panel (b) shows that when $h \to \infty$, the estimate of the regression function converges to the true linear function, unlike a kernel estimate shown in Figure 3.1, which approaches a constant function.

Figure 5.4 illustrates the estimation of a one-dimensional quadratic regression function. Panel (a) shows the true regression function and the data. The true regression function is the quadratic function $f(x) = x^2$. The simulated regression

[47] Note also that $\sum_{i=1}^n q_i(x)(X_i - x) = 0$.

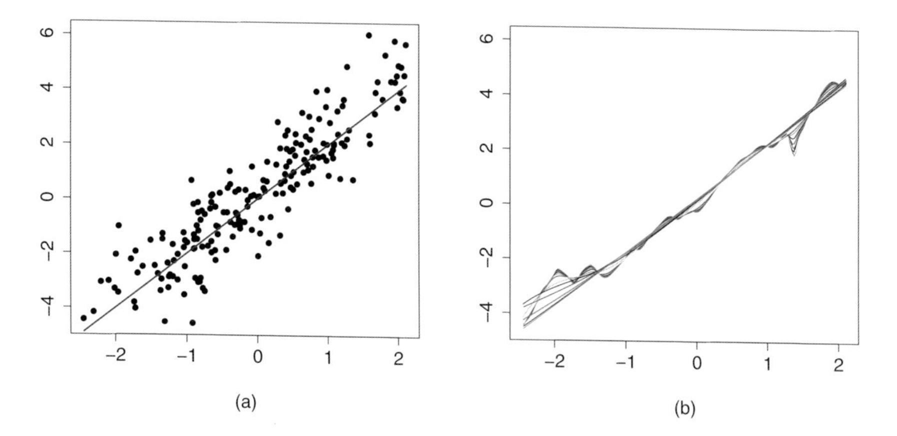

Figure 5.3 *Local linear estimates of a linear function.* (a) The data and the true regression function. (b) Local linear estimates of the regression function with smoothing parameters $h = 0.08, \ldots, 5$.

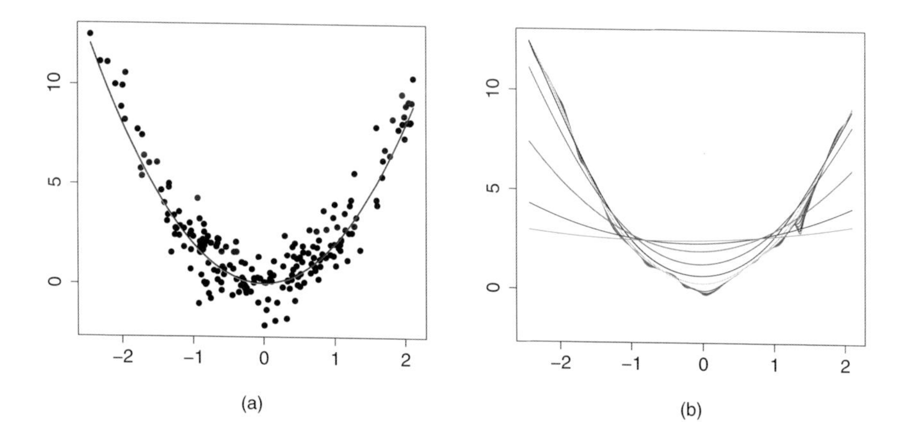

Figure 5.4 *Local linear estimates of a quadratic function.* (a) The data and the true regression function. (b) Local linear estimates of the regression function with smoothing parameters $h = 0.08, \ldots, 5$.

has sample size $n = 200$, $Y_i = f(X_i) + \epsilon_i$, where $\epsilon_i \sim N(0,1)$ and $X_i \sim N(0,1)$, where X_i and ϵ_i are uncorrelated. Panel (b) shows local linear estimates of f with smoothing parameters $h = 0.08, \ldots, 5$, when the kernel is the standard Gaussian density function. The kernel estimates in Figure 3.2 are not considerably different.

Figure 5.5 shows estimates of the derivative of a regression function. In panel (a), $f(x) = 2x$, so that $f'(x) \equiv 2$. In panel (b), $f(x) = x^2$, so that $f'(x) = 2x$. In

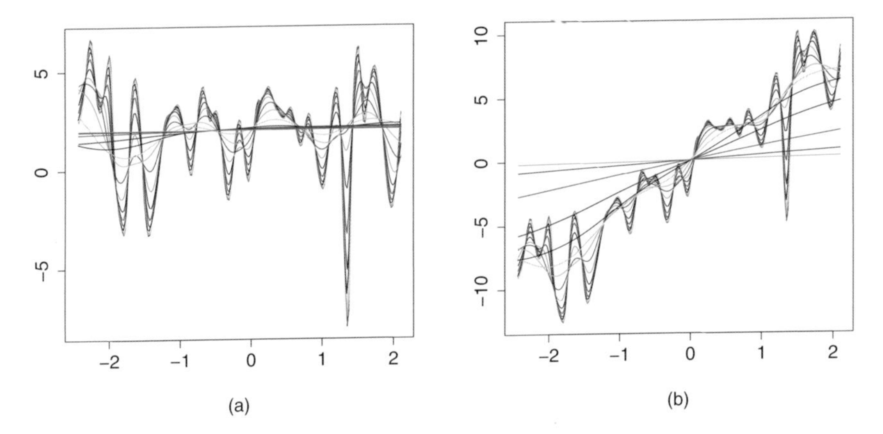

Figure 5.5 *Local linear estimates of derivative.* (a) Estimates of $f'(x) \equiv 2$. (b) Estimates of $f'(x) = 2x$. The smoothing parameters are $h = 0.08, \ldots, 5$.

both cases the sample size is $n = 200$, $Y_i = f(X_i) + \epsilon_i$, where $\epsilon_i \sim N(0,1)$ and $X_i \sim N(0,1)$, where X_i and ϵ_i are uncorrelated. The smoothing parameters are $h = 0.08, \ldots, 5$ and the kernel is the standard Gaussian density function. In both cases the derivative estimate converges to a constant when h increases.

Figure 5.6 illustrates the weights $q_i(x)$ in the case of one-dimensional explanatory variable X. We have used the standard Gaussian kernel and smoothing parameter $h = 0.2$. Panel (a) shows a perspective plot of the function $(x, X_i) \mapsto p_i(x)$, where $X_1, \ldots, X_n$ is a simulated sample of size $n = 200$ from the uniform distribution on $[-1, 1]$. Panel (b) shows the six functions $x \mapsto p_i(x)$ for the choices $X_i = -1, -0.5, \ldots, 1$. We can compare the local linear weights to the kernel weights in Figure 3.3. We can see that the weights differ in the boundaries $x = -1$ and $x = 1$. For $X_i = -1$ and $X_i = 1$ the local linear weights are larger at the boundaries. For $X_i = -0.5$ and $X_i = 0.6$ the local linear weights are negative at the boundaries.

Volatility Estimation with the Local Linear Estimator Let R_t be the S&P 500 return and let us choose

$$Y_t = R_t, \qquad X_t = R_{t-1}$$

and consider the estimation of the conditional variance with the local linear estimator, using the S&P 500 data described in Section 1.6.1. The function

$$\sigma^2(x) = E(Y_t^2 \mid X_t = x)$$

is now called the news impact curve.

Figure 5.7 shows a local linear estimator of the conditional volatility with the smoothing parameter $h = 0.04$ and the standard normal kernel.

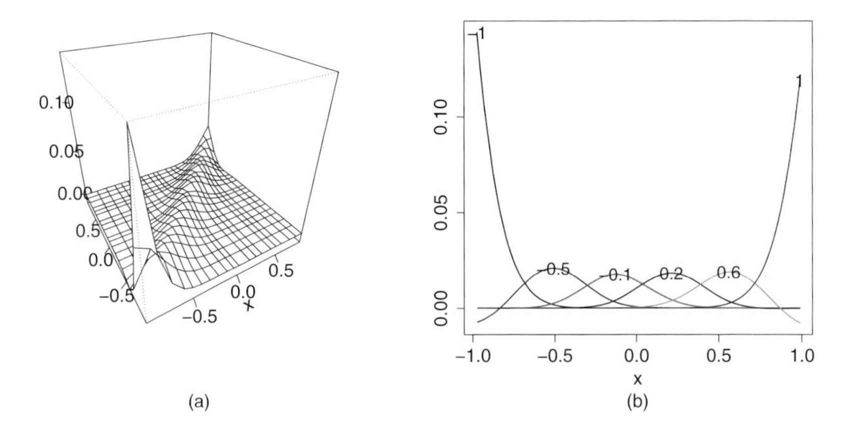

Figure 5.6 *Weights in local linear regression.* (a) The function $(x, X_i) \mapsto q_i(x)$. (b) The six slices $x \mapsto p_i(x)$ for the choices $X_i = -1, -0.5, \ldots, 1$.

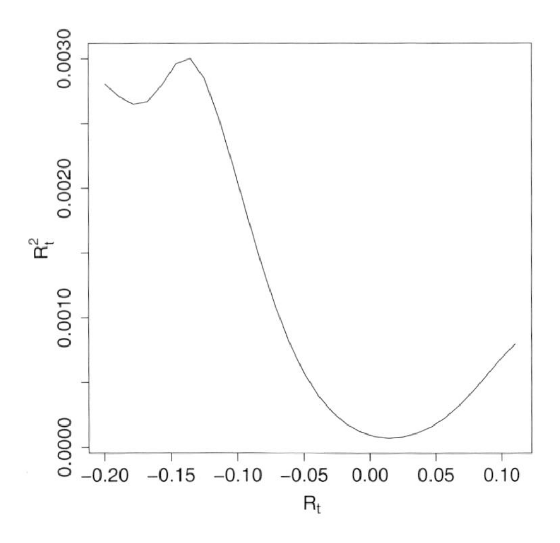

Figure 5.7 *S&P 500 volatility: News impact curve.* A local linear estimate of the conditional variance.

Kernel estimators for the news impact curve were shown in Figure 3.28. An ARCH(∞) model for the estimation of the news impact curve is mentioned in (3.64). A local polynomial volatility estimator was studied in Härdle & Tsybakov (1997), who developed a theory for the conditional heteroskedastic autoregressive nonlinear model (CHARN)

$$Y_t = f(Y_{t-1}) + \sigma(Y_{t-1})\epsilon_t.$$

See also Härdle et al. (2004, Section 4.4.2) for an example of a local linear estimate of the volatility of the DM/USD exchange rate. In the estimation of the conditional expectation $E(Y^2 \mid X = x)$ of Y^2 we can use the empirical risk

$$\sum_{i=1}^{n} p_i(x) \left[Y_i^2 - \psi(\alpha + \beta'(X_i - x))\right]^2,$$

where $\psi : \mathbf{R} \to \mathbf{R}$. Then the estimator of the conditional expectation is $\hat{f}(x) = \psi(\hat{\alpha}(x))$, where $(\hat{\alpha}(x), \hat{\beta}(x))$ minimizes the empirical risk. For example, $\psi(x) = \exp(x)$ has been used in Ziegelmann (2002) to estimate volatility functions. The use of the exponential function guarantees the positiveness of the estimator.

Local Polynomial Estimator We can define a local polynomial estimator of order $q \geq 1$. Minimize

$$\sum_{i=1}^{n} \left(Y_i - \alpha - \sum_{l=1}^{q} \sum_{|j|=l} (X_1 - x_1)^{j_1} \cdots (X_{id} - x_d)^{j_d} \beta_{lj} \right)^2 w_i(x),$$

where the summation $\sum_{|j|=l}$ is over the indexes $j = (j_1, \ldots, j_d) \in \{0, 1, \ldots\}^d$ with $|j| = j_1 + \cdots + j_d = l$ and we denote $x = (x_1, \ldots, x_d)$, $X_i = (X_{i1}, \ldots, X_{id})$.

In the one dimensional case $d = 1$, we can define the local polynomial estimator by extending the definition (5.7). Let us denote by $\mathbf{X}$ the $n \times (q + 1)$ matrix

$$\mathbf{X} = \begin{bmatrix} 1 & X_1 - x & \cdots & (X_1 - x)^q \\ \vdots & \vdots & \vdots & \vdots \\ 1 & X_n - x & \cdots & (X_n - x)^q \end{bmatrix}.$$

As before, let $\mathbf{y} = (Y_1, \ldots, Y_n)'$ be the column vector of length n and let $W(x)$ be the $n \times n$ diagonal matrix with diagonal elements $w_i(x)$. Then,

$$\hat{f}(x) = e_1 \left[\mathbf{X}'W(x)\mathbf{X}\right]^{-1} \mathbf{X}'W(x)\mathbf{y}, \tag{5.11}$$

where $e_1 = (1, 0, \ldots, 0)'$ is a $(q + 1) \times 1$ vector.

Local Polynomial Moving Average Let as have a time series $Y_1, \ldots, Y_t$. We want to predict the next value Y_{t+1}. This can be done with moving averages, discussed in Section 3.2.4. The one-sided moving average was defined in (3.14) as

$$\hat{f}(t) = \sum_{i=1}^{t} p_i(t)\, Y_i,$$

where

$$p_i(t) = \frac{K((t-i)/h)}{\sum_{j=1}^{t} K((t-j)/h)}.$$

The kernel function K is zero on the negative real axis. We can take $K(x) = \exp(-x)I_{[0,\infty)}(x)$ or $K(x) = I_{[0,1]}(x)$, for example.

Local polynomial moving average of order $q \geq 1$ is similar to (5.11). Now we choose $X_i = i$, $x = t+1$, and denote by $\mathbf{X}$ the $t \times (p+1)$ matrix

$$\mathbf{X} = \begin{bmatrix} 1 & X_1 - t - 1 & \cdots & (X_1 - t - 1)^q \\ \vdots & \vdots & \vdots & \vdots \\ 1 & X_t - t - 1 & \cdots & (X_t - t - 1)^q \end{bmatrix}.$$

Let $\mathbf{y} = (Y_1, \ldots, Y_t)'$ be the column vector of length t and let $W(t)$ be the $t \times t$ diagonal matrix $\text{diag}(w_1(t), \ldots, w_t(t))$, where $w_i(t) = K((t-i)/h)$. Then,

$$\hat{f}(t) = e_1 \left[\mathbf{X}'W(t)\mathbf{X}\right]^{-1} \mathbf{X}'W(t)\mathbf{y},$$

where $e_1 = (1, 0, \ldots, 0)'$ is a $(q+1) \times 1$ vector. Smoothing parameter selection for local polynomial moving averages is discussed in Gijbels et al. (1999).

5.2.2 Local Likelihood Estimators

The local log-likelihood was defined in (1.67). Let us consider the locally constant likelihood estimator. We assume that the density function of Y is given by

$$f(y, \theta), \qquad y \in \mathbf{R}, \ \ \theta \in \mathbf{R}^k,$$

where θ is a k-dimensional vector of parameters. Let

$$\theta(x) = \text{argmin}_{\theta \in \mathbf{R}^k} \sum_{i=1}^{n} p_i(x) \log f(Y_i, \theta),$$

where the weights $p_i(x)$ depend on X_i. Again, the typical example are the kernel weights.

As an example, let us consider the model

$$Y_t = \sigma_t \epsilon_t,$$

where $\sigma_t^2 = E(Y_t^2 \,|\, \mathcal{F}_t)$, $\mathcal{F}_t$ is the sigma-algebra generated by $Y_{t-1}, Y_{t-2}, \ldots$, and ϵ_t is independent of $Y_t, Y_{t-1}, \ldots$. Let us observe $Y_1, \ldots, Y_T$. The likelihood was written in (2.80) under the Gaussian assumption $\epsilon_t \sim N(0,1)$. The likelihood, given the observations $Y_1, \ldots, Y_p$, is

$$\sum_{t=p+1}^{T} \left(\log_e(\sigma_t^2) + \frac{Y_t^2}{\sigma_t^2} \right). \tag{5.12}$$

Let us find the local constant estimator. Define the local likelihood

$$l_T(\sigma^2) = \sum_{t=p+1}^{T} p_t(T) \left(\log_e(\sigma^2) + \frac{Y_t^2}{\sigma^2} \right),$$

where $p_t(T)$ are the weights. The maximizer of the local likelihood is

$$\sum_{t=1}^{T} p_t(T) Y_t^2.$$

We can choose p_t as the kernel weights $p_t(T) = K((T-t)/h) / \sum_{i=1}^{T} K((T-i)/h)$, where $K : [0, \infty) \to \infty$. This leads us to the moving average estimator of the variance. The moving average was defined in (3.14). The special case of the exponential moving average is obtained by choosing $K(x) = \exp(-x) I_{[0,\infty)}(x)$ and is defined for variance estimation in (3.79).

Fan & Gu (2003) present a local likelihood method for the estimation of the conditional variance of stock returns. Let S_t be the asset price at time t and let

$$Y_t = \log \frac{S_t}{S_{t-1}}$$

be the logarithmic return. Let us consider the model

$$Y_t = \theta_t S_{t-1}^{\beta_t} \epsilon_t,$$

where the time varying variance is $\sigma_t^2 = \theta_t^2 S_{t-1}^{2\beta_t}$ and $\epsilon_t \sim N(0, 1)$. We get the likelihood, similarly as in (5.12), as

$$\sum_{t=p+1}^{T} \left(\log_e(\theta_t^2 S_{t-1}^{2\beta_t}) + \frac{Y_t^2}{\theta_t^2 S_{t-1}^{2\beta_t}} \right).$$

Instead of giving θ_t and β_t a parametric form as in ARCH or in GARCH, we can use the local likelihood

$$l_T(\theta, \beta) = \sum_{t=p+1}^{T} p_t(T) \left(\log_e(\theta^2 S_{t-1}^{2\beta}) + \frac{Y_t^2}{\theta^2 S_{t-1}^{2\beta}} \right).$$

Define

$$(\hat{\theta}_T, \hat{\beta}_T) = \mathrm{argmax}_{\theta,\beta} l_T(\theta, \beta).$$

For a given β the maximizer is

$$\hat{\theta}_T^2(\beta) = \sum_{t=1}^{T} p_t(T) Y_t^2 S_{t-1}^{-2\beta}.$$

Thus we need to solve numerically

$$\hat{\beta}_T = \text{argmax}_\beta l_T(\hat{\theta}_T(\beta), \beta).$$

Finally we get the estimator of the conditional variance

$$\hat{\sigma}_t^2 = \hat{\theta}_t^2 S_{t-1}^{2\hat{\beta}_t}.$$

Yu & Jones (2004) proposed the following approach in the fixed design regression model $Y_i = f(x_i) + \sigma(x_i)\epsilon_i$. If $\epsilon_i \sim N(0,1)$, then the log-likelihood is

$$\sum_{i=1}^n \left(\frac{(Y_i - f(x_i))^2}{\sigma^2(x_i)} + \log \sigma^2(x_i) \right).$$

One can use the linear local likelihood and minimize

$$\sum_{i=1}^n p_i(x) \left(\frac{(Y_i - a_0 - a_1(x_i - x))^2}{\exp\{-v_0 - v_1(x_i - x)\}} + v_0 + v_1(x_i - x) \right)$$

over $a_0, a_1, v_0, v_1 \in \mathbf{R}$, where $p_i(x) = K_h(x - x_i)$, $K_h(x) = K(x/h)/h$, $K : \mathbf{R} \to \mathbf{R}$, and $h > 0$.

5.3 SUPPORT VECTOR MACHINES

Let $(X_1, Y_1), \ldots, (X_n, Y_n)$ be regression data for the estimation of the conditional expectation $f(x) = E(Y \mid X = x)$. Let $\mathcal{K} : \mathbf{R}^d \times \mathbf{R}^d \to \mathbf{R}$ be a positive definite kernel.[48] We make the eigenvalue decomposition

$$\mathcal{K}(x, z) = \sum_{i=1}^\infty \delta_i \, g_i(x) g_i(z),$$

where $\delta_i > 0$ are the eigenvalues and $g_i : \mathbf{R}^d \to \mathbf{R}$ are the eigenfunctions. The support vector machine estimator is $\hat{f}(x) = f(x, \hat{w})$, where $\hat{w}$ minimizes

$$\sum_{i=1}^n \gamma\left(Y_i, f(X_i, w)\right) + \lambda \sum_{i=1}^\infty \frac{w_i^2}{\delta_i}$$

over $w = (w_1, w_2, \ldots)$, where $\lambda \geq 0$ is a regularization parameter and

$$f(x, w) = \sum_{i=1}^\infty w_i g_i(x).$$

[48]A positive definite kernel satisfies that for all $a_1, \ldots a_k \in \mathbf{R}$ and $x_1, \ldots, x_k \in \mathbf{R}^d$, $\sum_{i=1}^k \sum_{j=1}^k a_i a_j \mathcal{K}(x_i, x_j) > 0$.

The solution can be written as

$$\hat{f}(x) = \sum_{i=1}^{n} \hat{\pi}_i \mathcal{K}(x, X_i).$$

The set $\{X_i : i = 1, \ldots, n, \ \ \hat{\pi}_i \neq 0\}$ of those observations for which the weight $\hat{\pi}_i$ is nonzero is called the set of support vectors. An equivalent procedure is obtained by defining

$$f(x, \pi) = \sum_{i=1}^{n} \pi_i \mathcal{K}(x, X_i),$$

where $\pi = (\pi_1, \ldots, \pi_n)$. Now we minimize

$$\sum_{i=1}^{n} \gamma\left(Y_i, f(X_i, \pi)\right) + \lambda \sum_{i,j=1}^{n} \pi_i \pi_j \mathcal{K}(X_i, X_j)$$

over π. The estimator is again $\hat{f}(x) = \sum_{i=1}^{n} \hat{\pi}_i \mathcal{K}(x, X_i)$, where $\hat{\pi} = (\hat{\pi}_1, \ldots, \hat{\pi}_n)$ is the minimizer. The useful kernels include the linear kernel $\mathcal{K}(x, z) = x'z$, the polynomial kernels $\mathcal{K}(x, z) = (x'z + a)^p$, the sigmoid $\mathcal{K}(x, z) = \tanh(x'z + a)$, and the radial kernel $\mathcal{K}(x, z) = \exp\{-b\|x - z\|^2\}$, where $a \in \mathbf{R}$, $p \geq 1$ is an integer, and $b > 0$. Vapnik (1995) and Vapnik, Golowich & Smola (1997) define support vector machines for regression function estimation. Wahba (1990) contains information about reproducing Hilbert spaces. A review of support vector machines is given in Hastie et al. (2001, Chapters 5.8 and 12.3.7).

Classification was introduced in Section 1.4. Let us consider the two-class classification problem with classification data $(X_1, Y_1), \ldots, (X_n, Y_n)$, where $X_i \in \mathbf{R}^d$ and $Y_i \in \{-1, 1\}$. We want to find a classifier $g : \mathbf{R}^d \to \{-1, 1\}$. A binary classifier is obtained from a real-valued classifier $h : \mathbf{R}^d \to \mathbf{R}$ by using $g = \operatorname{sign}(h)$. Let us consider functions

$$h(x, w) = \sum_{i=1}^{\infty} w_i h_i(x),$$

where $w = (w_1, w_2 \ldots)$ and $h_1, h_2, \ldots$ are eigenfunctions $\mathbf{R}^d \to \mathbf{R}$ of a positive definite kernel. Define the real-valued classifier as $h(x, \hat{w})$, where $\hat{w}$ is the minimizer of the penalized empirical risk

$$\sum_{i=1}^{n} \gamma\left(Y_i, h(X_i, w)\right) + \lambda \sum_{i=1}^{\infty} \frac{w_i^2}{\delta_i},$$

where δ_i are the eigenvalues of the positive definite kernel and $\lambda \geq 0$ is a regularization parameter. The empirical risk is defined as

$$\gamma(Y_i, h(X_i)) = \phi\left(Y_i h(X_i)\right),$$

where ϕ is the hinge loss $\phi(u) = \max\{0, 1 - u\}$. More generally, $\phi : \mathbf{R} \to (0, \infty)$ is a convex nonincreasing function with $\phi(u) \geq I_{(-\infty,0]}(u)$ for $u \in \mathbf{R}$. In addition to the hinge loss, we can take the exponential loss $\phi(u) = \exp\{-u\}$ or the logit loss $\phi(u) = \log_2(1 + e^{-u})$. The minimizers of the empirical risk over $h \in \mathcal{H}$ are called by Koltchinskii (2008) the large margin classifiers. The product $Yh(X)$ is the margin on (X, Y). If $h(X)Y \geq 0$, then (X, Y) is correctly classified by h; otherwise (X, Y) is misclassified.

5.4 STAGEWISE METHODS

In Section 5.4.1 we present general forward stagewise modeling algorithms. These algorithms are modified to get an algorithm for stagewise fitting of additive models, presented in Section 5.4.2, and to get an algorithm for projection pursuit regression, presented in Section 5.4.3.

5.4.1 Forward Stagewise Modeling

Forward stagewise additive modeling constructs an estimator that is a sum of basis functions. The estimator can be written as

$$\hat{f}(x) = \sum_{m=1}^{M} \hat{w}_m g(x, \hat{\theta}_m),$$

where the basis functions $g(\cdot, \theta) : \mathbf{R}^d \to \mathbf{R}$ are multivariate functions parameterized with the parameter θ, and the weights $\hat{w}_m$ and the parameters $\hat{\theta}_m$ are calculated using regression data $(X_1, Y_1), \ldots, (X_n, Y_n)$. Forward stagewise additive modeling is called sometimes boosting. Tukey (1977) used the name twicing for a forward stagewise additive modeling with $M = 2$ and with the squared error contrast function.

We write below three algorithms. The first algorithm is a general stagewise empirical risk minimization algorithm. The second algorithm specializes to the squared error contrast function. The specialization to the squared error loss helps us to obtain the third algorithm, where empirical risk minimization is replaced by a closed form estimator, like kernel regression estimator.

Algorithm for Stagewise Additive Modeling The estimator is constructed by adding new terms to the expansion so that an error criterion is minimized at each step. The error criterion is defined in terms of a contrast function $\gamma : \mathbf{R} \times \mathbf{R} \to \mathbf{R}$, see Section 5.1.1 for examples of contrast functions. Contrast functions can also be called loss functions. The following algorithm is the algorithm 10.2 from Hastie et al. (2001).

1. Initialize $f_0 = 0$.

2. For $m = 1$ to M:

   (a) Let $(\hat{w}_m, \hat{\theta}_m) = \text{argmin}_{w,\theta} \sum_{i=1}^{n} \gamma\left(Y_i, f_{m-1}(X_i) + wg(X_i, \theta)\right)$.

   (b) Let $f_m(x) = f_{m-1}(x) + \hat{w}_m g(x, \hat{\theta}_m)$.

3. Let the final estimate be $\hat{f}(x) = f_M(x)$.

An example of basis function $g(x, \theta)$ is a piecewise constant function that has a tree representation; and in this example, parameter θ determines the split points, the split variables, and the values of the function. The simplest example is the case of a regressogram with a two set partition, when the partition is made parallel to a coordinate axes. Such a regressogram is called a stump. Regressograms with a tree representation are discussed in Section 5.5.1.

Algorithm for Stagewise Additive Modeling with the Squared Error Contrast Function The squared error contrast function is $\gamma(y, f(x)) = (y - f(x))^2$. For this contrast function we can write the algorithm for forward stagewise additive estimation in the following way.

1. Initialize $f_0 = 0$.
2. For $m = 1$ to M:
   (a) Calculate the residuals $r_i = Y_i - f_{m-1}(X_i)$, $i = 1, \ldots, n$.
   (b) Let $(\hat{w}_m, \hat{\theta}_m) = \text{argmin}_{w,\theta} \sum_{i=1}^n (r_i - wg(X_i, \theta))^2$.
   (c) Let $f_m(x) = f_{m-1}(x) + \hat{w}_m g(x, \hat{\theta}_m)$.
3. Let the final estimate be $\hat{f}(x) = f_M(x)$.

Algorithm for Stagewise Additive Modeling with Nonparametric Base Learners We can generalize the previous algorithm so that at each step an arbitrary regression estimator is fitted to such regression data where the residual error is the response variable. This generalization allows to use kernel regression estimators in stagewise fitting. Instead of using empirical risk minimization to choose new additive components we can use a kernel estimator. We assume to have a method for regression function estimation which produces an estimator $\hat{g}$, based on regression data (X_i, r_i), $i = 1, \ldots, n$, where $r_i \in \mathbf{R}$ and $X_i \in \mathbf{R}^d$.

1. Initialize $f_0 = 0$.
2. For $m = 1$ to M:
   (a) Calculate the residuals $r_i = Y_i - f_{m-1}(X_i)$, $i = 1, \ldots, n$.
   (b) Find estimate $\hat{g}_m$ using the data (X_i, r_i), $i = 1, \ldots, n$.
   (c) Let $f_m = f_{m-1} + \hat{g}_m$.
3. The final estimator is $\hat{f} = f_M = \sum_{m=1}^M \hat{g}_m$.

The above algorithm can be modified to obtain stagewise fitting of additive models.

5.4.2 Stagewise Fitting of Additive Models

We have defined additive models in Section 4.2, where a backfitting algorithm was described. Algorithms in Section 5.4.1 lead to an estimate $\hat{f}$ which has the additive structure when the base learners are a function of one variable. We say that a regression function estimate $\hat{f} : \mathbf{R}^d \to \mathbf{R}$ has the additive structure if it can be written as

$$\hat{f}(x) = \sum_{k=1}^{d} \hat{f}_k(x_k), \qquad x \in \mathbf{R}^d, \tag{5.13}$$

for some univariate functions $\hat{f}_k : \mathbf{R} \to \mathbf{R}$. An example of a base learner depending only on one variable is a stump: a regressogram with a two-set partition, when the partition is made by a split parallel to a coordinate axis. The difference from the additive estimate obtained by backfitting in Section 4.2 is that the additive components are obtained by adding new terms to the previous component, instead of replacing the previous component. We write a further algorithm to construct an estimator with the additive structure.

Algorithm for Stagewise Additive Modeling with Univariate Base Learners The next algorithm was proposed by Bühlmann & Yu (2003), who used univariate smoothing splines as base learners. We can use also a one-dimensional kernel estimators as base learners.

1. Initialize $f_0 = 0$.
2. For $m = 1$ to M:
   (a) Compute the residuals $r_i = Y_i - f_{m-1}(X_i)$, $i = 1, \ldots, n$.
   (b) Calculate one-dimensional kernel estimators $\hat{g}_m^{(k)}$, $k = 1, \ldots, d$, using regression data (X_{ik}, r_i), $i = 1, \ldots, n$, where we write $X_i = (X_{i1}, \ldots, X_{id})$.
   (c) Define $\hat{d}_m = \mathrm{argmin}_{k=1,\ldots,d} \sum_{i=1}^{n} \left(r_i - \hat{g}_m^{(k)}(X_{ik}) \right)^2$.
   (d) Define $\hat{g}_m = \hat{g}_m^{(\hat{d}_m)}$.
   (e) Let $f_m = f_{m-1} + \hat{g}_m$.
3. The final estimator is $\hat{f} = f_M = \sum_{m=1}^{M} \hat{g}_m$.

It is useful to implement the algorithm in such a way that when we run the M steps of the algorithm we save the sequence $\hat{d}_1, \ldots, \hat{d}_M$ of the variable indexes, and save the sequence $r^{(1)}, \ldots, r^{(M)}$ of the residual vectors of length n. We get as a byproduct the evaluations of the estimator at the observations $X_1, \ldots, X_n$.

Now we can get the evaluation at an arbitrary point $x \in \mathbf{R}^d$. First we obtain the points $x_{\hat{d}_m}$ and the sequences $Z^{(m)} = (X_{1,\hat{d}_m}, \ldots, X_{n,\hat{d}_m})$, for $m = 1, \ldots, M$. Then we evaluate the kernel estimator at the point $x_{\hat{d}_m}$, when the kernel estimators is constructed using regression data $(Z_i^{(m)}, r_i^{(m)})$, $i = 1, \ldots, n$. Finally we sum those evaluations over $m = 1, \ldots, M$.

5.4.3 Projection Pursuit Regression

The projection pursuit regression finds an estimate of the form

$$\hat{f}(x) = \sum_{m=1}^{M} \hat{h}_m(\hat{\theta}'_m x), \qquad x \in \mathbf{R}^d, \tag{5.14}$$

where $\hat{h}_m : \mathbf{R} \to \mathbf{R}$ are univariate functions and $\hat{\theta}_m \in \mathbf{R}^d$ are projection vectors with $\|\hat{\theta}_m\| = 1$. Functions $x \mapsto \hat{h}_m(\theta'_m x)$ are called ridge functions. The algorithms of stagewise additive modeling of Section 5.4.1 can be applied and modified to construct projection pursuit estimators. In these algorithms, new terms $h_m(\theta'_m x)$ are added in a forward stagewise manner. General projection pursuit dates back to Friedman & Tukey (1974), and projection pursuit regression was studied in Friedman & Stuetzle (1981).

The estimator of the type (5.14) can be considered as a generalization of single-index modeling, discussed in Section 4.1. In a single-index model the estimator is $\hat{f}(x) = \hat{g}(\hat{\theta}' x)$, but now we have a sum of such of functions. The estimator of the type (5.14) can also be considered as a generalization of an estimator of the additive structure as in (5.13). Indeed, we can write an estimator with the additive structure as

$$\hat{f}(x) = \sum_{k=1}^{d} \hat{f}_k(e'_k x), \qquad x \in \mathbf{R}^d,$$

where $e_k \in \mathbf{R}^d$ is the vector whose kth component is one, and the other components are zero. Vectors e_k project to the coordinate axes. The estimator of the projection pursuit type (5.14) is obtained when the vectors e_k are replaced by arbitrary projection vectors and it is allowed to include more (or less) than d terms to the sum.

Algorithm for Projection Pursuit Regression The algorithm is similar to the stagewise additive modeling algorithms of Section 5.4.1.

1. Initialize $f_0 = 0$.
2. For $m = 1$ to M:
   (a) Compute the residuals $r_i = Y_i - f_{m-1}(X_i)$, $i = 1, \ldots, n$.
   (b) Estimate $\hat{\theta}_m$ using single-index estimation and regression data (X_i, r_i), $i = 1, \ldots, n$.
   (c) Let $Z_i = \hat{\theta}'_m X_i$ and calculate one dimensional regression function $\hat{h}_m$ using regression data (Z_i, r_i), $i = 1, \ldots, n$.
   (d) Define $\hat{g}_m(x) = \hat{h}_m(\hat{\theta}'_m x)$.
   (e) Let $f_m = f_{m-1} + \hat{g}_m$.
3. The final estimator is

$$\hat{f}(x) = f_M(x) = \sum_{m=1}^{M} \hat{g}_m(x) = \sum_{m=1}^{M} \hat{h}_m(\hat{\theta}'_m x).$$

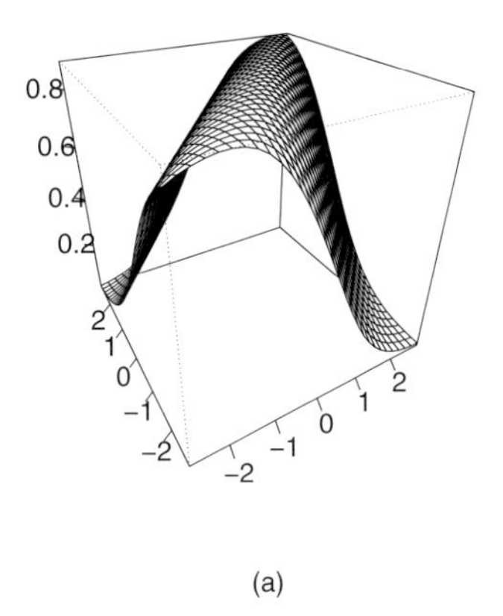

(a)

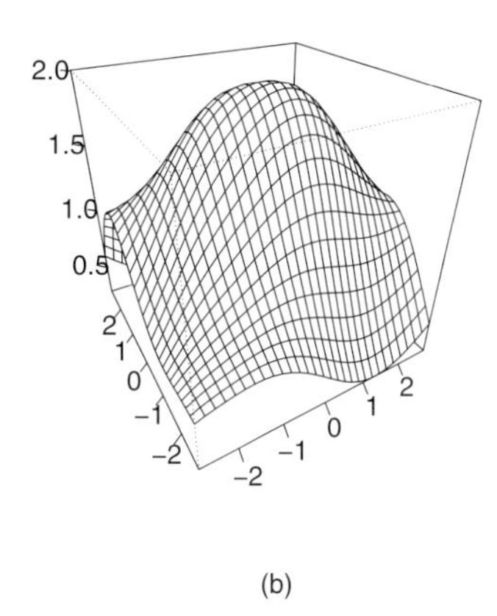

(b)

Figure 5.8 *Projection pursuit estimates.* Shown are projection pursuit estimates for step numbers (a) $M = 1$ and (b) $M = 2$.

The algorithm uses single-index estimation at each stage to estimate the projection vectors $\hat{\theta}_m$. Several methods to estimate $\hat{\theta}_m$ were presented in Section 4.1.2: minimization estimators, derivative, and average derivative estimators. The algorithm uses univariate regression at each stage, and we can choose kernel regression estimator, for example.

It is useful to implement the algorithm in such a way that during the execution of the M steps of the algorithm we save the sequence $\hat{\theta}_1, \ldots, \hat{\theta}_M$ of the projection vectors, and save the sequence $r^{(1)}, \ldots, r^{(M)}$ of the residual vectors of length n. We get as a byproduct the evaluations of the estimator at the observations $X_1, \ldots, X_n$.

Now we can get the evaluation at an arbitrary point $x \in \mathbf{R}^d$. First we calculate the points $\hat{\theta}_m' x$ and calculate the vectors $Z^{(m)} = \mathbf{X}\hat{\theta}_m$ of length n, for $m = 1, \ldots, M$, where $\mathbf{X}$ is the $n \times d$ matrix of the observed values of the explanatory variables. Then we evaluate the kernel estimators at the points $\hat{\theta}_m' x$, when the kernel estimators are constructed using regression data $(Z_i^{(m)}, r_i^{(m)})$, $i = 1, \ldots, n$. Finally we sum the evaluations over $m = 1, \ldots, M$.

Figure 5.8 shows projection pursuit estimates for step numbers $M = 1$ and $M = 2$. The regression function is the standard two-dimensional normal density function. The vector of explanatory variables $X = (X_1, X_2)$ has the standard normal distribution, and $Y = f(X) + \epsilon$, where $\epsilon \sim N(0, 0.1^2)$. We have $n = 1000$ observations. We use the projection pursuit regression, where the derivative method as in (4.8) is used to find the direction vectors. In the derivative method it is useful to choose the point where the derivative is estimated randomly from the observed values of the explanatory variables. For $M = 1$ the estimate has the single-index structure, but for $M = 2$ the estimate is already close to a spherically symmetric function.

5.5 ADAPTIVE REGRESSOGRAMS

A regressogram is a piecewise constant regression function estimate, defined as in Section 3.1. A regressogram is constructed by first finding a partition in the space of the explanatory variables, and then the value of the estimate is taken to be the mean of the y-values in each set of the partition. Adaptive regressograms are regressograms where the partition of the space of explanatory variables is chosen empirically. We consider mainly partitions that are found by a recursive partition scheme.

Let $(X_1, Y_1), \ldots, (X_n, Y_n)$ be identically distributed regression data. A regressogram is

$$\hat{f}(x, \mathcal{P}) = \sum_{R \in \mathcal{P}} \hat{Y}_R I_R(x), \qquad x \in \mathbf{R}^d, \tag{5.15}$$

where $\mathcal{P}$ is a partition of $\mathbf{R}^d$ and is the average of the response variables whose corresponding explanatory variable is in R:

$$\hat{Y}_R = \frac{1}{\#\{X_i \in R\}} \sum_{i: X_i \in R} Y_i. \tag{5.16}$$

We consider in Section 5.5.1 greedy regressograms, which are such that the partition of the space of explanatory variables is found by a myopic stepwise algorithm, where the space is splitted recursively to finer sets. We cover two estimation problems: the local estimation problem, where we find the estimate of the regression function at one point, and the global estimation problem, where we find the estimate at all points simultaneously. The recursive splitting method is the same both in local and global problems, but the stopping rule for the splitting is typically different. Thus the collection of all local estimates leads typically to an estimate different from the global estimate.

Section 5.5.2 describes the CART method. The CART method starts with a construction of a greedy regressogram but the final choice of the partition is made from a sequence of pruned subtrees. Section 5.5.3 describes the dyadic CART method. In this method the choice of the partition is made from a deterministically defined sequence of partitions. Section 5.5.4 describes bootstrap aggregation.

5.5.1 Greedy Regressograms

A greedy partition is a partition of the space of the explanatory variables which is found by a stepwise algorithm, which recursively splits the space to finer sets. This algorithm is called greedy, or stepwise, because we do not try to find a global minimum for the optimization problem but find the optimizer one step at a time. Morgan & Sonquist (1963) presented this type of algorithm, although they did not restrict themselves to binary splits but allowed a large number of splits to be made simultaneously.

First we define the split points over which we search the best splits. The splits are made parallel to the coordinate axes, and thus we have to define a grid of possible split points for each direction. Let us denote the sets of possible split points by

$$\mathcal{G}_1, \ldots, \mathcal{G}_d, \tag{5.17}$$

where $\mathcal{G}_k \subset \mathbf{R}$ is a finite grid of split points in direction k. A natural possibility for choosing $\mathcal{G}_k$ is to take it to be the collection of the midpoints of the coordinates of the observations: $\mathcal{G}_k = \{Z_1^k, \ldots, Z_{n-1}^k\}$, $k = 1, \ldots, d$, where Z_i^k is the midpoint of $X_{(i)}^k$ and $X_{(i+1)}^k$:

$$Z_i^k = \frac{1}{2}\left(X_{(i)}^k + X_{(i+1)}^k\right),$$

where $X_{(1)}^k, \ldots, X_{(n)}^k$ is the order statistic of the kth coordinate of the observations $X_1, \ldots, X_n$. This choice of possible split points guarantees that all the cells, even at the finest resolution level, contain observations.

When rectangle $R \subset \mathbf{R}^d$ is splitted through the point $s \in \mathbf{R}$ in direction $k = 1, \ldots, d$, then we obtain sets

$$R_{k,s}^{(0)} = \{(x_1, \ldots, x_d) \in R : x_k \le s\} \tag{5.18}$$

and

$$R_{k,s}^{(1)} = \{(x_1, \ldots, x_d) \in R : x_k > s\}. \tag{5.19}$$

The split point s satisfies

$$s \in S_{R,k} \stackrel{def}{=} \mathcal{G}_k \cap \text{proj}_k(R), \tag{5.20}$$

where $\text{proj}_k(R) = R_k$, when $R = R_1 \times \cdots \times R_d$.

Pointwise Estimator Now we consider the case of estimating the regression function only at one point $x \in \mathbf{R}^d$. We want to find a rectangle $R_x \subset \mathbf{R}^d$ such that $x \in R_x$ and use estimate

$$\hat{f}(x) = \hat{Y}_{R_x},$$

where $\hat{Y}_{R_x}$ is defined in (5.16). The neighborhood R_x is chosen as one of the rectangles in the sequence of rectangles found by the following procedure.

- Start with the whole space and make splits as long as the obtained rectangle contains a sufficient number of observations.
- Split at each step the rectangle so that the empirical risk of the corresponding regressogram is minimized. The empirical risk is the sum of squared residuals of the regressogram.
- At each step the minimization is done over all directions and over all split points in the given direction.

We describe the splitting procedure more precisely in the next definition. The splitting is restricted so that we do not split rectangles which contain less observations than a given minimal number m of observations.

Let us define a sequence of greedy neighborhoods $\mathbf{R}^d \supset R_0 \supset R_1 \supset \cdots \supset R_M$, with the minimal observation number $m \geq 1$. The sequence is defined recursively by the following rules. Start with set $R_0 = \mathbf{R}^d$. Assume that we have found sets $R_0, \ldots, R_L$, where $L \geq 0$.

1. If $\#\{X_i \in R_L\} \leq m$, then we choose $M = L$ and stop splitting.

2. If $\#\{X_i \in R_L\} > m$, then we split R_L in the following way. Let

$$I_{R_L} = \{(k, s) : k = 1, \ldots, d,\ s \in S_{R_L,k}\},$$

where $S_{R,k}$ is the set of split points defined in (5.20). We construct new sets $R^{(0)}_{\hat{k},\hat{s}}$ and $R^{(1)}_{\hat{k},\hat{s}}$, where we use the notation defined in (5.18) and (5.19), and

$$\left(\hat{k}, \hat{s}\right) = \text{argmin}_{(k,s)\in I_{R_L}} ERR\left(R_L, k, s\right), \tag{5.21}$$

where

$$\text{ERR}(R, k, s) = \sum_{i:X_i \in R^{(0)}_{k,s}} \left(Y_i - \hat{Y}_{R^{(0)}_{k,s}}\right)^2 + \sum_{i:X_i \in R^{(1)}_{k,s}} \left(Y_i - \hat{Y}_{R^{(1)}_{k,s}}\right)^2 \tag{5.22}$$

and $\hat{Y}_R$ is defined in (5.16). Finally, the new set is chosen as $R_{L+1} = R^{(0)}_{\hat{k},\hat{s}}$ if $x \in R^{(0)}_{\hat{k},\hat{s}}$ and as $R_{L+1} = R^{(1)}_{\hat{k},\hat{s}}$ otherwise.

The greedy pointwise regressogram is defined for $x \in \mathbf{R}^d$ by choosing a greedy neighborhood

$$R_x \in \{R_0, \ldots, R_M\},$$

where $R_0, \ldots, R_M$ is a sequence of greedy neighborhoods. The greedy regressogram is defined as

$$\hat{f}(x) = \frac{1}{\#\{X_i \in R_x\}} \sum_{i:X_i \in R_x} Y_i, \qquad x \in \mathbf{R}^d.$$

Global Estimator Now we consider the case of estimating the regression function globally at every point $x \in \mathbf{R}^d$. We want to find a partition $\mathcal{P}$ of $\mathbf{R}^d$ and use the regressogram $\hat{f}(x, \mathcal{P})$. A greedy partition $\mathcal{P}$ is one of the partitions in the sequence of partitions found by the following procedure.

- Start with the partition $\{\mathbf{R}^d\}$ and split the rectangles of the partition as long as some rectangle contains a sufficient number of observations.

- Make splits so that the empirical risk of the corresponding regressogram is minimized. The minimization is done over all rectangles in the current partition, over all directions, and over all split points in the given rectangle and in the given direction.

We describe the procedure more precisely in the following. We restrict the growing of the partition so that we do not split rectangles which contain less observations than a given threshold. The partition is grown by minimizing an empirical risk of the estimator, which is typically defined as the sum of the squared errors of the estimator $\hat{f}$. We say that partition $\mathcal{P}$ is grown if it is replaced by partition

$$\mathcal{P}_{R,k,s} = \mathcal{P} \setminus \{R\} \cup \left\{ R_{k,s}^{(0)}, R_{k,s}^{(1)} \right\}, \tag{5.23}$$

where rectangle $R \in \mathcal{P}$ is splitted in direction $k = 1, \ldots, d$ through the point $s \in S_{R,k}$.

A sequence of greedy partitions $\mathcal{P}_1, \ldots, \mathcal{P}_M$, with minimal observation number $m \geq 1$, is defined recursively by the following rules. Start with the partition $\mathcal{P}_1 = \{R\}$, where $R = \mathbf{R}^d$. Assume that we have constructed partitions $\mathcal{P}_1, \ldots, \mathcal{P}_L$, where $L \geq 1$.

1. If all $R \in \mathcal{P}_L$ satisfy $\#\{X_i \in R\} \leq m$, then partition $\mathcal{P}_L$ is the final partition.

2. Otherwise, we construct next partition $\mathcal{P}_{\hat{R},\hat{k},\hat{s}}$, where

$$\left(\hat{R}, \hat{k}, \hat{s}\right) = \text{argmin}_{(R,k,s)\in I} \sum_{i=1}^{n} \left(Y_i - \hat{f}(X_i, \mathcal{P}_{R,k,s})\right)^2, \tag{5.24}$$

where

$$I = \{(R, k, s) : R \in \mathcal{P}_L, \ \#\{X_i \in R\} \geq m, \ k = 1, \ldots, d, \ s \in S_{R,k}\},$$

$S_{R,k}$ is the set of split points defined in (5.20), $\mathcal{P}_{R,k,s}$ is the partition defined in (5.23), and $\hat{f}(\cdot, \mathcal{P})$ is the regressogram defined in (3.4).

Let

$$\hat{\mathcal{P}} \in \{\mathcal{P}_1, \ldots, \mathcal{P}_M\},$$

be a greedy partition, where $\mathcal{P}_1, \ldots, \mathcal{P}_M$ is a sequence of greedy partitions. The greedy regressogram is defined by

$$\hat{f} = \hat{f}\left(\cdot, \hat{\mathcal{P}}\right),$$

where $\hat{f}$ is defined in (3.4). We can use sample splitting to find a good partition $\hat{\mathcal{P}}$ and thus a good regressogram. Let $n^* = [n/2]$ and use the data (X_i, Y_i), $i = 1, \ldots, n^*$, to construct the sequence $\mathcal{P}_1, \ldots, \mathcal{P}_M$ and the corresponding sequence of estimators $\hat{f}_1, \ldots, \hat{f}_M$. Then we calculate for each estimate the sum of squared residuals using the second part of the data:

$$\text{SSR}_m = \sum_{i=n^*+1}^{n} \left(Y_i - \hat{f}_m(X_i)\right)^2, \qquad m = 1, \ldots, M.$$

The final estimate is $\hat{f}_{\hat{m}}$, where $\hat{m} = \text{argmin}_{m=1,\ldots,M} \text{SSR}_m$.

5.5.2 CART

The CART (classification and regression tress) procedure was introduced in Breiman et al. (1984). In Section 5.5.1 a sequence of partitions was constructed in a stepwise manner and then one partition was selected from this sequence, using sample splitting. CART constructs the sequence of partitions in a different way. First a fine partition is grown with stepwise optimization and then the sequence of partitions is found by a complexity penalized pruning. The new way of constructing the sequence opens up the possibility for using cross validation to choose the final partition, instead of sample splitting. Also, the complexity penalized pruning may increase the quality of partitions in the sequence. In contrast to dyadic CART, the large partition $\mathcal{P}^*$ is now data-dependent. Otherwise the final estimate is obtained analogously as in dyadic CART, by minimizing a complexity penalized sum of squared residuals. The CART sequence is found by the following steps.

1. Choose a large partition $\mathcal{P}^*$. This partition is the largest partition $\mathcal{P}_M$ from the sequence of greedy partitions defined in Section 5.5.1 for the global estimator.

2. For $\alpha \geq 0$, let

$$\mathcal{P}_\alpha = \mathrm{argmin}_{\mathcal{P} \subset \mathcal{P}^*} \left[\sum_{i=1}^{n} \left(Y_i - \hat{f}(X_i, \mathcal{P}) \right)^2 + \alpha \cdot \#\mathcal{P} \right], \tag{5.25}$$

   where $\hat{f}$ denotes a regressogram as defined in (3.4). For $\alpha = 0$ we have $\mathcal{P}_\alpha = \mathcal{P}^*$, and for large enough α we obtain $\mathcal{P}_\alpha = \{\mathbf{R}^d\}$. Since there are a finite number of subsets of $\mathcal{P}^*$, there are a finite number of values $0 = \alpha_1 < \cdots < \alpha_M$ such that

$$\mathcal{P}_\alpha = \mathcal{P}_{\alpha_i}, \text{ when } \alpha_i \leq \alpha < \alpha_{i+1}, \tag{5.26}$$

   for $i = 1, \ldots, M$, and we denote $\alpha_{M+1} = \infty$. Now $\mathcal{P}_{\alpha_1} = \mathcal{P}^*$ and $\mathcal{P}_{\alpha_M} = \{\mathbf{R}^d\}$.

A sequence of CART partitions

$$\mathcal{P}_1, \ldots, \mathcal{P}_M \tag{5.27}$$

is defined, with an abuse of notation, by $\mathcal{P}_i = \mathcal{P}_{\alpha_i}$, $i = 1, \ldots, M$, where $\alpha_1, \ldots, \alpha_M$ is defined by (5.26).

We can use cross validation to find a good partition and thus a good regressogram. In the case of greedy partitions we had to use sample splitting, that is, twofold cross validation, but in the case of CART partitions the penalization parameter α can be used to connect different partitions and we can use K-fold cross validation for $2 \leq K \leq n$. Let us denote by $I_1, \ldots, I_K$ a partition of the index set $\{1, \ldots, n\}$, where $2 \leq K \leq n$. For example, we can partition observations into $K = 10$ subsets (ten-fold cross validation). We can partition the observations at most to n subsets and at least to two subsets. Observations (X_i, Y_i), $i \notin I_k$, are used to construct sequence

$\hat{f}_{\alpha_{1,k}}, \ldots, \hat{f}_{\alpha_{M_k,k}}$, where $\alpha_{1,k} < \cdots < \alpha_{M_k,k}$, $k = 1, \ldots, K$. For each estimate in the sequence we calculate the average of squared residuals (ASR) using (X_i, Y_i), $i \in I_k$:

$$\text{ASR}_{j,k} = \frac{1}{\# I_k} \sum_{i \in I_k} \left(Y_i - \hat{f}_{\alpha_{j,k}}(X_i) \right)^2, \qquad j = 1, \ldots, M_k, \; k = 1, \ldots, K.$$

Finally we use the complete data to find a sequence $\hat{f}_{\alpha_1}, \ldots, \hat{f}_{\alpha_M}$ and a grid $\alpha_1, \ldots, \alpha_M$. We make a partition of $(0, \infty) = U_{m=1}^{M} A_m$, where $\alpha_m \in A_m$ and estimate

$$\text{ASR}_{\alpha_m} = \frac{\sum \{ SSR_{j,k} : \alpha_{j,k} \in A_m \}}{\# \{ (j,k) : \alpha_{j,k} \in A_m \}}, \qquad m = 1, \ldots, M.$$

The final estimate is $\hat{f}_{\alpha_{\hat{m}}}$, where $\hat{m} = \text{argmin}_{m=1,\ldots,M} \text{ASR}_{\alpha_m}$.

We need two algorithms to find the sequence $\mathcal{P}_1, \ldots, \mathcal{P}_M$: a growing algorithm for growing the large partition $\mathcal{P}^*$ and a pruning algorithm for producing the sequence from this large partition. Both algorithms use the fact that the partitions which we consider can be represented as binary trees, where the rectangles of the partition are the nodes of the tree. The representation as a binary tree follows from the stepwise splitting procedure. We take the whole space to be the root of the tree. After that, when a node (a rectangle) is splitted, the two obtained rectangles are taken to be the child nodes of the splitted node.

We choose $\mathcal{P}^*$ as the largest partition $\mathcal{P}_M$ from the sequence of greedy partitions defined in Section 5.5.1 for the global estimator. We can now use a faster algorithm to obtain $\mathcal{P}^*$ than the algorithm described in Section 5.5.1, since this algorithm uses unnecessary time to optimize the order in which the partition is grown, and we are interested only in the final partition and not in the intermediate partitions. Thus we can use an algorithm based on the following recursion. Let the minimal observation number be $m \geq 1$.

1. Start with the partition $\mathcal{P} = \{\mathbf{R}^d\}$. The rectangle $\mathbf{R}^d$ is taken to be the root node of the initial binary tree.

2. Assume that we have constructed partition $\mathcal{P}$. This partition is interpreted as a binary tree.

   (a) If all child nodes $R \in \mathcal{P}$ satisfy $\#\{X_i \in R\} \leq m$, then we finish the splitting.

   (b) Otherwise, choose a child node $R \in \mathcal{P}$ with $\#\{X_i \in R\} > m$. Construct new partition $\mathcal{P}_{R,\hat{k},\hat{s}}$, where

$$\left(\hat{k}, \hat{s} \right) = \text{argmin}_{(k,s) \in I_R} \sum_{i=1}^{n} \left(Y_i - \hat{f}(X_i, \mathcal{P}_{R,k,s}) \right)^2,$$

   where $I_R = \{(k, s) : k = 1, \ldots, d, \; s \in S_{R,k}\}$, $S_{R,k}$ is the set of split points defined in (5.20), $\mathcal{P}_{R,k,s}$ is the partition defined in (5.23), and

$\hat{f}(\cdot, \mathcal{P})$ is the regressogram defined in (3.4). Partition $\mathcal{P}_{R,\hat{k},\hat{s}}$ is interpreted as a binary tree, where rectangle $R^{(0)}_{\hat{k},\hat{s}}$ is the left child node of node R and rectangle $R^{(1)}_{\hat{k},\hat{s}}$ is the right child node of node R, where we use the notation of (5.23).

After growing the large partition $\mathcal{P}^*$ we need an algorithm to find the CART sequence of (5.27). To solve for a given α the complexity penalized minimization problem (5.25), we can use a dynamic programming algorithm which starts at the leaves of the binary tree T^* corresponding to $\mathcal{P}^*$. If t is a node of T^*, denote the sum of squared residuals associated with this node by

$$\mathrm{ssr}(t) = \sum_{i:X_i \in R_t} \left(Y_i - \bar{Y}_{R_t}\right)^2,$$

where R_t is the rectangle associated with node t. Denote with $Q(t)$ the sum of $\mathrm{ssr}(t')$ over the leafs t' of the subtree T_t whose root is t. Starting at the leaf nodes, we compare at each node t whether

$$Q(t) + \alpha \cdot \#T_t < \mathrm{ssr}(t) + \alpha, \tag{5.28}$$

where $\#T_t$ is the number of leaves in the subtree T_t. If this holds, then the subtree whose root is t should be kept, because the complexity penalized error is smaller than obtained by making t a leaf node. Otherwise, the tree is pruned at node t, and t is made a leaf node. The value $Q(t)$ can be calculated during the pruning process.

To extend this idea to find the complete CART sequence and the corresponding values $\alpha_1, \ldots, \alpha_M$, note that we have for every nonterminal node t of T^* that $Q(t) < \mathrm{ssr}(t)$. As long as (5.28) holds, branch T_t has a smaller error-complexity than the single node $\{t\}$; but at some critical value of α, the two error-complexities become equal. At this point the subbranch $\{t\}$ is smaller than T_t, has the same error-complexity, and is therefore preferable. To find this α, solve (5.28) to get

$$\alpha < \frac{\mathrm{ssr}(t) - Q(t)}{|T_t| - 1}.$$

The algorithm is based on finding the "weakest links," which are the nodes minimizing

$$g_k(t) = \begin{cases} \frac{\mathrm{ssr}(t) - Q(t)}{|T_t| - 1}, & t \text{ is not a leaf in } T_k, \\ \infty, & t \text{ is a leaf in } T_k, \end{cases} \tag{5.29}$$

$k = 1, \ldots, K$. Let $t_1 = \mathrm{argmin}_{t \in T_0} g_0(t)$, $T_0 = T^*$. Then t_1 is the root node and $\alpha_1 = g_0(t_1) = 0$. Let T_1 be the subtree of T^* obtained by making t_1 a leaf node. We continue in this way: $t_k = \mathrm{argmin}_{t \in T_{k-1}} g_{k-1}(t)$ and $\alpha_k = g_{k-1}(t_k)$ for $k = 1, \ldots, M$.[49] We get a sequence where $\alpha_1 = 0$, and for $\alpha_k \le \alpha < \alpha_{k+1}$ the

[49] If at any stage there is a multiplicity of weakest links– for example, $g_{k-1}(t_k) = g_{k-1}(t'_k)$–then define $T_k = T_{k-1} - T_{t_k} - T_{t'_k}$.

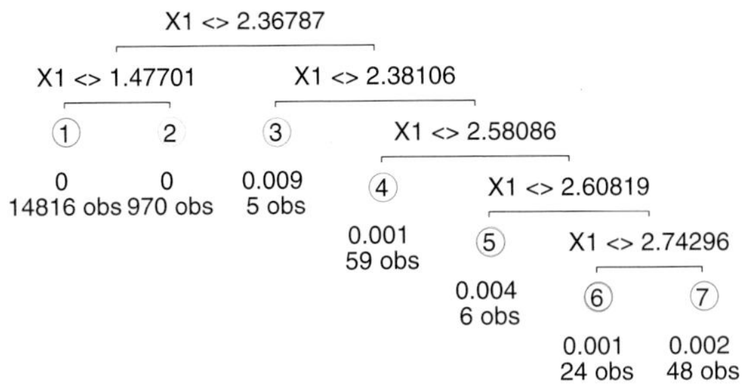

X1 <> 2.36787
X1 <> 1.47701
X1 <> 2.38106
1
2
3
X1 <> 2.58086
0
14816 obs
0
970 obs
0.009
5 obs
4
X1 <> 2.60819
0.001
59 obs
5
X1 <> 2.74296
0.004
6 obs
6
7
0.001
24 obs
0.002
48 obs

Figure 5.9 *A CART estimate in volatility prediction.* A tree plot of a CART estimate when the response variable is the squared S&P 500 return and there are two explanatory variables.

corresponding partition $\mathcal{P}_{\alpha_k}$ is the collection of rectangles associated with the leaf nodes of tree T_k.

We illustrate CART estimator with a two-dimensional volatility prediction estimator. Let us define

$$Y_t = R_t^2, \qquad X_t = \left(\sum_{i=t-k_1}^{t-1} R_i^2, \sum_{i=t-k_1-k_2}^{t-k_1-1} R_i^2 \right),$$

where R_t is the net S&P 500 return and $k_1 = k_2 = 5$. We make the copula transform to the standard Gaussian margins of the data X_t. The S&P 500 data are described in Section 1.6.1.

Figure 5.9 shows a CART estimate based on the regression data (X_t, Y_t).[50] The estimate has made splits only with respect X_1. The numbers below the terminal nodes show the value of the estimate and the number of observations occurring in that terminal node. Figure 3.25 shows a perspective plot of the kernel density estimate, defined in (3.86), for the same regression data.

5.5.3 Dyadic CART

Dyadic CART was introduced in Donoho (1997), in the two-dimensional case $d = 2$, for the fixed equidistant design. Let $f : [0, 1]^2 \to \mathbf{R}$ and

$$Y_i = \bar{f}(i) + \sigma\, \epsilon_i, \tag{5.30}$$

[50] We have used function "tree" from R-package "tree." For drawing we have used function "draw.tree" from R-package "maptree."

where $i = (i_1, i_2)$ are fixed equispaced design points, $i_1, i_2 = 0, \ldots, m-1$, m is dyadic (an integral power of 2), $\bar{f}(i)$ is the cell average over the cell C_i: $\bar{f}(i) = \int_{C_i} f/\text{volume}(C_i)$, with $C_i = [i_1/m, (i_1+1)/m) \times [i_2/m, (i_2+1)/m)$. Furthermore, ϵ_i are identically distributed with mean zero and unit variance, and $\sigma > 0$. The number of observations is $n = m^2$.

Dyadic CART can be defined in two steps.

1. Let $\mathcal{P}^*$ be the largest possible dyadic partition. A dyadic partition is a partition that is obtained by midpoint splits of $[0,1]^2$. When the side length of a rectangle is m^{-1}, then this side is not allowed to be split. Thus the largest dyadic partition consists of the rectangles with volume m^{-2}.

2. Let $\mathcal{P}_\alpha$ be the partition of $[0,1]^2$ that minimizes

$$\sum_{i_1, i_2 = 1}^{m} \left(Y_i - \hat{f}_n(x_i, \mathcal{P}) \right)^2 + \alpha \cdot \#\mathcal{P}$$

among all dyadic subpartitions of $\mathcal{P}^*$, where x_i is the midpoint of cell C_i, $\hat{f}(\,\cdot\,, \mathcal{P})$ is the regressogram with partition $\mathcal{P}$, $\alpha > 0$ and $\#\mathcal{P}$ is the cardinality of partition $\mathcal{P}$. Define the dyadic CART estimator by

$$\hat{f}_n(x) = \hat{f}_n(x, \mathcal{P}_\alpha).$$

Donoho (1997) proposes an algorithm with $O(n)$ steps for the calculation of the optimal partition, where $n = m^2$ is the number of observations. The dyadic CART estimator has the optimal convergence rate in anisotropic Nikolskii smoothness classes, as proved in Donoho (1997) when ϵ_i are independent and identically distributed Gaussian random variables. A regression estimate for random design regression based on similar ideas than the Dyadic CART estimate but using piecewise polynomials was analyzed for univariate data in Kohler (1999).

5.5.4 Bootstrap Aggregation

Adaptive regressograms (greedy and CART regressograms) give piecewise constant estimates of low granularity. It is possible to increase the granularity of these estimates by bootstrap aggregation. In bootstrap aggregation B bootstrap samples are generated from the original sample $(X_1, Y_1), \ldots, (X_n, Y_n)$. For example, we can use one of the following methods.

1. In *n-out-of-n with replacement bootstrap* we take B bootstrap samples of size n with replacement from the original sample $(X_1, Y_1), \ldots, (X_n, Y_n)$. Some observations may appear more than once in the bootstrap sample and some observations may be missing.

2. In *$n/2$-out-of-n without replacement bootstrap* we take B bootstrap samples of size $[n/2]$ without replacement. Every observation appears at most once in the bootstrap sample.

We call the following procedure the bootstrap aggregation.

1. Generate B bootstrap samples from the original sample.

2. Calculate adaptive regressogram $\hat{f}_j$, $j = 1, \ldots, B$, based on each bootstrap sample. We may choose the adaptive regressograms to be greedy regressograms defined in Section 5.5.1 or CART regressograms defined in Section 5.5.2.

3. Define the regression function estimator $\hat{f}$ as the arithmetic mean of the estimators $\hat{f}_j$:
$$\hat{f}(x) = \frac{1}{B} \sum_{j=1}^{B} \hat{f}_j(x), \qquad x \in \mathbf{R}^d.$$

The original proposal was made in in Breiman (1996), where n-out-of-n with replacement bootstrap was suggested. Breiman (2001) considers a class of procedures which includes (a) random split selection from the set of best splits and (b) a random perturbation of observations. Bootstrap aggregation has been interpreted as a method of decreasing variance of an unstable estimator, like an adaptive regressogram.

PART II

VISUALIZATION

CHAPTER 6

VISUALIZATION OF DATA

Section 6.1 discusses scatter plots. Scatter plots are a natural way to visualize two-dimensional data. Scatter plots can be used in the one-dimensional, three-dimensional, and higher-dimensional cases by transforming data to a two-dimensional data.

Section 6.2 illustrates the use of kernel and histogram estimates to visualize one-dimensional data. Kernel and histogram estimates are based on smoothing, and they do not visualize the raw data like one-dimensional scatter plots. In fact, smoothing is useful also in drawing of the two-dimensional scatter plots when the number of observations is large.

Section 6.3 defines projection pursuit and multidimensional scaling. Both of these methods can be seen as methods of reducing the dimension of the data. Projection pursuit searches for an optimal linear projection. Multidimensional scaling searches for a such configuration of the data in two dimension that the distances of the data points are equal to the original distances.

Section 6.4 introduces graphical matrices, parallel coordinate plots, Andrews' curves, and faces. All of these visualization methods represent observations as graphical objects. In the case of graphical matrices the data matrix is represented as a matrix of graphical objects. In the case of parallel coordinate plots the observations are represented as piecewise linear functions. Andrews' curves represent the observations as functions and faces represent observations as schematical human faces.

6.1 SCATTER PLOTS

We define first two-dimensional scatter plots and only after that we define one-dimensional scatter plots. We use this order of definition because two-dimensional scatter plots are easier to define and interpret than one-dimensional scatter plots. Also, there is not an established way to define one-dimensional scatter plots, and we give two useful definitions: tail plots and QQ plots.

6.1.1 Two-Dimensional Scatter Plots

A two-dimensional scatter a plot is plot of a two-dimensional point cloud $\{x_1, \ldots, x_n\} \subset \mathbf{R}^2$ in the Cartesian coordinate system. We have shown several scatter plots; see, for example, Figure 1.1(a) and Figure 1.4(b).

When the sample size is large, then the scatter plot is mostly black, so the visuality of density of the points in different regions is obscured. In this case it is possible to use smoothing to visualize the density.

Figure 6.1 shows scatter plots of the data (X_t, Y_t), where

$$X_t = \log_e |R_{t-1}|, \qquad Y_t = \log_e |R_t|,$$

and $R_t = (S_t - S_{t-1})/S_{t-1}$ are the daily net returns of the S&P 500 index. The data are introduced in Section 1.6.1. There are more than 15,000 observations. Panel (a) shows a scatter plot. Panel (b) shows a histogram with 80 bins. The bins are colored with a gray scale.[51] Histogram is defined in (3.49). See Carr, Littlefield, Nicholson & Littlefield (1987) for a study of histogram plotting.

6.1.2 One-Dimensional Scatter Plots

Let $x_1, \ldots, x_n \in \mathbf{R}$ be a sequence of real numbers which we want to visulize. One-dimensional scatter plots are two-dimensional scatter plots of the points

$$(x_i, \text{level}(x_i)), \qquad i = 1, \ldots, n, \tag{6.1}$$

where level : $\{x_1, \ldots, x_n\} \to \mathbf{R}$ is a mapping that attaches a real value to each data point. Thus, a one-dimensional scatter plot visualuzes data by lifting each observation in a suitable way.

[51] First we have took square roots $f_i = \sqrt{n_i}$ of the bin counts n_i and then we have defined $g_i = 1 - (f_i - \min_i(f_i) + 0.5)/(\max_i(f_i) - \min_i(f_i) + 0.5)$. Now $g_i \in [0, 1]$. Values g_i close to one are shown in light gray, and values g_i close to zero are shown in dark gray.

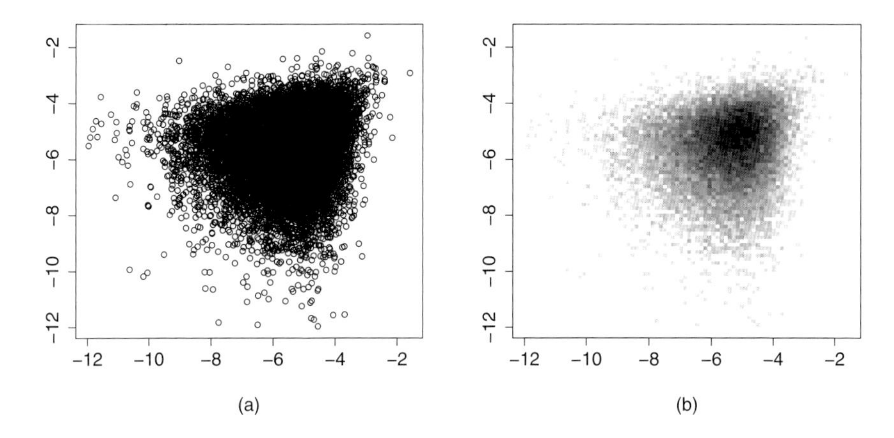

Figure 6.1 *Scatter plots.* (a) A two dimensional scatter plot. (b) A gray scale histogram.

Left and Right Tail Plots Left and right *tail plots* can be used to visualize the heaviness of the tails of the underlying distribution. Left and right tail plots of GARCH(1,1) and EWMA residuals were shown in Figure 3.11 and in Figure 3.18.

We divide the data into the left tail and the right tail, and we visualize separately the two tails. The right tail is

$$\mathcal{R} = \{x_i : x_i > \text{med}_n,\ i = 1, \ldots, n\},$$

where $\text{med}_n = \text{median}(x_1, \ldots, x_n)$ is the sample median, defined in (1.10). We choose the level

$$\text{level}(x_i) = \#\{x_j : x_j \geq x_i, x_j \in \mathcal{R}\}, \qquad x_i \in \mathcal{R}.$$

The level of x_i is the number of observations larger or equal to x_i. Thus, the largest observation has level one, the second largest observation has level 2, and so on. The right tail plot is the two-dimensional scatter plot of the points $(x_i, \text{level}(x_i))$, $x_i \in \mathcal{R}$. A right tail plot visualizes the heaviness of the right tail. The left tail is

$$\mathcal{L} = \{x_i : x_i < \text{med}_n,\ i = 1, \ldots, n\}.$$

For the left tail plot we choose the level

$$\text{level}(x_i) = \#\{x_j : x_j \leq x_i, x_j \in \mathcal{L}\}, \qquad x_i \in \mathcal{L}.$$

Thus, the smallest observation has level one, the second smallest observation has level two, and so on. It is useful to use a logarithmic axis for the y-axis in both right and left tail plots. Tail plots have been applied in Mandelbrot (1963), Bouchaud & Potters (2003), and Sornette (2003).

Figure 6.2 shows tail plots of S&P 500 returns and of utility transformed S&P 500 returns. The S&P 500 index data are described in Section 1.6.1. Panel (a)

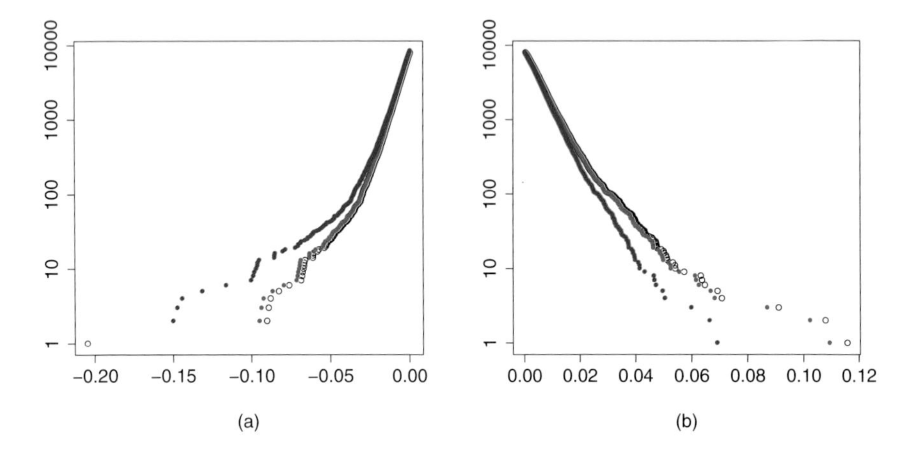

Figure 6.2 *Left and right tail plots.* (a) Left tail plots of the utility transformed S&P 500 returns, and (b) right tail plots of the utility transformed S&P 500 returns. The data are the S&P 500 returns: net returns (black), logarithmic returns (red), and the power utility transformed returns with risk aversion $\gamma = 10$ (blue).

shows left tail plots and panel (b) shows right tail plots, The black points show the net returns $X_t = (S_t - S_{t-1})/S_{t-1}$. The red points show the logarithmic returns $Y_t = \log(S_t/S_{t-1})$. The blue points show the power utility transformed returns $Z_t = (S_t/S_{t-1})^{1-\gamma}/(1-\gamma)$, where $\gamma = 10$. The use of utility functions in portfolio selection is introduced in (1.95). We use a logarithmic scale for the y-axis. We do not show two outliers that are in the left tail plots (a red and a blue point). Note that the left tail for blue points is heavier than the left tail for black and red points, but for the right tail the relation reverses.

QQ Plots A QQ plot is a way to compare the heaviness of tails of two distributions. QQ plots are called also quantile plots or quantile–quantile plots. QQ plots of GARCH(1,1) residuals were shown in Figure 3.12.

A QQ plot of data $x_1, \ldots, x_n \in \mathbf{R}$ and data $y_1, \ldots, y_n \in \mathbf{R}$ is a 2D scatter plot of the points

$$\left(x_{(i)}, y_{(i)}\right), \qquad i = 1, \ldots, n,$$

where $x_{(1)} \le \cdots \le x_{(n)}$ and $y_{(1)} \le \cdots \le y_{(n)}$ are the ordered samples. A QQ plot is a special case of a 1D scatter plot as defined in (6.1). In a QQ plot level$(x_{(i)}) = y_{(i)}$.

The empirical distribution function $\hat{F}_X$ of data $x_1, \ldots, x_n$ was defined in (1.25). As noted in (1.27), it holds that

$$\hat{F}_X^{-1}(p) = \inf\{t \in \mathbf{R} : \hat{F}_X(t) \ge p\} = \begin{cases} x_{(1)}, & 0 < p \le 1/n, \\ x_{(2)}, & 1/n < p \le 2/n, \\ \vdots & \end{cases}$$

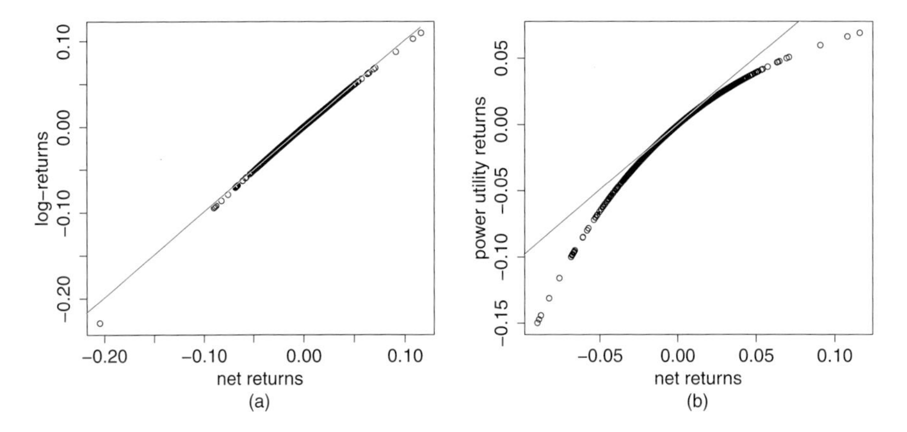

Figure 6.3 *QQ plots.* (a) The x-observations are the net returns and the y-observations are the logarithmic returns. (b) The x-observations are the net returns and the y observations are the power utility transformed returns with risk aversion $\gamma = 10$. The graph of line $y = x$ is shown as red.

so that the QQ plot is a plot of the points

$$\left(\hat{F}_X^{-1}(p_i), \hat{F}_Y^{-1}(p_i)\right), \qquad i = 1, \ldots, n,$$

where $p_i = (i - 1/2)/n$ and $\hat{F}_Y$ is the empirical distribution function of data $y_1, \ldots, y_n$. We can define more generally a QQ plot of data $x_1, \ldots, x_n \in \mathbf{R}$, associated with a reference distribution function $F : \mathbf{R} \to [0, 1]$, as a 2D scatter plot of the points

$$\left(x_{(i)}, F^{-1}(p_i)\right), \qquad i = 1, \ldots, n.$$

The general version of a QQ plot is a special case of a 1D scatter plot as defined in (6.1), and now $\text{level}(x_{(i)}) = F^{-1}(p_i)$.

Figure 6.3 shows QQ plots of S&P 500 returns and utility transformed S&P 500 returns. We continue with the data shown in Figure 6.2. The S&P 500 index data is described in Section 1.6.1. Panel (a) shows a QQ plot where x-observations are the net returns and the y-observations are the logarithmic returns. Panel (b) shows a QQ plot where x-observations are again the net returns and the y observations are the power utility transformed returns with risk aversion $\gamma = 10$. The graph of the line $y = x$ ($x_1 = x_2$) is also included in the plots. We see that the points in both QQ plots are below the red line $y = x$. This means that in both cases the left tail of y-observations is heavier than the left tail of x-observations and the right tail of y-observations is lighter than the right tail of the x-observations. One outlier is removed from the left tail in panel (b).

6.1.3 Three- and Higher-Dimensional Scatter Plots

If the dimension d of data $x_1, \ldots, x_n \in \mathbf{R}^d$ is three or higher, we can project data to two dimensions and then apply two-dimensional scatter plots. A projection $g : \mathbf{R}^d \to \mathbf{R}^2$ can be defined by $g(x) = Ax$, where A is a $2 \times d$ matrix satisfying $AA' = I_2$.

A projection can be constructed by first making a rotation $(x_1, \ldots, x_d) \mapsto (y_1, \ldots, y_d)$ of the data and then projecting $(y_1, \ldots, y_d) \mapsto (y_1, y_2)$. Rotation matrices are orthogonal with determinant 1. All two-dimensional rotation matrices are given by

$$\begin{bmatrix} \cos\theta & -\sin\theta \\ \sin\theta & \cos\theta \end{bmatrix}.$$

All three dimensional rotation matrices are given by $R = R_x(\theta_1)R_y(\theta_2)R_z(\theta_3)$, where

$$R_x(\theta) = \begin{bmatrix} 1 & 0 & 0 \\ 0 & \cos\theta & -\sin\theta \\ 0 & \sin\theta & \cos\theta \end{bmatrix}, \quad R_y(\theta) = \begin{bmatrix} \cos\theta & 0 & \sin\theta \\ 0 & 1 & 0 \\ -\sin\theta & 0 & \cos\theta \end{bmatrix},$$

and

$$R_z(\theta) = \begin{bmatrix} \cos\theta & -\sin\theta & 0 \\ \sin\theta & \cos\theta & 0 \\ 0 & 0 & 1 \end{bmatrix}.$$

After projecting data to two dimensions, we can use histogram smoothing as in Figure 6.1. This amounts to estimating a two-dimensional marginal density.

A second possibility is to estimate a high-dimensional regular histogram as defined in (3.49). Then we make a data set from the histogram by considering the center points of the bins to be the new data points. The new data set can be projected to two dimensions. Each new data point is associated with a bin count. However, we cannot make a histogram plot of the new data set as in Figure 6.1, where the dimension of the data was originally two. The two-dimensional data set obtained from a projection of high-dimensional bn counts is such that many data points are masked by other data points. Thus we have to plot the new data in such a way that the points in the foreground mask the points in the background, and not conversely.

Figure 6.4 shows scatter plots of the data (X_t, Y_t, Z_t), where

$$X_t = \log_e |R_{t-2}|, \qquad Y_t = \log_e |R_{t-1}|, \qquad Z_t = \log_e |R_t|,$$

and $R_t = (S_t - S_{t-1})/S_{t-1}$ are the daily net returns of the S&P 500 index. The data are introduced in Section 1.6.1. There are more than 15,000 observations. We calculate a three-dimensional histogram with $80^3 = 512,000$ bins. There are 11 810 nonempty bins, and this is the number of new observations. Panel (a) shows the rotation $R_x(\theta_1)R_y(\theta_2)R(\theta_3)$, where $\theta_1 = \theta_2 = \pi$ and $\theta_3 = 0$, followed by the projection $(x, y, z) \mapsto (x, y)$. Panel (b) shows the slice where only 6000 observations in the background are shown. Now we can see high-density regions in the center of the point cloud.

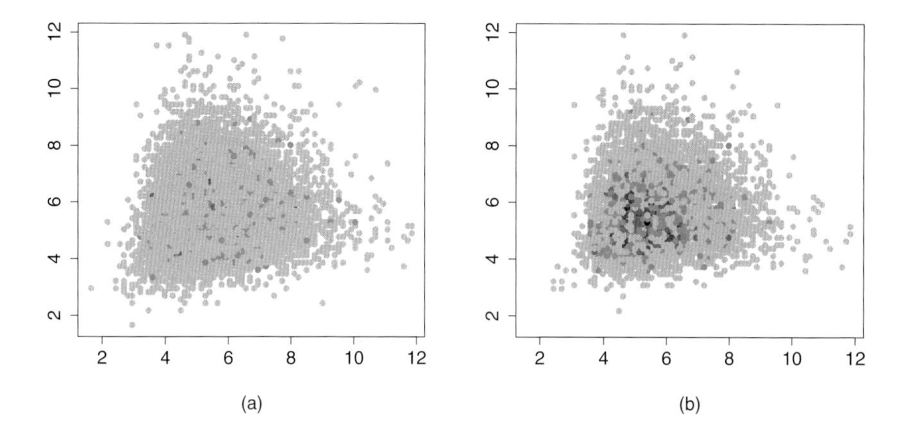

Figure 6.4 *3D scatter plot.* (a) A gray scale 3D histogram projected to two dimensions. (b) A slice where the observations in the background are shown.

6.2 HISTOGRAM AND KERNEL DENSITY ESTIMATOR

Histograms and kernel density estimates are a useful way to visualize one-dimensional data, as an alternative to one-dimensional scatter plots discussed in Section 6.1.2. One-dimensional scatter plots, like tail plots and QQ plots, are useful to visualize the tails of the distribution, but they are not good in visualizing multimodality appearing in the central areas of the distribution. The kernel density estimator is defined in (3.39). The histogram is defined in (3.49).

Figure 6.5 shows a sequence of kernel density estimates. The data consist of portfolio returns. Let

$$X_t^{(b)} = bR_t^{bond} + (1-b)R_t^{sp500},$$

where R_t^{bond} is the U.S. Treasury 10-year bond monthly return, R_t^{sp500} is the S&P 500 monthly return, and $b \in [0,1]$ is the weight of the bond in the portfolio. The returns are annualized. We have monthly data during the period 1953-05–2011-12, which makes 704 observations.[52] We use smoothing parameter $h = 0.08$ and the standard normal kernel in the kernel density estimator. The weight b of the bond is in the range $b = 0, 0.2, \ldots, 1$. We see that for small b the distribution has larger variance (the density estimate is is more flat). For large b the density estimates are more peaked: the distribution has smaller variance.

Figure 6.6 shows a histogram which is calculated from the pay-offs of a call option. For these data it is more natural to use a histogram instead of a kernel density estimator, because the data come from a distribution which is not a continuous distribution and

[52]The bond data are from the Federal Reserve Bank of St. Louis and the S&P 500 data are from Yahoo. The yields y_t are transformd to annualized returns R_t^{bond} with the formula $R_t^{bond} = 120(y_{t-1} - y_t) + y_t$.

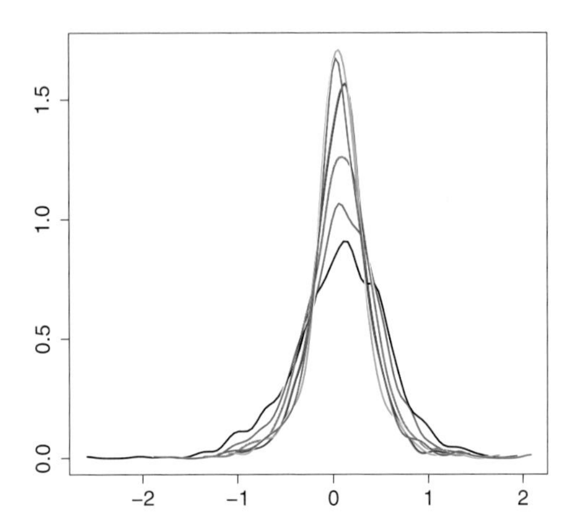

Figure 6.5 *Kernel density estimates.* We show kernel density estimates of the distribution of the portfolio returns when the weight b of the bond is in the range $b = 0, 0.2, \ldots, 1$ and the weight of the S&P 500 is $1 - b$.

not a discrete distribution. The data are the monthly S&P 500 returns during 1953-05–2011-12, Which makes 704 observations. An option strategy which buys at the beginning of each month a call option with the strike equal to the current price gives the pay-offs

$$X_t = \max\{S_t - S_{t-1}, 0\},$$

where S_t is the value of S&P index at time t and the option premium is not taken into account. The histogram is not normalized to integrate to one: The y-axis shows the frequencies of the bins. The number of bins changes the histogram considerably. Panel (a) shows a histogram with 20 bins and panel (b) shows a histogram with 150 bins. The number $\#\{X_t = 0\}$ of zero expirations is 288. With a small number of bins the first bin on the left does not give information of the number of zero expirations. When there are a large number of bins the frequency of first bin is close to the number of zero expirations but the rest of the distribution is not shown accurately.

6.3 DIMENSION REDUCTION

6.3.1 Projection Pursuit

Projection pursuit tries to find interesting low-dimensional projections of high-dimensional data by maximizing a projection index. In the theoretical version of projection pursuit we find a $d \times d$ projection matrix A that maximizes $Q(A'X)$,

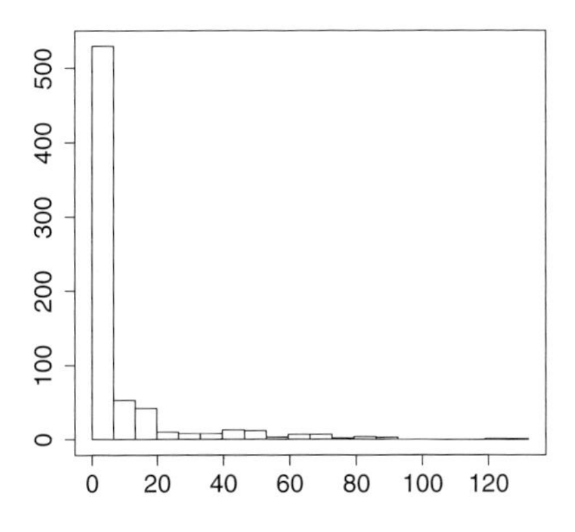

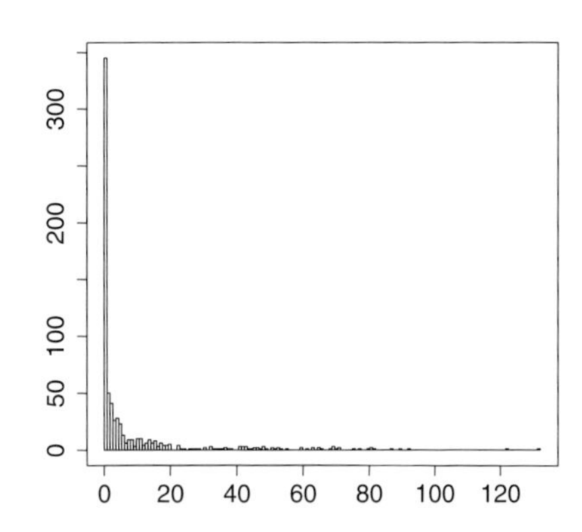

Figure 6.6 *Histogram.* (a) A histogram estimate of the call pay-offs when the number of bins is 20. (b) Number of bins is 150.

where $X \in \mathbf{R}^d$ is a random vector and Q is a projection index. A projection matrix is a matrix A which is symmetric and idempotent: $A' = A$ and $A^2 = A$.

We can proceed in the following sequential way. Let $Q(Y)$ be a projection index, defined for univariate random variables Y. Let us find a projection vector $a_1 \in \mathbf{R}^d$ maximizing the projection index:

$$a_1 = \text{argmax}_{a:\|a\|=1} Q(a'X).$$

For $k = 2; \ldots, d_0$, let $a_k \in \mathbf{R}^d$ be a projection vector, perpendicular to $a_1, \ldots, a_{k-1}$ maximizing the projection index:

$$a_k = \text{argmax}_{a:\|a\|=1, a \perp a_1, \ldots, a \perp a_{k-1}} Q(a'X).$$

We obtain column vectors $a_1, \ldots, a_d$ and define the projection matrix $A = (a_1, \ldots, a_d)$.

Huber (1985) defines the following projection indexes for one-dimensional random variables Y. In all cases it can be shown that $Q(Y) \geq 0$ and $Q(Y) = 0$ if Y is normal.

1. *Cumulant* Let

$$Q(X) = |c_m(X)|/c_2(X)^{m/2}, \qquad m > 2,$$

where c_m is the mth cumulant:

$$c_m = \frac{1}{i^m} \frac{\partial^m}{\partial t^m} \log E \exp\{itX\} \bigg|_{t=0}, \qquad m \geq 2, \tag{6.2}$$

where i is the imaginary unit. For $m = 3$ we obtain the absolute skewness and for $m = 4$ the absolute kurtosis.

2. *Fisher Information* Let

$$Q(X) = \mathrm{Var}(X)I(X) - 1,$$

where $I(X)$ is the Fisher information

$$I(X) = \int_{-\infty}^{\infty} \frac{(f'_X)^2}{f_X},$$

where f_X is the density of X.

3. *Shannon Entropy* Let

$$Q(X) = -S(x) + \log\left(\mathrm{std}(X)\sqrt{2\pi e}\right),$$

where $S(X)$ is the Shannon entropy

$$S(X) = -\int_{-\infty}^{\infty} f_X \log f_X,$$

where f_X is the density of X.

We can also take the variance as a projection index:

$$V(X) = \mathrm{Var}(Y).$$

In this case, A is the matrix of the eigenvectors of the covariance matrix $\Sigma = \mathrm{Cov}(X) = E[(X - EX)(X - EX)']$ of X. This leads to principl components analysis where the spectral representation, or the eigendecomposition, of Σ is

$$\Sigma = A\Lambda A',$$

where Λ is the $d \times d$ diagonal matrix of the eigenvalues of Σ and A is the orthogonal $d \times d$ matrix of the eigenvectors, or the principal components, of Σ. The columns of A are the eigenvectors. Principal component transformation is defined by

$$A'(X - EX) \in \mathbf{R}^d. \tag{6.3}$$

Huber (1985) gives a review of projection pursuit. The term "projection pursuit" was coined by Friedman & Tukey (1974). Further studies of projection pursuit include Cook, Buja & Cabrera (1993) and Cook, Buja, Cabrera & Hurley (1995).

6.3.2 Multidimensional Scaling

Multidimensional scaling makes a nonlinear mapping of data $X_1, \ldots, X_n \in \mathbf{R}^d$ to $\mathbf{R}^2$, or to any space $\mathbf{R}^k$ with $2 \le k < d$. We can define the mapping $Q : \{X_1, \ldots, X_n\} \to \mathbf{R}^2$ of multidimensional scaling in two steps:

1. Reduce the information in the data by calculating the pairwise distances $\|X_i - X_j\|, j \neq i$.

2. Find a set $Q(X_1), \ldots, Q(X_n) \in \mathbf{R}^2$ so that $\|Q(X_i) - Q(X_j)\| = \|X_i - X_j\|$ for $i \neq j$.

We can use also some other distance than the Euclidean distance. In practice, we may not be able to find a mapping that preserves the distances exactly, but we find a mapping $Q : \{X_1, \ldots, X_n\} \to \mathbf{R}^2$ so that the stress functional

$$\sum_{1 \le i < j \le n} \left(\|X_i - X_j\| - \|Q(X_i) - Q(X_j)\| \right)^2$$

is minimized. Sammon's mapping uses the stress functional

$$\sum_{1 \le i < j \le n} \frac{\left(\|X_i - X_j\| - \|Q(X_i) - Q(X_j)\| \right)^2}{\|X_i - X_j\|}.$$

This stress functional emphasizes small distances. Numerical minimization is used to solve the minimization problems.

Multidimensional scaling can be used to visualize correlations between different time series. Let $x_i = (x_{i1}, \ldots, x_{iT})$ be the time series of returns of company i, where $i = 1, \ldots, n$. When we normalize the time series of returns so that the vector of returns has sample mean zero and sample variance one, then the Euclidean distance is equivalent to using the correlation distance. Indeed, let

$$y_i = \frac{x_i - \bar{x}_i}{\mathrm{s}(x_i)},$$

where $\bar{x}_i = T^{-1} \sum_{t=1}^T x_{it}$ and $\mathrm{s}^2(x_i) = T^{-1} \sum_{t=1}^T x_{it}^2 - \bar{x}_i^2$. Now

$$\frac{1}{T} \|y_i - y_j\|^2 = 2[1 - \rho(x_i, x_j)],$$

where $\rho(x_i, x_j) = \gamma(x_i, x_j)/[s(x_i)s(x_j)]$ with $\gamma(x_i, x_j) = T^{-1} \sum_{t=1}^T x_{it} x_{jt} - \bar{x}_i \bar{x}_j$. Thus, we apply the multidimensional scaling for the norm

$$\|y_i - y_j\|_{2,T}^2 = \frac{1}{T} \|y_i - y_j\|^2 = \frac{1}{T} \sum_{t=1}^T (y_{it} - y_{jt})^2,$$

which is obtained by dividing the Euclidean norm by $\sqrt{T}$. Since $-1 \le \rho(x_i, x_j) \le 1$, we have that $0 \le \|y_i - y_j\|_{2,T} \le 2$. Zero correlation gives $\|y_i - y_j\|_{2,T} = \sqrt{2}$, positive correlations give $0 \le \|y_i - y_j\|_{2,T} < \sqrt{2}$, and negative correlations give $\sqrt{2} < \|y_i - y_j\|_{2,T} \le 2$.

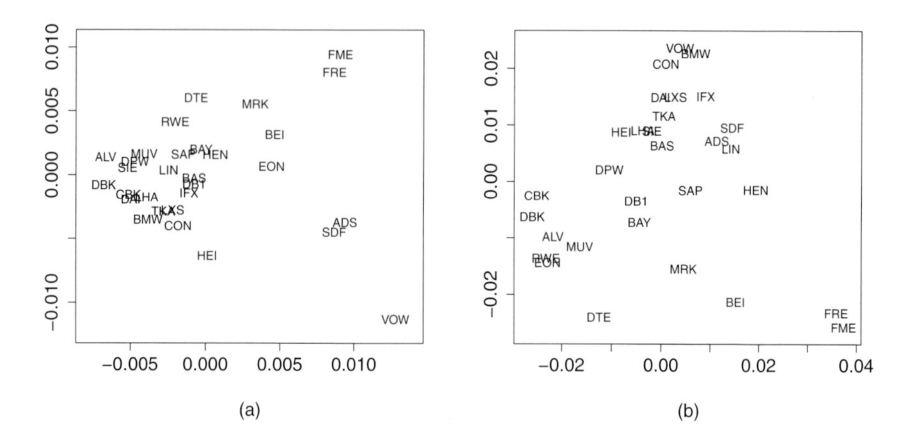

Figure 6.7 *Multidimensional scaling*: *Time series correlations.* The distances between the DAX 30 companies shown in the scatter plots are inversely proportional to the correlations between the return time series. Panel (a) shows the period starting at 2003-01-02 and ending at 2013-02-08, and panel (b) shows the period starting at 2010-12-01 and ending at 2011-11-22.

We study the daily returns of DAX 30 companies during the period starting at 2003-01-02 and ending at 2013-02-08. We have together 2568 observations. The data are obtained from Yahoo.[53]

Figure 6.7 shows multidimensional scaling. Panel (a) shows the complete period starting at 2003-01-02 and ending at 2013-02-08. Panel (b) shows the period of about one year starting at 2010-12-01 and ending at 2011-11-22. We can see, for example, that the Volkswagen (VOW) is uncorrelated during the longer period, but for the one-year period it is highly correlated with BMW. Fresenius medical care (FME) and Fresenius (FRE) are highly correlated during the both time periods.

6.4 OBSERVATIONS AS OBJECTS

In a graphical matrix, each observation was represented with a graphical element of a matrix. In a parallel coordinate plot, each observation was represented with a partially linear function. We can find other ways to represent observations with objects. These include Andrew's curves and faces.

[53]We use the following symbols for the DAX 30 companies: ADS Adidas, ALV Allianz, BAS Basf, BAY Bayer, BEI Beiersdorf, BMW Bayerische Motoren Werke, CBK Commerzbank, CON Continental, DAI Daimler, DB1 Deutsche Boerse, DBK Deutsche Bank, DPW Deutsche Post, DTE Deutsche Telekom, EON E.On, FME Fresenius Medical Care, FRE Fresenius, HEI HeidelbergCement, HEN Henkel, IFX Infineon Technologies, LHA Deutsche Lufthansa, LIN Linde, LXS Lanxess, MRK Merck, MUV Münchener Rückversicherungs AG, RWE RWE, SAP Sap, SDF K+S, SIE Siemens, TKA ThyssenKrupp, and VOW Volkswagen.

6.4.1 Graphical Matrices

A data matrix is an $n \times d$ matrix of real numbers, where n is the number of sampled objects and d is the number of variables; the number in the ith row and in the jth column gives the measurement of the ith object with respect to the jth variable.

A graphical matrix is an $n \times d$ matrix of graphical elements; an $n \times d$ data matrix of real numbers is transformed by representing each real number by a graphical element. Graphical matrices were studied by Bertin (1967, 1981). A related method to visualize data is the Data Image, developed in Minnotte & West (1999).

When the data consist only of binary values, one may represent these values of the data matrix with white and black rectangles. Otherwise, one may use gray scale, colors, or varying-sized objects to represent measured values. For example, real values can be represented with bars or lines, whose length is proportional to the value. Bertin (1981) mentions seven visual variables that can be used to code numerical values: form, position, size, value, texture, color, and orientation.

In order to make a graphical matrix useful, the rows and the columns of the matrix have to be permutated so that it is possible to see significant patterns. In the case of regression data, when one of the variables is a response variable and the other variables are explanatory variables, there are two natural ways to search for a useful permutation of the rows (observations).

1. The first possibility is to order the observations according to the value of the response variable.

2. The second possibility is to (a) cluster the observations using only the explanatory variables and (b) order the rows so that the rows corresponding to the observations in the same cluster are together. This rule does not specify the order of the observations inside the cluster.

Figure 6.8 shows a graphical matrix of the absolute returns of S&P 500 returns. S&P 500 index data are described in Section 1.6.1. We make a data matrix with four variables: The absolute value of the net returns is taken as the response variable, and the three lags of the absolute values of the returns are the explanatory variables:

$$Y_t = |R_t|, \qquad X_{t,1} = |R_{t-1}|, \qquad X_{t,2} = |R_{t-2}|, \qquad X_{t,3} = |R_{t-3}|,$$

where $R_t = (S_t - S_{t-1})/S_{t-1}$ is the net return and S_t is the price of the S&P 500 index. We have used k-means clustering to find $k = 5$ clusters from the three-dimensional data $(X_{t,1}, X_{t,2}, X_{t,3})$, $t = 1, \ldots, T$. The k-means clustering is explained in Klemelä (2009, Section 8.2), for example. The black cluster is the low-volatility cluster and the red cluster is the high-volatility cluster. The green cluster is the high-volatility cluster for lag 1, the turquoise cluster is the high-volatility cluster for lag 2, and the blue cluster is the high-volatility cluster for lag 3. The volatility of the response variable seems to be higher in the turquoise cluster than in the green class.

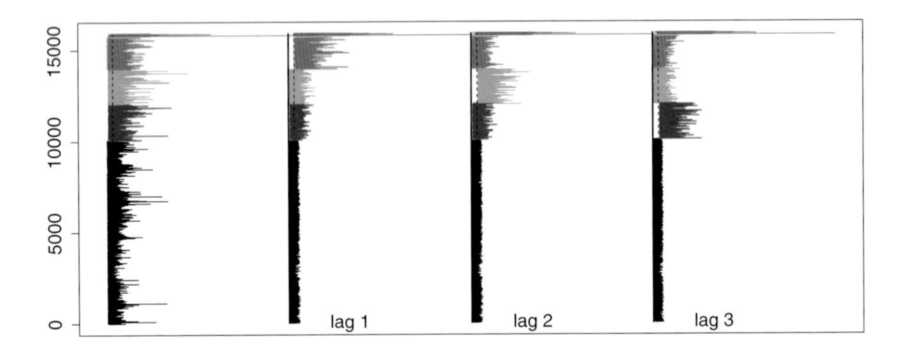

Figure 6.8 *Graphical matrix.* The leftmost column shows the response variable, which is the absolute value of the S&P 500 return. The next three columns show the three explanatory variables, which are three lags of the absolute value of the S&P 500 return. The colors show the five clusters found by the k-means algorithm from the data of the three explanatory variables.

6.4.2 Parallel Coordinate Plots

In a parallel coordinate plot, data points $x_1, \ldots, x_n \in \mathbf{R}^d$ are drawn as piecewise linear curves. Observation $x_i = (x_i^1, \ldots, x_i^d) \in \mathbf{R}^d$, $i = 1, \ldots, n$, is represented as the curve that linearly interpolates the points

$$(1, x_i^1), (2, x_i^2), \ldots, (d, x_i^d).$$

Thus in a parallel coordinate plot the number of the coordinate is mapped onto the horizontal axis and the value is mapped onto the vertical axis, and then these points are linearly interpolated. Parallel coordinate plots were introduced by Inselberg (1985). See also Inselberg & Dimsdale (1990), Wegman (1990), and Inselberg (1997).

Figure 6.9 illustrates the definition of a parallel coordinate plot. Panel (a) shows a scatter plot, and panel (b) shows a corresponding parallel coordinate plot. The sample size is $n = 30$ and dimension is $d = 2$.[54]

Figure 6.10 points out that a typical plot of a vector time series is identical with a parallel coordinate plot. This example illustrates the case where parallel coordinate plots perform well: The number of cases is small, the number of variables is large, and the variables have a natural ordering. We show daily prices of stock indexes DAX 30 (black), MDAX 50 (red), FTSE 100 (blue), and CAC 40 (green) during the time period 1990-11-26 to 2013-06-07. The days are the variables and the stocks are

[54]The data are the daily returns of stock indexes DAX 30 and MDAX 50 during the period of 30 trading days between 2009-04-22 and 2009-06-05. We have made the copula preserving transform with the standard Gaussian marginals, as explained in Section 1.7.2.

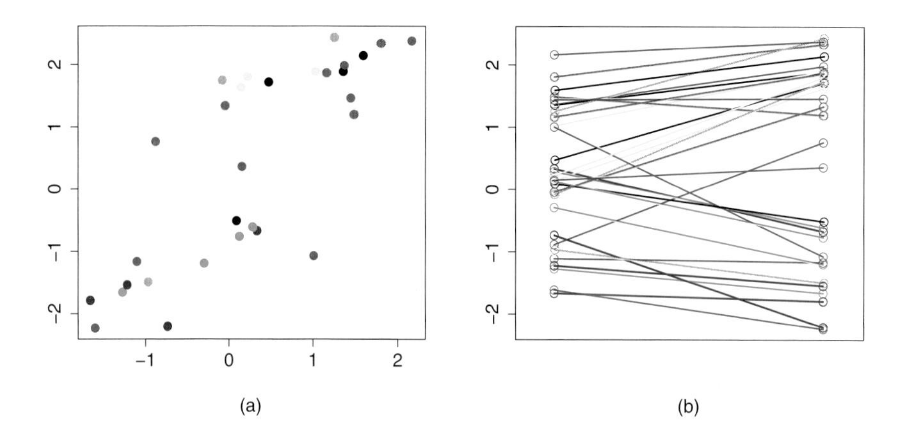

Figure 6.9 *Parallel coordinate plot: Illustration.* (a) A scatter plot. (b) A corresponding parallel coordinate plot.

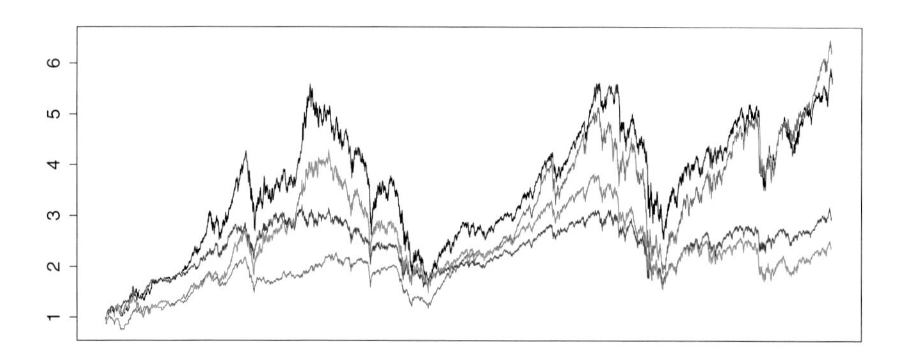

Figure 6.10 *Parallel coordinate plot of stock prices.* Prices of four stock indexes during a period from 1990-11-26 to 2013-06-07; DAX 30 (black), MDAX 50 (red), FTSE 100 (blue), and CAC 40 (green).

the cases, so that $n = 4$ and $d = 4602$. The prices are normalized to have value one at the first day.

Figure 6.11 illustrates the use of a parallel coordinate plot to visualize points in a nearest neigborhood. Nearest-neighbor regression estimate was defined in (3.29) as a weighted average:

$$\hat{f}(x) = \sum_{i=1}^{n} p_i(x)\, Y_i. \tag{6.4}$$

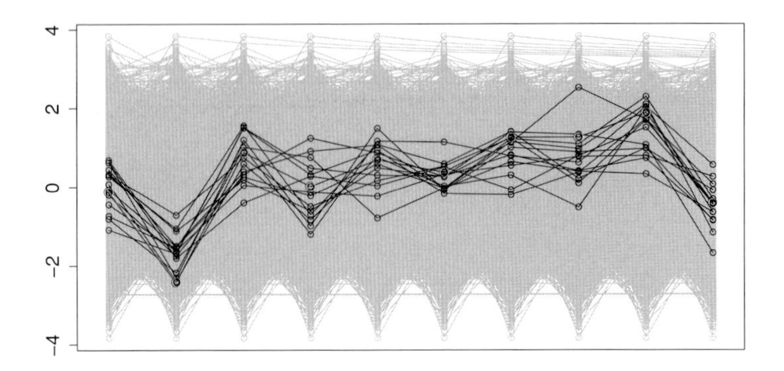

Figure 6.11 *Parallel coordinate plot of a nearest neighborhood.* A parallel coordinate plot of a nearest neighborhood of 15 points. The central point is colored with red.

where $\sum_{i=1}^{n} p_i(x) = 1$ and either $p_i(x) = 0$ or $p_i(x) = 1$. It is of interest to visualize the neighborhood

$$\{X_i : p_i(x) = 1, \;\; i = 1, \ldots, n\},$$

where $x \in \mathbf{R}^d$ is a given point where we want to make a prediction. The data consist of consecutive squared returns $(R_t^2, \ldots, R_{t-9}^2)$, where R_t is the net return of the the S&P 500 index. The data are transformed with the copula transform with the standard normal marginals. S&P 500 index data are explained in Setion 1.6.1. Now we have $n = 15,929$ observations and $d = 10$ dimensions. We look at the neigborhood of $k = 15$ points. The center x is shown with the red color, the 15 points in the nearest neigborhood are shown with the black color, and the rest of the data are shown with the gray color.

Figure 6.12 illustrates the use of a parallel coordinate plot to visualize points in a neigborhood defined by kernel weights. The kernel regression estimator was defined in (3.6). Kernel regression estimator is a weighted average as in (6.4). The weights $p_i(x)$ satisfy $\sum_{i=1}^{n} p_i(x) = 1$ and $p_i(x) \geq 0$. The weights $p_i(x)$ give more weight to points Y_i, which are such that X_i is near x. If we use a kernel function which is everywhere positive, like a Gaussian kernel, then all weights are positive. Thus we cannot apply parallel coordinate plots straightforwardly as in the case of a nearest-neighbor estimate, where weights are either 0 or 1. We make a parallel coordinate plot, where the lines corresponding to observations are shown in a gray scale. The lines corresponding to the observations that are closest to the x value are shown darkest. We use a gray scale coding where white is 1 and black is 0. Thus we make a gray scale from the weights $p_i(x)$ by associating every data point X_i by the value

$$1 - \frac{p_i(x)}{\max_{i=1,\ldots,n} p_i(x)}.$$

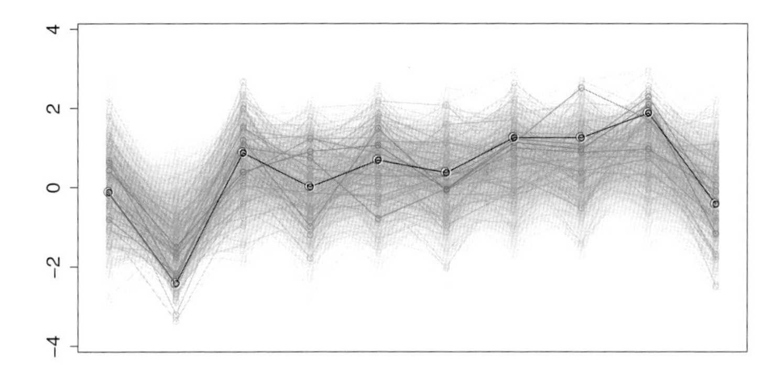

Figure 6.12 *Parallel coordinate plot of a kernel neighborhood.* A parallel coordinate plot where the observations with more weight are colored darker. The central point is colored with red.

The data is the same as in Figure 6.11. The smoothing parameter is $h = 1.5$ and the standard normal kernel is applied.

6.4.3 Other Methods

Andrew's curves Andrew's curves, introduced in Andrews (1972), represent observations as one-dimensional curves. The curves are Fourier series whose coefficients are equal to the the observation values. The definition of the ith Andrew's curve is

$$f_i(t) = \begin{cases} 2^{-1/2}x_{i1} + x_{i2}\sin(t), & \text{when } d = 2, \\ 2^{-1/2}x_{i1} + x_{i2}\sin(t) + x_{i3}\cos(t), & \text{when } d = 3, \\ 2^{-1/2}x_{i1} + x_{i2}\sin(t) + x_{i3}\cos(t) + x_{i4}\sin(2t), & \text{when } d = 4, \\ \vdots & \end{cases}$$

for $i = 1, \ldots, n$, where $t \in [-\pi, \pi]$ and we write the observations as $x_i = (x_{i1}, \ldots, x_{id})$. Note that $\int_{-\pi}^{\pi} \sin^2(t)\, dt = \int_{-\pi}^{\pi} \cos^2(t)\, dt = \pi$.

As in the case of graphical matrices and parallel coordinate plots, the ordering of the variables affects the visualization: The last variables will have only a small contribution to the visualization. Andrews (1972) suggests to use the ordering of the variables given by the principal component analysis; see Section 6.3.1 for the definition of the principal component analysis.

Faces Faces are graphical representations of a data matrix where the size of face elements are assigned to variables (the columns of the data matrix). Chernoff (1973) introduced faces, and one often uses the term "Chernoff faces." Flury & Riedwyl

(1981) and Flury & Riedwyl (1988) have further developed the face technique and defined the following characteristics: right eye size, right pupil size, position of right pupil, right eye slant, horizontal position of right eye, curvature of right eyebrow, density of right eyebrow, horizontal position of right eyebrow, vertical position of right eyebrow, right upper hair line, right lower hair line, right face line, darkness of right hair, right hair slant, right nose line, right size of mouth, right curvature of mouth, and the same for the left side of the face.

Each observation is represented by one face. We have defined together 36 characteristics of a face. Thus we may visualize data up to 36 variables, but we are not able to visualize many observations with faces. With faces we may find clusters from data by looking groups of similar looking faces. Härdle & Simar (2003) discuss parallel coordinate plots, Andrews's curves, and Chernoff faces.

CHAPTER 7

VISUALIZATION OF FUNCTIONS

Section 7.1 defines a slice of a multivariate function. A one-dimensional slice is obtained by choosing one free variable and fixing the values of the other variables. A two-dimensional slice is obtained by choosing two free variables and fixing the values of the other variables.

Section 7.2 defines a partial dependence function obtained from a regression function. The partial dependence function can be called the marginal effect, because its value at a point is obtained by averaging over the distribution of the other than one or two explanatory variables. (Note that we use the term partial effect to refer to a partial derivative of the regression function.) A partial dependence function is a visualization method analogous to projecting the data to one or two dimensions.

Section 7.3 discusses methods for the reconstruction of a set from a finite number of points. We mention three reconstruction methods: using an equispaced grid, using a union of balls, and using a union of polyhedrons. These reconstruction methods are useful when we turn into the construction of level set trees.

Section 7.4 presents level set tree-based methods for the visualization of multivariate functions. A level set tree is a tree structure of level sets of a function. By using the volumes of the level sets, we can derive a univariate function from the tree structure. We call this univariate function the volume function. Using the locations of the level sets, we can draw the level set tree in such a way that it expresses the location information. We call the plot showing the location information the barycenter plot.

The calculation of a level set tree requires approximation of the function and its level sets. The use of an equispaced grid leads to an accurate level set tree; but when the dimension increases, the size of the grid increases exponentially. The equispaced grid can be replaced by a union of balls, centered at the observations. This leads to a faster construction of a level set tree, but the method is more inaccurate and the calculation of the volumes of the level sets poses additional problems.

Level set tree-based visualizations can be applied by visualizing density estimates and regression function estimates. Visualization of density estimates is natural with level set trees, because level set trees represent the structure of the local maximums of the function. In the case of regression functions, we are interested in both the local minima and the local maxima. The local minima need to be visualized with a separate level set tree. In the case of a regression function visualization, we are interested in the partial effects of the explanatory variables, and these can be visualized with the help of level set trees of the estimates of the partial derivatives of the regression function.

7.1 SLICES

A slice of function $f : \mathbf{R}^d \to \mathbf{R}$ is obtained by fixing some variables to have a constant value and keeping the rest of the variables as free variables. For example, when $d \geq 3$, a two-dimensional slice is

$$g(x_1, x_2) = f(x_1, x_2, x_{30}, \ldots, x_{d0}), \qquad (x_!, x_2) \in \mathbf{R}^2,$$

where $(X_{30}, \ldots, x_{d0}) \in \mathbf{R}^{d-2}$ is a fixed point.

A problem with slices is that there are a large number of possible slices. It is reasonable to look at one- or two-dimensional slices. Two-dimensional slices can be drawn with contour plots or perspective plots. To go through all two dimensional slices, we have to first choose two variables, and this can be done in $d(d-1)/2$ ways. After choosing the two variables, we have to make a grid on $\mathbf{R}^{d-2}$ and draw for each grid point a two-dimensional function.

Figure 7.1 shows a regression function and its kernel estimate. We have simulated $n = 100$ observations (X, Y), where $Y = f(X) + \sigma\epsilon$, X is uniformly distributed on $[-2, 6]^2$, $\epsilon \sim N(0, \sigma^2)$, and $\sigma = 0.01$. We have used smoothing parameter $h = 0.5$ and the standard Gaussian kernel. Figure 7.2 shows one-dimensional slices of the estimate whose perspective plot is shown in Figure 7.1(b). Panel (a) shows 80 slices $x_1 \mapsto f(x_1, x_{20})$, where the first coordinate x_1 is the free variable and the second coordinate x_{20} is fixed to 80 different values. Panel (b) shows 80 slices $x_2 \mapsto f(x_{10}, x_2)$.

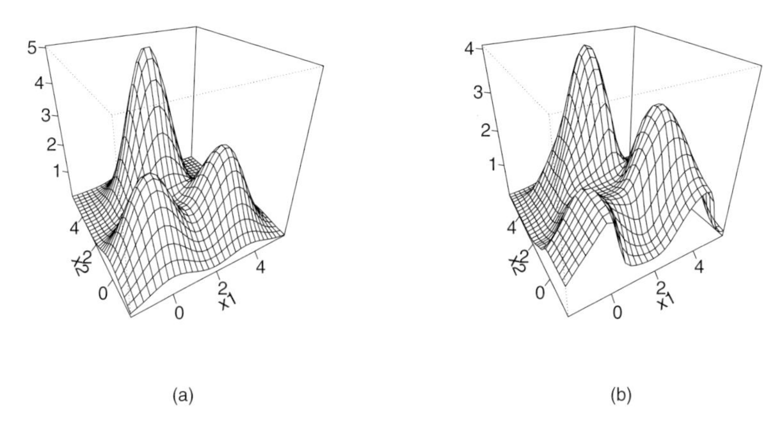

Figure 7.1 *Regression function and its estimate.* (a) A perspective plot of a regression function. (b) A kernel estimate of the regression function.

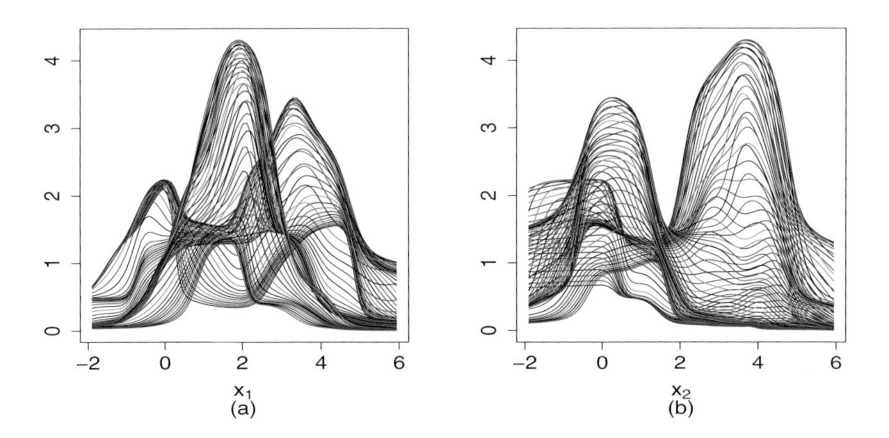

Figure 7.2 *One dimensional slices.* (a) Slices $x_1 \mapsto f(x_1, x_{20})$ for 80 different values of x_{20}. (b) Slices $x_2 \mapsto f(x_{10}, x_2)$. The slices are from the estimate in Figure 7.1(b).

7.2 PARTIAL DEPENDENCE FUNCTIONS

A partial dependence function can be used to visualize a regression function. The partial dependence function is defined as marginal averages of a regression function. A partial dependence function is analogous to a marginal density function of a multivariate density function. A partial dependence function is sometimes called the marginal effect or the average dependence function. A partial dependence plot is a plot of a partial dependence function.

Let $f(x) = E(Y \mid X = x)$ be a regression function with continuous regressors $X = (X_1, \ldots, X_d)$. The partial dependence function of X_1 is

$$\begin{aligned} g_{X_1}(x_1) &= Ef(x_1, X_2, \ldots, X_d) \\ &= \int_{\mathbf{R}^d} f(x_1, x_2, \ldots, x_d)\, f_{X_2,\ldots,X_d}(x_2, \ldots, x_d)\, dx_2 \cdots dx_d, \end{aligned} \quad (7.1)$$

where $f_{X_2,\ldots,X_d} : \mathbf{R}^{d-1} \to \mathbf{R}$ is the joint density of the regressors $X_2, \ldots, X_d$. We can define similarly the partial dependence function of (X_1, X_2) as

$$g_{X_1,X_2}(x_1, x_2) = Ef(x_1, x_2, X_3, \ldots, X_d).$$

Partial dependence functions can be defined similarly for any subset of $\{X_1, \ldots, X_d\}$.

A partial dependence function of a regression function is analogous to a marginal density. If $f_X : \mathbf{R}^d \to \mathbf{R}$ is a density function of the distribution of random vector $X = (X_1, \ldots, X_d)$, then a marginal density is obtained by integrating out some of the variables. For example, a two dimensional marginal density is

$$f_{X_1,X_2}(x_1, x_2) = \int f_X(x_1, \ldots, x_d)\, dx_3 \cdots dx_d,$$

where we assume that $d \geq 3$.

Note that for the conditional expectation we have

$$\bar{f}_{X_1}(x_1) = E\left[f(X_1, \ldots, X_d) \mid X_1 = x_1\right] = E(Y \mid X_1 = x_1).$$

Thus the conditional expectation ignores the effects of $X_2, \ldots, X_d$, unlike the partial dependence function which takes into account the average effects of the variables $X_2, \ldots, X_d$. Partial dependence plots are discussed in Hastie et al. (2001, Section 10.3.2).

If $\hat{f} : \mathbf{R}^d \to \mathbf{R}$ is a regression function estimator based on regression data $(X_1, Y_1) \ldots, (X_n, Y_n)$, then we get an estimator of the partial dependence function as

$$\hat{g}_{X_1}(x_1) = \frac{1}{n} \sum_{i=1}^{n} \hat{f}(x_1, X_{i,2}, \ldots, X_{i,2}). \quad (7.2)$$

For example, if $\hat{f}$ is a kernel regression estimator, defined in (3.6), the estimator of the two dimensional partial dependence function is

$$\hat{g}_{X_1}(x_1) = \sum_{i=1}^{n} q_i(x)\, Y_i,$$

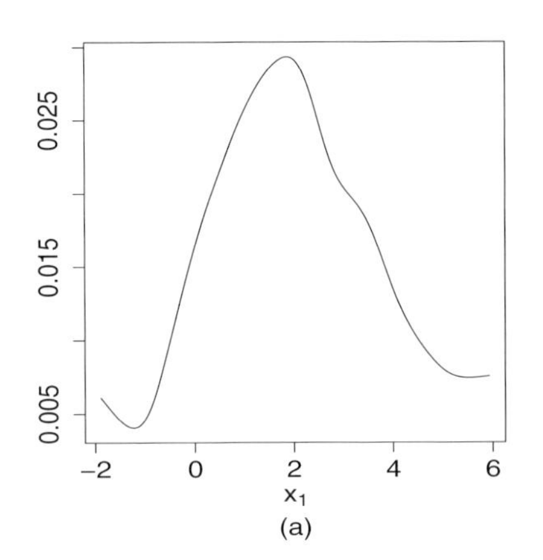

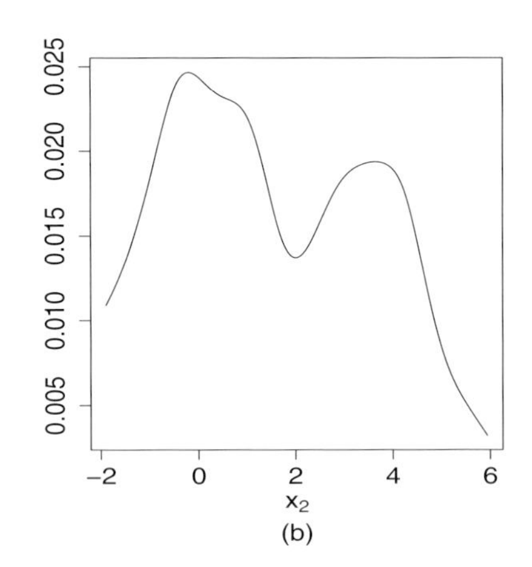

Figure 7.3 *One-dimensional partial dependency functions.* (a) An estimate of $x_1 \mapsto Ef(x_1, X_2)$. (b) An estimate $x_2 \mapsto Ef(X_1, x_2)$.

where

$$q_i(x) = \frac{1}{n} \sum_{j=1}^{n} p_i(x_1, X_{j,2}, \ldots, X_{j,d}),$$

where $p_i(x)$ are the kernel regression weights. The estimator in (7.2) is almost the marginal integration estimator, defined in (4.15) for the estimation in additive models. In the case of the additive model

$$E(Y \mid X = x) = c + f_1(X_1) + \cdots + f_d(X_d)$$

the partial dependence function g_{X_1} of X_1 is $g_{X_1}(x_1) = c + f_1(x_1)$.

Figure 7.3 shows estimates of the one-dimensional partial dependency functions of the two-dimensional regression function whose perspective plot is shown in Figure 7.1(a). Panel (a) shows an estimate of function $x_1 \mapsto Ef(x_1, X_2)$, and panel (b) shows an estimate of function $x_2 \mapsto Ef(X_1, x_2)$. We have used the same simulated data that is used in the estimate of Figure 7.1(b). We have used the kernel regression estimator with smoothing parameter $h = 0.5$ and the standard Gaussian kernel.

7.3 RECONSTRUCTION OF SETS

We consider the problem of reconstructing set $\mathbf{A} \subset \mathbf{R}^d$ using a finite number of points $x_1, \ldots, x_n \in \mathbf{A}$. We are mainly interested in the approximation of a level set of a function, which is discussed in Section 7.3.1, but we describe also point cloud data in Section 7.3.2.

7.3.1 Estimation of Level Sets of a Function

Let $f : \mathbf{R}^d \to \mathbf{R}$ be a function. We want to approximate the level set

$$\Lambda(f, \lambda) = \{x \in \mathbf{R}^d : f(x) \geq \lambda\},$$

where $\lambda \in \mathbf{R}$. We are interested in the case where f is either a regression function estimate or a density estimate. We describe three approximations of $\Lambda(f, \lambda)$. The first uses a partition to rectangles, the second uses a collection of balls, and the third uses a collection of simplexes.

Approximation Using a Grid The first approximation is made using a grid. We assume that we can find a rectangle $R \subset \mathbf{R}^d$, whose sides are parallel to the coordinate axis and which is such that $\Lambda(f, \lambda) \subset R$. Let $R_1, \ldots, R_M$ be a partition of R to small rectangles, and let μ_i be the center of R_i, $i = 1, \ldots, M$. Define the approximation of the level set as

$$\tilde{\Lambda}(f, \lambda) = \bigcup \{R_i : f(\mu_i) \geq \lambda,\ i = 1, \ldots, M\}.$$

This method of approximation suffers from the curse of dimensionality. The number M of the small rectangles needed to partition R grows exponentially as a function of dimension d; see Section 3.2.6.

Figure 7.4 introduces the example which we use to illustrate the methods of level set approximation. In this example we have a sample $X_1, \ldots, X_n \in \mathbf{R}^2$ of size $n = 200$ from a distribution whose density has three modes. Panel (a) shows a scatter plot of the data. Panel (b) shows a perspective plot of a kernel density estimate with smoothing parameter $h = 0.4$ and standard Gaussian kernel.

Figure 7.5 shows estimates of level sets using the example introduced in Figure 7.4. We want to have an approximation of the level set $\tilde{\Lambda}(\hat{f}, \lambda)$, where $\hat{f}$ is the kernel density estimate. We use a grid on the support of $\hat{f}$ which has 100^2 points. Panel (a) shows a contour plot of the kernel density estimate, and panel (b) shows the level set with $\lambda = 0.04$ of the kernel density estimate.

Approximation Using Balls A second possibility to approximate level set $\Lambda(f, \lambda)$ is to use unions of balls. Let $\mathcal{X} = \{x_1, \ldots, x_n\} \subset \mathbf{R}^d$. Let $\rho > 0$ be a radius and define the approximation of $\Lambda(f, \lambda)$ as

$$\tilde{\Lambda}(f, \lambda) = \bigcup \{B_\rho(x_i) : f(x_i) \geq \lambda,\ i = 1, \ldots, n\}, \tag{7.3}$$

where $B_\rho(x) \subset \mathbf{R}^d$ is the ball with radius ρ, centered at $x \in \mathbf{R}^d$.

Figure 7.6 illustrates the approximation using unions of balls with the example of Figure 7.4. We choose $\lambda = 0.04$. Panel (a) shows the case where the radius is $\rho = 0.2$. Panel (b) shows the case where the radius is $\rho = 0.4$.

The use of a union of balls can be computationally attractive because the number of balls is at most n, which does not grow with the dimension. However, the method introduces additional bias because the balls are typically extending out of the boundaries of the level set. Furthermore, the radius ρ is an additional smoothing parameter whose choice affects the estimates. For large dimension d, all balls tend to be separated until a very large radius ρ is chosen.

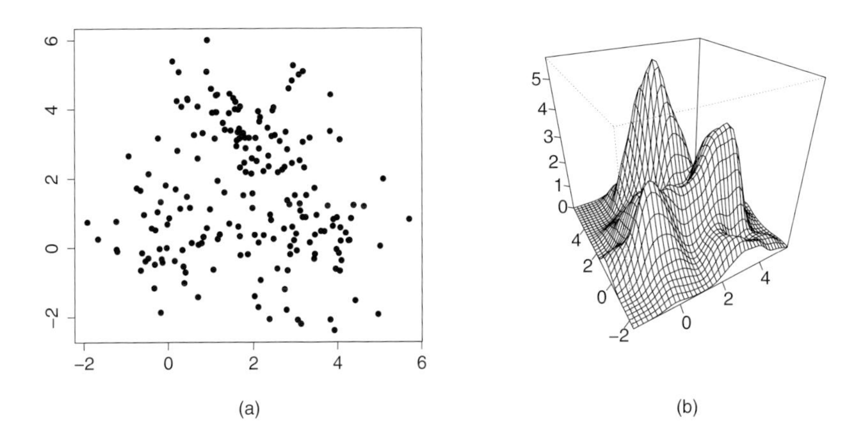

Figure 7.4 *Estimation of level sets: Data and density estimate.* (a) A data of size $n = 200$ sampled from a distribution with three modes. (b) A perspective plot of a kernel density estimate.

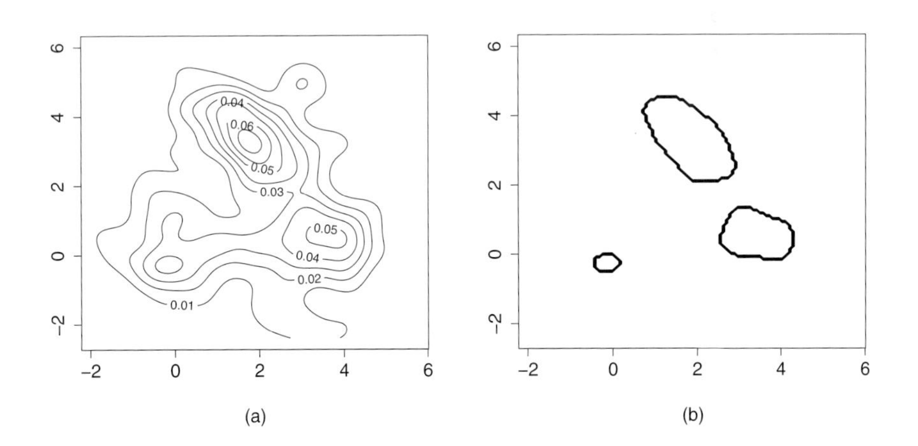

Figure 7.5 *Estimation of level sets: Grid.* (a) A contour plot of a kernel density estimate. (b) A level set of the density estimate with level $\lambda = 0.04$.

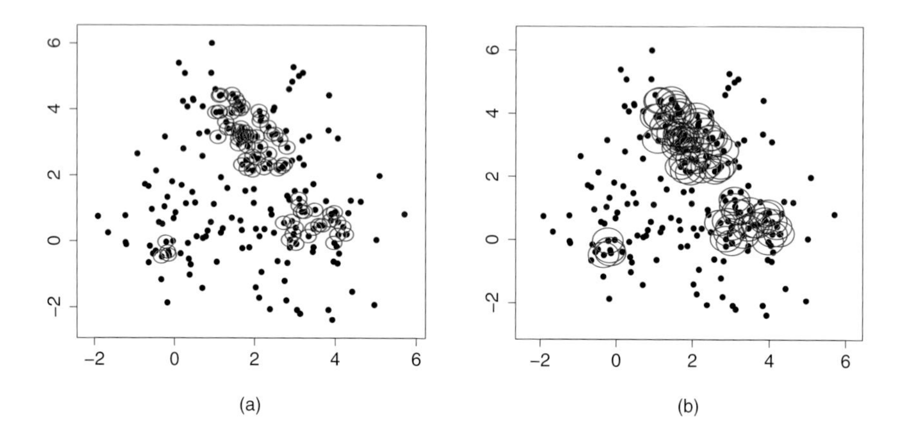

Figure 7.6 *Estimation of Level sets: Union of circles.* (a) A union of circles with radius $\rho = 0.2$. (b) $\rho = 0.4$.

Approximation Using Simplexes A third possibility is to consider suitable unions of convex hulls of simplexes. Let $\mathcal{X} = \{x_!, \ldots, x_n\} \subset \mathbf{R}^d$. Let us call set $\sigma \subset \mathcal{X}$ a simplex of dimension d if it is a collection of $d+1$ points. Let $\sigma = (x_{i_1}, \ldots, x_{i_{d+1}})$ be a simplex of dimension d. The points $x_{i_1}, \ldots, x_{i_{d+1}}$ are called the vertices of σ. Let us denote by $\mathcal{U}_\rho(\mathcal{X})$ the collection of all simplexes of dimension $d+1$ whose vertices are in $\mathcal{X}$ and which are such that the maximum distance between the vertices is at most $\rho > 0$:

$$\mathcal{U}_\rho(\mathcal{X}) = \left\{\sigma = (x_{i_1}, \ldots, x_{i_{d+1}}) : \|x_{i_j} - x_{i_k}\| \le \rho, \; x_{i_1}, \ldots, x_{i_{d+1}} \in \mathcal{X}\right\}.$$

Now we can define an approximation of $\Lambda(f, \lambda)$ as

$$\tilde{\Lambda}(f, \lambda) = \bigcup \left\{\mathcal{H}(\sigma) : \sigma \in \mathcal{U}_\rho(\mathcal{X}), \; \text{mean}(f(\sigma)) \ge \lambda\right\}, \tag{7.4}$$

where $\mathcal{H}(\sigma)$ is the convex hull of σ and $\text{mean}(f(\sigma)) = \sum\{f(x) : x \in \sigma\}/(d+1)$. The use of random polyhderons in density support estimation has been analyzed in Aaron (2013).

Figure 7.7 illustrates the approximation of a level set using unions of triangles with the example of Figure 7.4. We have $\lambda = 0.04$. Panel (a) shows the case where the maximum length of the edge of a triangle is $\rho = 0.5$. Panel (b) shows the case where the maximum length is $\rho = 1$.

Using a union of simplexes to approximate a level set is computationally not as attractive as using balls, because the number of simplexes used in the construction can be very large. On the other hand the approximation can be more accurate.

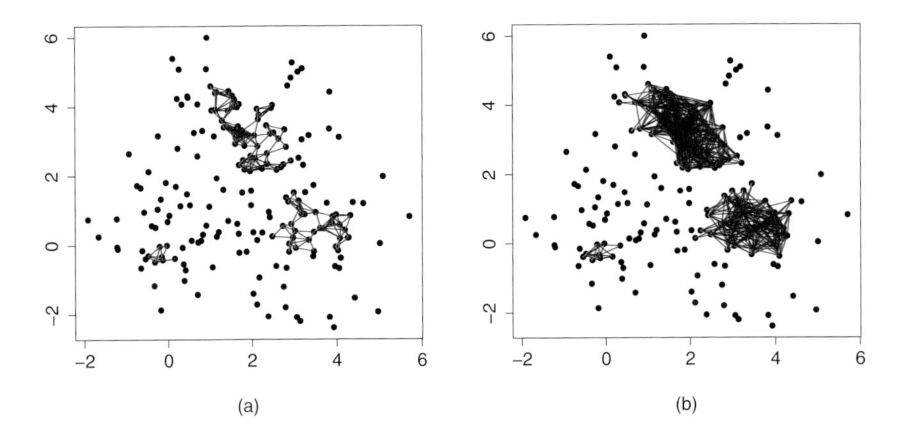

Figure 7.7 *Estimation of level sets: Union of triangles.* (a) A union of triangles with maximum side length $\rho = 0.4$. (b) $\rho = 1$.

7.3.2 Point Cloud Data

Point cloud data are a sample from a uniform distribution on a set $A \subset \mathbf{R}^d$. More generally, the data may be a sample from a distribution whose support is approximately equal to A. Point cloud data can be used as a data structure to represent and store a solid object to a computer.

Point cloud data are analyzed in topological data analysis. The aim is to infer topological properties of set A using these data. The topological properties of set A are, for example, the number of connected components of A (Betti number of order 0), the number of holes in A (Betti number of order 1), and so on; see Carlsson (2009).

Point cloud data may be analyzed by reconstructing a solid object $\hat{A} \subset \mathbf{R}^d$ from the sample $x_1, \ldots, x_n \in A$. The construction can be a union of the sets in a simplicial complex. Let us call a simplicial complex of $\mathcal{X} = \{x_1, \ldots, x_n\} \subset \mathbf{R}^d$ a collection K of sets $\sigma \subset \mathcal{X}$, which is such that if $\sigma \in K$ and $\tau \subset \sigma$, then $\tau \in K$.

We call sets $\sigma \in K$ simplexes. Let $\#\sigma$ be the cardinality of σ. If $\#\sigma = 1$, then we call σ a vertex, if $\#\sigma = 2$, then we call σ an edge, if $\#\sigma = 3$, then we call σ a triangle, and if $\#\sigma = 4$, then we call σ a tetrahedron. If $\#\sigma = k + 1$, then we say that the dimension of σ is k. The $k + 1$ points of a k-dimensional simplex are called the vertices of the simplex.

Let us denote the Vietoris–Rips complex of $\mathcal{X}$ at scale $\rho > 0$ by $\mathcal{V}_\rho(\mathcal{X})$. The Vietoris–Rips complex is such a simplicial complex that for every $\sigma \in \mathcal{V}_\rho(\mathcal{X})$, the vertices of σ are pairwise within distance at most ρ:

$$\mathcal{V}_\rho(\mathcal{X}) = \{\sigma \subset \mathcal{X} : \|u - v\| \leq \rho \text{ for all } u, v \in \sigma\}.$$

Of a special interest is a subcomplex of $\mathcal{V}_\rho(\mathcal{X})$ which is called a k-skeleton. A k-skeleton $\mathcal{U}_\rho(\mathcal{X})$ is such a simplicial complex that the vertices of σ are pairwise within distance at most ρ and the simplexes have all dimension at most k. For example, in $\mathbf{R}^2$ a two-skeleton of $\mathcal{V}_\rho(\mathcal{X})$ leads to a collection of triangles such that the edges of the triangles have length at most ρ. We have used essentially a d-skeleton in (7.4) to approximate a level set. See Zomorodian (2010) and Zomorodian (2012) for definitions and algorithms.

7.4 LEVEL SET TREES

First we define level set trees, second we discuss algorithms for the calculation of level set trees, third we define a volume function for a useful presentation of a level set tree, fourth we define the barycenter plot to show the location information, and fifth we illustrate regression function visualization with an example of the estimation of a news impact curve in volatility prediction.

7.4.1 Definition and Illustrations

The *level set* $\Lambda(f, \lambda)$ of function $f : \mathbf{R}^d \to \mathbf{R}$ at level $\lambda \in \mathbf{R}$ is defined as the set of those points where the function is greater than or equal to the value λ:

$$\Lambda(f, \lambda) = \{x \in \mathbf{R}^d : f(x) \geq \lambda\}.$$

A *level set tree* of a function is a tree whose nodes are the disconnected components of some level sets of the function.

To construct a level set tree we choose a finite number of levels $\lambda_1 < \cdots < \lambda_L$. We assume that each level set $\Lambda(f, \lambda_l)$, $l = 1, \ldots, L$, is a connected set or, alternatively can be decomposed into a finite number of connected subsets, which are separated from each other. Then we can construct a level set tree with a finite number of nodes. The root of a level set tree is the level set with the lowest level; but if this level set has many disconnected components, then the level set tree has many roots. Given a node of a level set tree, the child nodes of this node are among the disconnected parts of the level set that is at one step higher level than the given node. The parent–child relation holds when the set associated with a child node is a subset of the set associated with the parent node.

A *level set tree* of function $f : \mathbf{R}^d \to \mathbf{R}$, associated with set of levels $\mathcal{L} = \{\lambda_1 < \cdots < \lambda_L\}$, where $\lambda_L \leq \sup_{x \in \mathbf{R}^d} f(x)$, is a tree whose nodes are associated with subsets of $\mathbf{R}^d$ and levels in $\mathcal{L}$ in the following way:

1. Write the lowest level set as

$$\Lambda(f, \lambda_1) = A_1 \cup \cdots \cup A_K,$$

where sets A_i are pairwise separated, and each is connected. The level set tree has K root nodes that are associated with sets A_i, $i = 1, \ldots, K$, and each root node is associated with the same level λ_1.

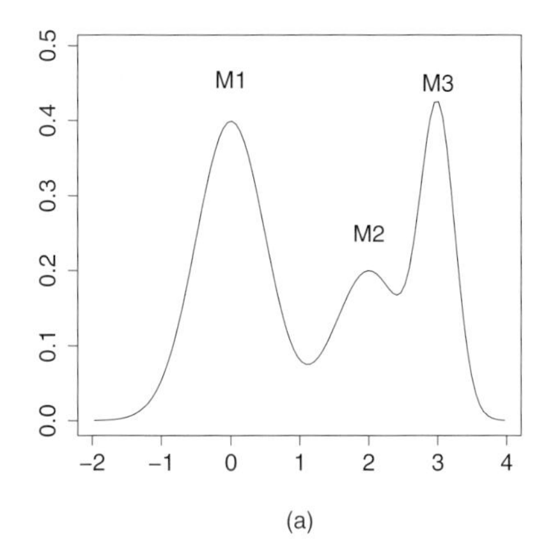

(a)

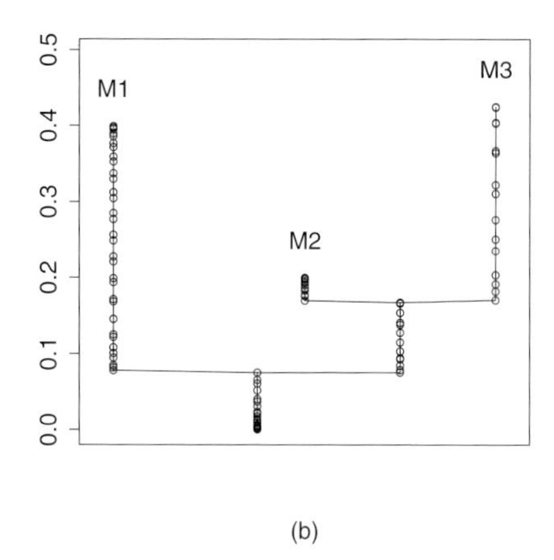

(b)

Figure 7.8 *Level set tree of a 1D density function.* (a) A 1D density function with three local maxima, labeled with M1, M2, M3. (b) A level set tree of the density function.

2. Let node m be associated with set $B \subset \mathbf{R}^d$ and level $\lambda_l \in \mathcal{L}$, $1 \leq l < L$.

   (a) If $B \cap \Lambda(f, \lambda_{l+1}) = \emptyset$, then node m is a leaf node.

   (b) Otherwise, write

   $$B \cap \Lambda(f, \lambda_{l+1}) = C_1 \cup \cdots \cup C_M,$$

   where sets C_i are pairwise separated, and each is connected. Then node m has M children, which are associated with sets C_i, $i = 1, \ldots, M$, and each child is associated with the same level λ_{l+1}.

We have defined the level set tree as in Klemelä (2004). The concept of a level set tree has its origins in Reeb graphs, defined originally in Reeb (1946). A level curve of function $f : \mathbf{R}^d \to \mathbf{R}$ at level $\lambda \in \mathbf{R}$ is $\Gamma(f, \lambda) = \{x \in \mathbf{R}^d : f(x) = \lambda\}$. A graph of level curves is called a Reeb graph, or a level curve graph.

A level set tree is a concept well-suited to represent and visualize the modes of a density function, when by modes we mean the local maxima of a function. Figure 7.8 shows a 1D density function with three modes. Panel (a) shows a graph of the function, and panel (b) shows a level set tree of the function. Figure 7.9 shows a 2D density function with three modes. Panel (a) shows a perspective plot of the function, and panel (b) shows a level set tree of the function.

In addition to representing and visualizing local maxima of a function, we need also to represent and visualize the local minima of a function. This can be done with *lower level set trees*, which are trees whose nodes are disconnected parts of lower level sets. The *lower level set* of function $f : \mathbf{R}^d \to \mathbf{R}$ at level $\lambda \in \mathbf{R}$ is defined as

$$\Lambda(f, \lambda) = \{x \in \mathbf{R}^d : f(x) \leq \lambda\}.$$

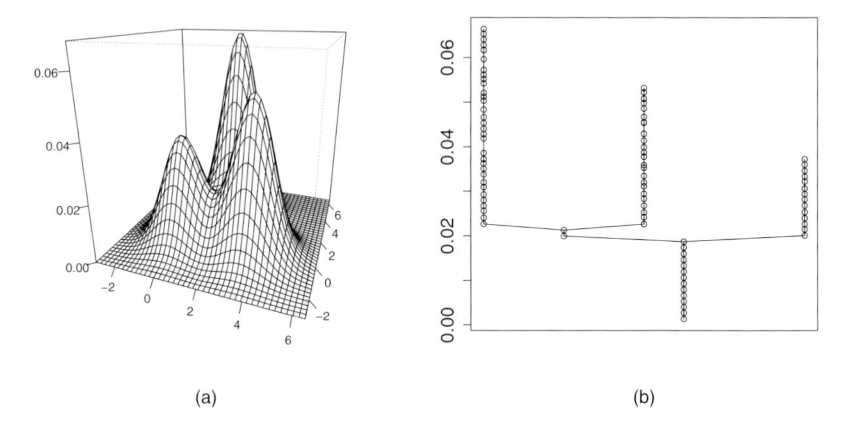

Figure 7.9 *Level set tree of a 2D density function.* (a) A 2D density function with three local maxima. (b) A level set tree of the density function.

The definition of a lower level set tree is analogous to the definition of a level set tree. The root node of a lower level set tree is the level set with the highest level. The child nodes of a node of a lower level set tree are disconnected parts of the level set that is at one step lower level than the parent set, and the child nodes are subsets of the parent set.

Figure 7.10 shows a 1D function with three local maxima and three local minima. Panel (a) shows a graph of the function, panel (b) shows a level set tree of the function, and panel (c) shows a lower level set tree of the function. Figure 7.11 shows a 2D function with three local maxima and two local minima. Panel (a) shows a perspective plot of the function, panel (b) shows a level set tree of the function, and panel (c) shows a lower level set tree of the function.

In the 2D case a single perspective plot may not be able to show all local maxima or minima due to the fact that local extremes may be hidden behind the surface filling the foreground. Thus, even in the 2D case level set trees and lower level set trees can give a useful summary of the local maxima and minima, although the main use of level set trees is in the higher-dimensional cases.

Figure 7.10 and Figure 7.11 show that level set trees or lower level set trees are not always intuitive in visualizing the local extremes of a function. However, in many cases it is natural to decompose a function to the positive and to the negative part and use level set trees to visualize separately the negative and the positive part. We use this technique in Section 7.4.5 to visualize the partial derivatives of a regression function. The positive part and the negative part of a partial derivative of a regression function have a natural interpretation. The positive part of a partial derivative of a regression function shows the regions where the dependence of an explanatory variable is positive and the negative part of a partial derivative of a regression function shows the regions where the dependence of an explanatory variable is negative. The

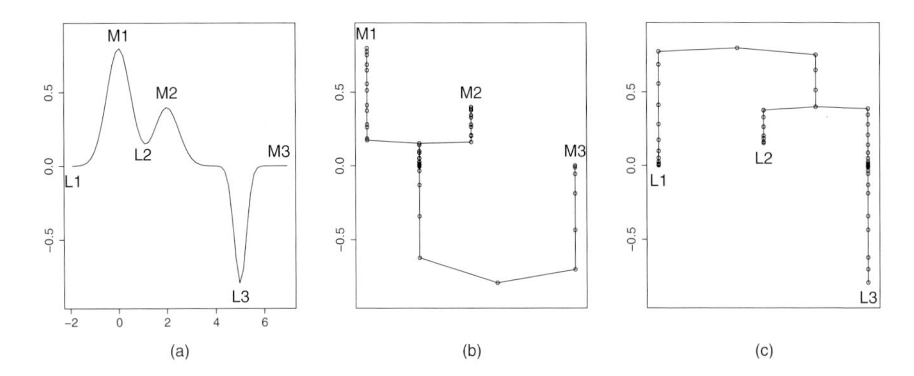

Figure 7.10 *A level set tree and a lower level set tree in 1D.* (a) A 1D function with three local maxima and three local minima, labeled with M1, M2, M3 and L1, L2, L3. (b) A level set tree of the function. (c) A lower level set tree of the function.

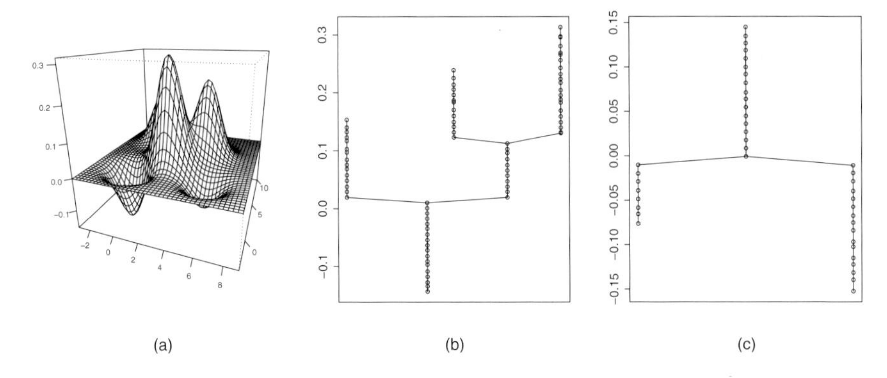

Figure 7.11 *A level set tree and a lower level set tree in 2D.* (a) A 2D function with three local maxima and two local minima. (b) A level set tree of the function. (c) A lower level set tree of the function.

positive part of a function $f : \mathbf{R}^d \to \mathbf{R}$ is defined as $f_+(x) = \max\{f(x), 0\}$, and the negative part is defined as $f_-(x) = -\min\{f(x), 0\}$. It holds that $f = f_+ - f_-$.

7.4.2 Calculation of Level Set Trees

We want to calculate a level set tree of function $f : \mathbf{R}^d \to \mathbf{R}$, for levels $\lambda_1 < \cdots < \lambda_L$. We make an assumption about the level sets $\Lambda(f, \lambda_l)$, $l = 1, \ldots, L$: We assume that there is a collection of sets $A_1, \ldots, A_L \subset \mathbf{R}^d$ so that the level sets can be written

$$\Lambda(f, \lambda_l) = \bigcup_{j=l}^{L} A_j. \tag{7.5}$$

Property (7.5) says that there exists a collection $A_1, \ldots, A_L$ of "elementary sets" so that all level sets are unions of these elementary sets. The lowest level set is a union of all the elementary sets and higher level sets are unions of subsets of the elementary sets. If the level sets of function f do not satisfy property (7.5), then it can be possible to obtain a function through an approximation whose level sets satisfy the property. We give two examples of such approximation methods: (1) approximation using a partition and (2) approximation using level sets. Both methods give a piecewise constant approximation. We discuss the interpolation methods more precisely after presenting the algorithm.

We call the algorithm for calculating a level set tree the Leafsfirst algorithm. We describe the algorithm for calculating upper level set trees, and the calculation of lower level set trees can be done analogously.

Algorithm Leafsfirst:

1. The input of the algorithm is the levels $\lambda_1, \ldots, \lambda_L$ and the sets $A_1, \ldots, A_L$ that appear in (7.5). First we have to order the levels in increasing order. From now on, we assume that $\lambda_1 < \cdots < \lambda_L$.

2. We start at the highest level. Let the first leaf node be associated with level λ_L and with set A_L. This node is now a "temporary root node," because the node has not yet a parent node.

3. For $l = L - 1$ to 1: Consider level λ_l and set A_l.

   Create a new node to the level set tree. The level of this node is λ_l. We have to find the set which is associated to this node and we have to find the child nodes of the new node. This is done in the following way.

   We find which sets associated with the temporary root nodes (those nodes which do not yet have a parent) touch set A_l. At this step we will apply the bounding box technique, which is explained below. We have two possible cases.

   (a) If set A_l touches sets $B_1, \ldots, B_M$ associated with the temporary root nodes, then create the connections between A_l and $B_1, \ldots, B_M$; the new node is the parent of the nodes that are touched. The set associated with this new parent node is $A_l \cup B_1 \cup \cdots \cup B_M$.

(b) If set A_l does not touch any sets associated with the temporary root nodes, then the new node is a leaf node of the tree and does not have children. The set associated with this node is A_l.

The new node is a new "temporary root node."

4. When we have gone through all the levels, the remaining "temporary root nodes" are the final root nodes.

The sorting at item 1 of the algorithm takes $O(L \log L)$ steps. In the worst case the item 3 of the algorithm requires the pairwise comparison of all sets $A_1, \ldots, A_L$ to find which sets touch, and this takes typically $O(dL^2)$ steps, since we need $O(d)$ steps to calculate whether two rectangles touch or weather two balls touch. Thus the worst-case complexity of the algorithm is

$$O\big(L \log L + dL^2\big) = O\big(dL^2\big), \tag{7.6}$$

where L is the number of elementary sets appearing in (7.5) and d is the dimension of function (dimension of the space where the function is defined).

We enhance the algorithm with the *bounding box technique*. In the bounding box technique we associate the nodes with the bounding box of the set associated with a node. The bounding box of set $A \subset \mathbf{R}^d$ is the smallest rectangle containing A, such that the sides of the rectangle are parallel to the coordinate axes. At item 3 of the algorithm we find which sets, associated with the current root nodes, are touched by set A_l. If set A_l does not touch the bounding box of those sets, then it does not touch any sets inside the bounding box. Only if it does touch the bounding box will we have to travel further toward the leaf nodes to find whether A_l touches any of the smaller bounding boxes. With the bounding box enhancement the worst-case complexity of item 3 is still $O(dL^2)$, but with this technique we achieve considerable improvements in typical cases.

The Leafsfirst algorithm was introduced in Klemelä (2006), and it was applied with a piecewise constant approximation with a regular equispaced grid.

Approximation Using a Partition Let $f_0 : \mathbf{R}^d \to \mathbf{R}$ be an initial function whose level set tree we want to calculate. Assume that the level sets of f_0 do not satisfy condition (7.5). We can use a partition to construct such an approximation of f_0 that property (7.5) holds.

First we choose a rectangle $R \subset \mathbf{R}^d$ containing the region where we approximate f_0. We define a partition $\{A_1, \ldots, A_L\}$ of R, where A_l are rectangles. This means that $\cup_{l=1}^{L} A_l = R$ and $A_l \cap A_m = \emptyset$ for $l \neq m$. Let x_l be the center point of A_l, $l = 1, \ldots, L$. The initial function $f_0 : \mathbf{R}^d \to \mathbf{R}$ is evaluated at points x_l and we define

$$f(x) = \sum_{l=1}^{L} f_0(x_l)\, I_{A_l}(x). \tag{7.7}$$

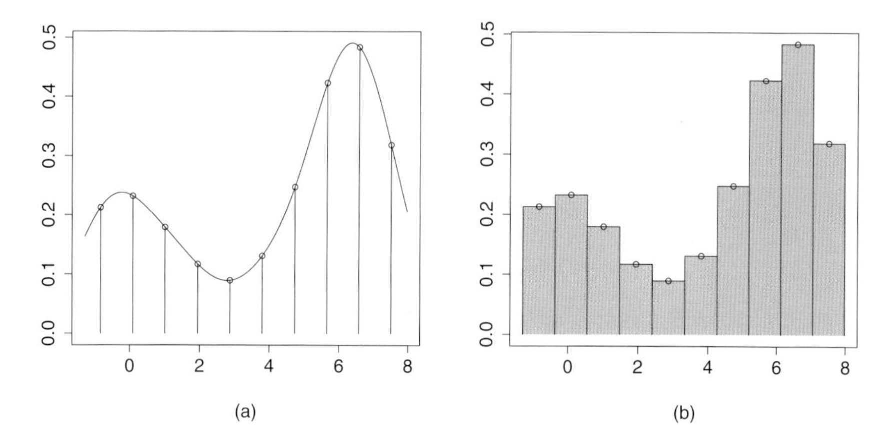

Figure 7.12 *Approximation using a partition.* (a) The function is evaluated at equispaced points. (b) The values are interpolated to obtain a piecewise constant function.

Now f satisfies property (7.5) when we choose $\lambda_l = f_0(x_l)$ and assume, without loosing generality, that $f_0(x_1) < \cdots < f_0(x_L)$.

For example, when f_0 is a density function, we choose rectangle R to contain the support of f_0; and when f_0 is a regression function, then we choose R to contain the support of the density f_X of the explanatory variables. If the support of f_0 or f_X is not bounded (the support is $\mathbf{R}^d$, for example), then we choose a suitable large part of the support (a large rectangle R containing the region where f_0 or f_X is larger than some small $\epsilon > 0$).

In fact, we do not have to use rectangles to make a partition of R, but a partition using any types of sets would lead to the property (7.5). However, rectangles are convenient because it is easy to check whether two rectangles touch each other, and performing this check is required in the algorithm for the calculation of a level set tree. Note also that it is not required that the partition to rectangles would be a regular partition, when by a regular partition we mean such a partition where all the sets in the partition have the same size and shape. Namely, we can use also such partitions that are obtained from greedy regressograms or from the CART procedure discussed in Section 5.5. These partitions are irregular in the sense that all the rectangles in the partition can have a different size and shape.

Figure 7.12 illustrates the approximation using a partition in the one-dimensional case. We use formula (7.7) to construct an approximation. Panel (a) shows a one-dimensional function and an equispaced grid of 10 points, where the function is evaluated. Panel (b) shows how the equispaced grid leads to a regular partition which is used in interpolation to give a piecewise constant function.

We noted in (7.6) that the the number of steps of the Leafsfirst algorithm is $O\left(dL^2\right)$, where L is the number of sets used in the approximation and d is the dimension of the range of of the function. In the case of approximation using a partition, the number L can be quite high. A regular equispaced partition of rectangle $R \subset \mathbf{R}^d$ leads to exponentially increasing number L as a function of d. For example, when we partition $[0,1]^d$ so that each direction has 10 parts, this makes the total of $L = 10^d$ sets in the partition. This is an example of the curse of dimensionality. Thus we are led to consider an alternative approximation method called approximation using level sets.

Approximation Using Level Sets In Section 7.3.1 we have defined three methods to approximate level sets. The first of them uses an approximation with a grid, and this is a special case of the partition method described before. Next we discuss the two other methods: approximation with unions of balls as in (7.3) and approximation with unions of polyhedrons as in (7.4).

Let $f_0 : \mathbf{R}^d \to \mathbf{R}$ be an initial function whose level set tree we want to calculate, and assume that the level sets of f_0 do not satisfy condition (7.5). We can construct an approximation of f_0 by defining directly the level sets of the approximating function so that property (7.5) is satisfied. We approximate a level set tree of a function $f_0 : \mathbf{R}^d \to \mathbf{R}$ using only the evaluations $f_0(X_1), \ldots, f(X_n)$, where $X_1, \ldots, X_n \in \mathbf{R}^d$. The basic examples are the case where f_0 is a regression function estimate calculated with data $(X_1, Y_1), \ldots, (X_n, Y_n)$ and the case where f_0 is a density function estimate calculated with data $X_1, \ldots, X_n$.

1. Let us denote $\lambda_1 = f_0(X_1), \ldots, \lambda_n = f_0(X_n)$. We assume that $\lambda_1 < \cdots < \lambda_L$. Let $f : \mathbf{R}^d \to \mathbf{R}$ be a function with finitely many different level sets

$$\Lambda(f, \lambda_j) = \bigcup_{i=j}^{n} B_\rho(X_i), \qquad j = 1, \ldots, n, \tag{7.8}$$

where $B_\rho(X_i)$ is the ball with radius ρ centered at X_i: $B_\rho(X_i) = \{x \in \mathbf{R}^d : \|x - X_i\| \le \rho\}$.

2. A d-dimensional simplex σ is a collection of $d+1$ points. Those points are called the vertices of σ. Denote $\mathcal{X} = \{X_1, \ldots, X_n\} \subset \mathbf{R}^d$. Let $\mathcal{U}_\rho(\mathcal{X})$ be the collection of all simplexes of dimension d whose vertices are in $\mathcal{X}$ and which are such that the maximum distance between the vertices is at most $\rho > 0$:

$$\mathcal{U}_\rho(\mathcal{X}) = \left\{\sigma = (x_{i_1}, \ldots, x_{i_{d+1}}) : \|x_{i_j} - x_{i_k}\| \le \rho,\ x_{i_1}, \ldots, x_{i_{d+1}} \in \mathcal{X}\right\}.$$

Let us define the level of $\sigma \in \mathcal{U}_\rho(\mathcal{X})$ to be the minimum value of f_0 at the vertices of σ:

$$\lambda(\sigma) = \min\{f_0(x) : x \in \sigma\}.$$

Let $\mathcal{U}_\rho(\mathcal{X}) = \{\sigma_1, \ldots, \sigma_L\}$. Denote $\lambda_i = \lambda(\sigma_i)$. We assume that $\lambda_1 < \cdots < \lambda_L$. Define

$$\Lambda(f, \lambda_l) = \bigcup_{i=l}^{L} \mathcal{H}(\sigma_i), \qquad l = 1, \ldots, L, \tag{7.9}$$

where $\mathcal{H}(\sigma_i)$ is the convex hull of σ_i.

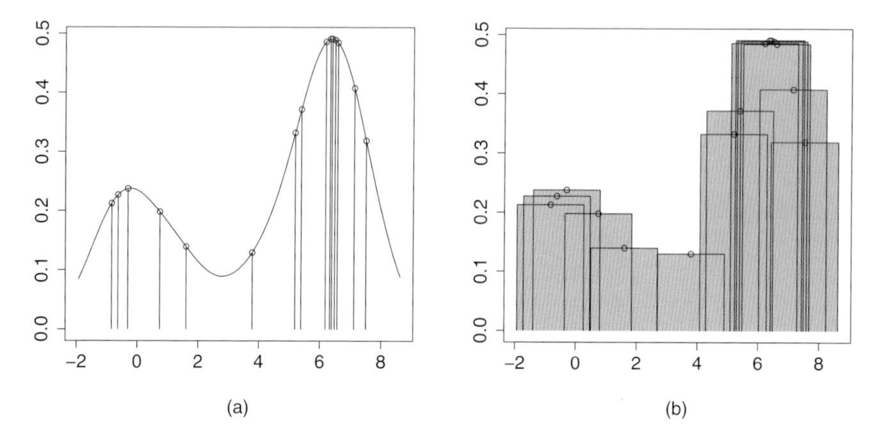

Figure 7.13 *Approximation using level sets in 1D.* (a) The function is evaluated at irregularly spaced points. (b) The values are interpolated to obtain a piecewise constant function.

Both approximations lead to functions satisfying the property (7.5), which is required to hold in order that algorithm Leafsfirst can be applied. Approximation of f_0 through f can be quite complex, but the level sets of f have a simple expression, given in (7.8) and (7.9). The level set tree of f gives us an approximation for the corresponding level set tree of f_0.

Figure 7.13 illustrates the approximation using level sets in the one-dimensional case. Now we define the function through its level sets and use the formula (7.8). Panel (a) shows the same one-dimensional function as in Figure 7.12(a), but now the function is evaluated at irregularly spaced points. In fact, the function is a kernel density estimate that is constructed using as data the same points that are used to evaluate the function. Panel (b) shows how the evaluations at irregularly spaced points are used in approximation to obtain a piecewise constant function. Each point is a center of an interval with length 2.2, and an interval associated with a higher value is overriding an interval with a lower value.

Figure 7.14 illustrates the approximation of a function through approximating its level sets in the two-dimensional case. Panel (a) shows an approximation of level sets with unions of balls. Panel (b) shows an approximation of level sets with unions of rectangles. The balls and rectangles are colored with a gray scale so that a darker color indicates a larger value of function f.

The Leafsfirst algorithm takes in the worst-case $O\left(dL^2\right)$ steps, as noted in (7.6), where L is the number of sets used in the approximation and d is the dimension of the range of of the function. In the case of approximation (7.8) using unions of balls, and when the underlying function f_0 is a density estimate or a regression function estimate, we have that $L = n$, where n is the sample size. Thus, in the case of approximation (7.8) the Leafsfirst algorithm leads to a computationally feasible procedure even in quite high-dimensional spaces.

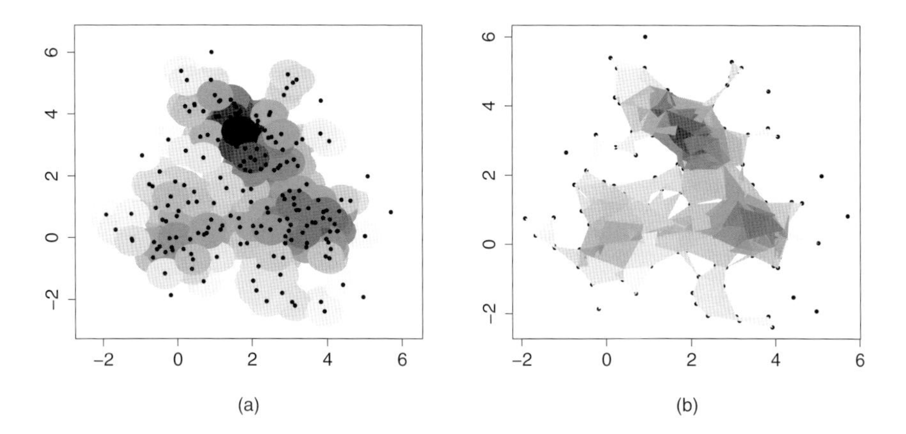

Figure 7.14 *Approximation using level sets in 2D.* (a) Level sets are estimated with unions of balls. (b) Level sets are estimated with unions of triangles.

The accuracy of the procedure (7.8) may not be as good as in the case of the approximation using a partition, or using the approximation with simplexes. Approximation using unions of balls leads to biased estimators of the functions, in the sense that the level sets are estimated to be too large. The estimation error can be seen in Figure 7.13(b), where the two modes are not as clearly separated as in the original function.

7.4.3 Volume Function

Definition of a Volume Function A volume function is constructed from a level set tree of a function. A volume function of a level set tree is a tool for the visualization of the level set tree and the underlying function. In addition to the tree structure, a volume function shows information about the volumes of the disconnected parts of the level sets. We explain the construction of a volume function in four steps.

First we note that a tree can always be drawn as a collection of nested sets, so that a parent–child relation is represented as a set inclusion. The root node is a set that contains all the sets. The child nodes of a parent node are smaller sets inside the set representing the parent set. Figure 7.15 shows an example of a representation of a tree with nested sets. Panel (a) shows a usual representation of a tree, where nodes are shown as small circles and a parent–child relation is expressed with a line joining two nodes. A parent node is always drawn below a child node. Panel (b) shows the same tree represented with nested vectors. To make the 1D representation useful, we have drawn the vectors so that a child node is drawn higher than its parent node. Otherwise it would be difficult to show clearly the set inclusion of 1D sets. Panel (c) shows the same tree represented with nested 2D circles.

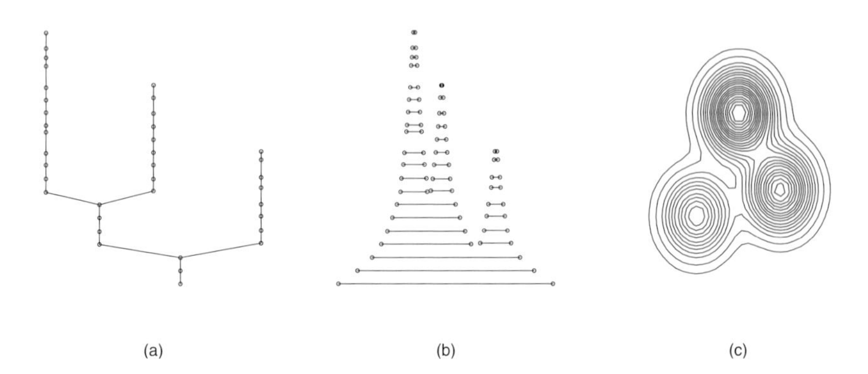

Figure 7.15 *Tree representations.* (a) A classical tree representation. (b) A representation of the tree with nested vectors. (c) A representation of the tree with nested circles.

Second we note that the representation of a tree with nested sets can be interpreted as a function. For example, Figure 7.15(c) shows the tree as nested sets and can be interpreted as a contour plot of a 2D function. Also, Figure 7.15(b) shows the tree as nested vectors; and it can be modified so that we obtain a graph of a 1D function by joining the endpoints of the vectors to the vector under it (and removing the vectors themselves). Thus we obtain a representation of a tree as a piecewise constant function.

The third step to obtain a volume function is to add information about the levels to the picture. Each node of a level set tree is associated with a level (i.e., the level of the level set from which the set associated with the node is a part). Thus we use the level of the node to draw the vectors in Figure 7.15(b) so that the height of the vector is equal to the level. Also, we choose the level of the contours in Figure 7.15(c) to be equal to the level of the node.

The fourth step to obtain a volume function is to add information about the volumes. This is done by choosing the nested sets to have the same volumes as the corresponding sets of the level set tree. We note that each node of a level set tree is associated with a set (this set is some disconnected region of a level set). We choose the vectors to have the length equal to the volume of the set associated with the node. Also, we choose the volume of a set in the contour plot representation to be equal to the volume of the set associated with the node.

We use the term *volume function* to denote the function obtained by the above transformation and we use the term *volume transformation* to denote the transformation itself. The term *volume plot* is used to denote a plot of a volume function.

Remarks about a Volume Function We have defined a volume function so that its dimension can be any dimension smaller than or equal to the dimension of the original function. In this book we use the transformation of multivariate functions to univariate functions, but there is also some interest to transform multivariate functions to 2D functions.

A given multivariate function has many level set trees, because a level set tree is chosen using a grid of levels and we can choose this grid in many ways. Furthermore, a given level set tree has many volume functions. We have the following three choices to be made when constructing a volume function: (1) A level set tree does not specify the ordering for the sibling nodes, and thus a volume function can have its local extremes in various permutations. (2) When we use set inclusion to represent a parent–child relation, the location of the child set is not specified exactly, but we require only that a child set is inside the parent set. In a 1D volume function it is natural to put the child sets symmetrically inside the parent set but, in a 2D volume function there are many natural ways to choose the location of the child sets. (3) We have not specified the location of the volume function. In the 1D volume functions we choose the location of the function so that the left end point of the support is equal to the origin.

Note that the definition of a level set tree does not require that the level sets would have a finite volume, but this assumption was made in the definition of a volume function. Typically this assumption is not restrictive. For example, a Gaussian density can well be approximated with a function with a bounded support.

Interpretation of a Volume Function The main advantage of a volume function is that we can represent a high-dimensional function with a lower-dimensional function, and this lower-dimensional function has similarities with the original multivariate function. In short, the original function and its volume function have the same structure of local extremes. We discuss in the following in detail the properties that remain invariant in the volume transformation.

First, the number of local extremes of the original function and that of its volume function are equal. The volume function from an upper level set tree has the same number of local maxima as the original function. The volume function from a lower level set tree has the same number of local minima as the original function.

Second, level set trees of the original function and its volume function have the same tree structure, when the trees are constructed using the same grid of levels. The same tree structure means that the trees have the same number of nodes and there is a mapping between the nodes so that the parent–child relations between the corresponding nodes match. The levels associated with the nodes of a level set tree of the original function and the levels associated with the corresponding nodes of the level set tree of the volume function are equal. However, the sets associated with the nodes are different, because the sets associated with the nodes of level set tree of the original function are in a higher-dimensional space than the nodes of the level set tree of the volume function.

Third, all the separated regions of level sets of a volume function have the same volumes as the corresponding separated regions of of the original multivariate function.

Fourth, as an implication of the previous similarities, it follows that the excess masses of the original function and its volume function are equal. That is,

$$\int (f - \lambda)_+ = \int (\mathrm{vf}(f) - \lambda)_+$$

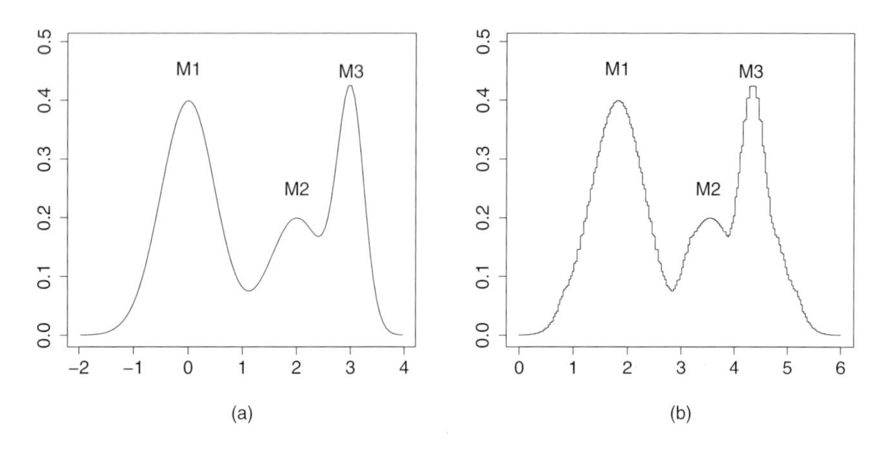

Figure 7.16 *Volume function in 1D.* (a) A 1D function with three local maxima, labeled with M1, M2, and M3. (b) Its volume function.

for all $\lambda \in \mathbf{R}$, where $\text{vf}(f)$ is a volume function of f and $(x)_+ = \max\{0, x\}$. In fact, a stronger property

$$\int_A (f - \lambda)_+ = \int_{\text{vf}(A)} (\text{vf}(f) - \lambda)_+$$

holds, where A is a set associated with a level set tree whose level is λ and $\text{vf}(A)$ is the set associated with the corresponding node of a level set tree of $\text{vf}(f)$.

Illustrations of a Volume Function Figure 7.16 shows a 1D function and its volume function. Panel (a) shows a graph of the function, and panel (b) shows a graph of a volume function. Note that the corresponding level set tree has been shown in Figure 7.8. It can be seen that the volume function is a piecewise constant function. We have used 100 level sets to make the volume function. By increasing the number of levels, we can obtain more smooth-looking volume functions. Note that the support of the original function is contained in $[-2, 4]$ but the support of the volume function is contained in $[0, 6]$, since we use volume functions with the left endpoint of the support is equal to the origin.

Figure 7.17 shows a 2D function and its volume function. Panel (a) shows a perspective plot of the function, and panel (b) shows a 1D volume function. Note that the corresponding level set tree has been shown in Figure 7.9.

Figure 7.18 shows a 2D standard Gaussian density function and its volume function. Panel (a) shows a perspective plot of the Gaussian density function, and panel (b) shows a volume function. The density is is unimodal but the volume function visualizes now the tail behavior of the function.

Figure 7.19 shows a 2D linear function and its volume function. Panel (a) shows a perspective plot of the function, and panel (b) shows a volume function. The function is unimodal on the box $[-3, 3] \times [-3, 3]$.

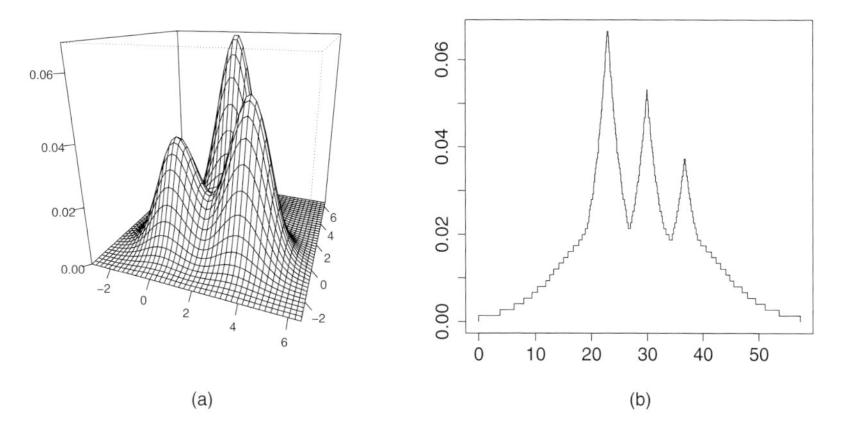

Figure 7.17 *Volume function in 2D.* (a) A 2D function with three local maxima. (b) A volume function of a level set tree of the function.

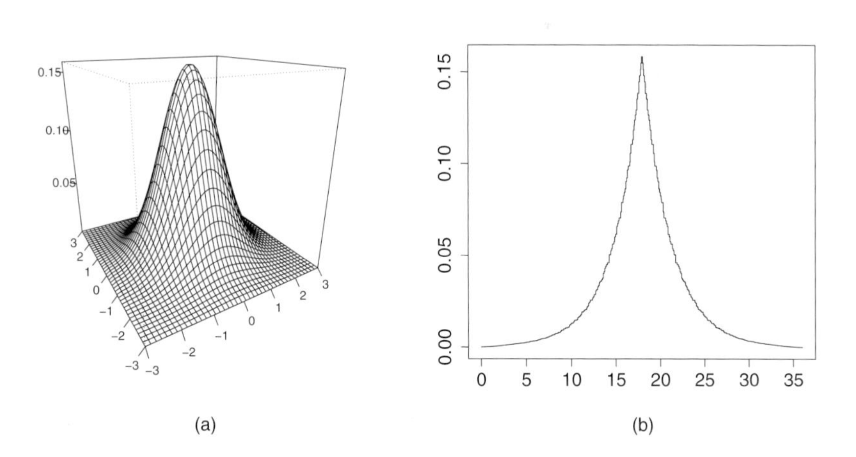

Figure 7.18 *Volume function of a Gaussian function.* (a) A perspective plot of the standard Gaussian density function. (b) A volume function of the standard Gaussian density function.

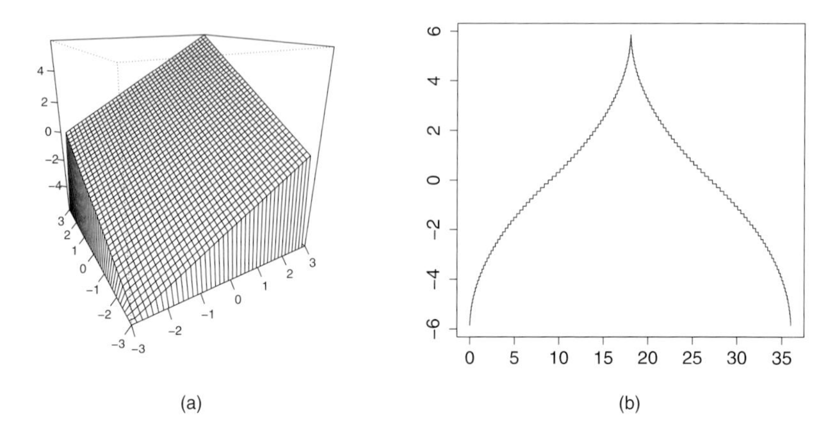

Figure 7.19 *Volume function of a linear function.* (a) A perspective plot of the linear function $f(x) = x_1 + x_2$. (b) A volume function of the linear function.

Calculation of a Volume Function A volume function is obtained from a level set tree with an otherwise rather easy construction, but the calculation of a volume function requires additionally the calculation of the volumes of the all separated parts of the level sets, and this can be computationally demanding.

To construct a level set tree, we have presented two approximation methods for functions: (1) approximation using a partition and (2) approximation using level sets. The calculation of the volumes of the separated regions of the level sets is different for these two methods of function approximation.

When we apply the approximation using a partition as in (7.7), then the calculation of the volumes of the disconnected parts of the level sets is simple, because each disconnected part of a level set is a union of disjoint rectangles. Thus the volume of a disconnected part of a level set is the sum of the volumes of the rectangles which make up this disconnected part.

When we apply approximation using level sets, so that the level sets of the approximating function are defined as in (7.8) or as in (7.9), then the calculation of the volumes of the disconnected parts of the level sets is complicated, and we will use numerical methods to calculate these volumes. We define below the volume approximation for the case of using unions of balls, but the case of using unions of polyhedrons is similar. We have to calculate the volumes of the sets of the type

$$A = \bigcup_{j=1}^{n_A} B_\rho(X_{i_j}), \tag{7.10}$$

where $i_1, \ldots, i_{n_A} \subset \{1, \ldots, n\}$ is a collection of indexes. These volumes can be calculated approximately using Monte Carlo integration and, in particular, the rejection method.[55] Monte Carlo integration can be computationally expensive in a high-dimensional space, but we can use a method where Monte Carlo integration is used only for a small number of disconnected parts of level sets, and for the rest of the disconnected parts of level sets, an interpolation method is used to approximate the volumes.

The interpolation method works in the following way. We choose levels $l_1, \ldots, l_m$ from the collection $\{\lambda_1, \ldots, \lambda_n\}$ of all levels of the level set tree. We find the disconnected parts of level sets $A_1, \ldots, A_m$ so that set A_i is the part of the level set with level l_i. We estimate the volumes of sets $A_1, \ldots, A_m$. Let us denote the corresponding estimates of the volumes with $v_1, \ldots, v_m$. Let A be a set which is associated with the node of a level set tree, so that A is as written in (7.10). Define

$$\kappa_A = \frac{\text{volume}(A)}{n_A \cdot \text{volume}(B_\rho(0))}.$$

Let λ be the level of the level set of which A is a part. We find levels l_i and l_{i+1} such that $l_i < \lambda < l_{i+1}$. We estimate κ_A with

$$\hat{\kappa}_A = \kappa_i + \frac{\lambda - l_i}{l_{i+1} - l_i}(\kappa_i - \kappa_{i+1}),$$

where

$$\kappa_i = \frac{v_i}{n_i \cdot \text{volume}(B_\rho(0))}, \qquad \kappa_{i+1} = \frac{v_i}{n_{i+1} \cdot \text{volume}(B_\rho(0))},$$

and n_i and n_{i+1} are the numbers from the representations $A_i = \bigcup_{j=1}^{n_i} B_\rho(X_{k_j})$, $A_{i+1} = \bigcup_{j=1}^{n_{i+1}} B_\rho(X_{l_j})$. Finally we estimate

$$\text{volume}(A) \approx \hat{\kappa}_A \cdot n_A \cdot \text{volume}(B_\rho(0)). \tag{7.11}$$

Figure 7.20 shows a level set tree and a volume function of a kernel density estimate of the three-modal function shown in Figure 7.17. We have estimated the density using the kernel density estimator with the smoothing parameter $h = 0.4$ and the standard Gaussian kernel. The sample has 200 observations.

Figure 7.21 illustrates the approximation of level sets with unions of balls. We approximate the kernel density estimate shown in Figure 7.20. Radius $\rho = 0.55$ is used in the formula (7.8). Panel (a) shows a level set tree of the density estimate. Panel (b) shows a volume function of the density. We have used the interpolation method (7.11) with $m = 20$ levels and with $M = 200$ numbers of observations in

[55] In the rejection method we first find the smallest ball $B_r(\mu)$ that contains the union of the small balls. Second, we generate a Monte Carlo sample of size M from the uniform distribution on the ball $B_r(\mu)$. (This sample can be generated by generating M random vectors $z_1, \ldots, z_M$ from the standard normal distribution, generating a sample $u_1, \ldots, u_M$ from the uniform distribution on $[0, 1]$, and the final sample is $\mu + r\sqrt{u_i} z_i / \|z_i\|$, $i = 1, \ldots, M$.) Third, we calculate the number n_{inside} of observations in the Monte Carlo sample which are inside of some small ball. The estimate of the volume of the union of small balls is equal to volume$(B_r(\mu)) \cdot n_{inside}/M$.

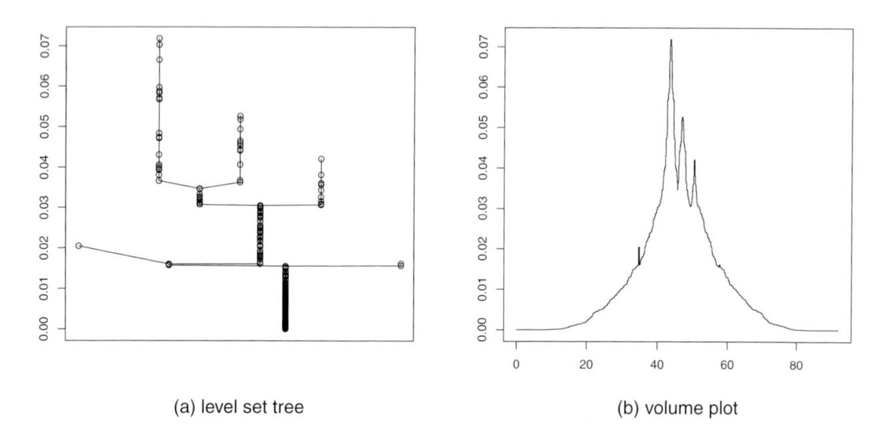

Figure 7.20 *Kernel density estimate.* (a) A level set tree of a kernel density estimate. (b) A volume function of the kernel density estimate.

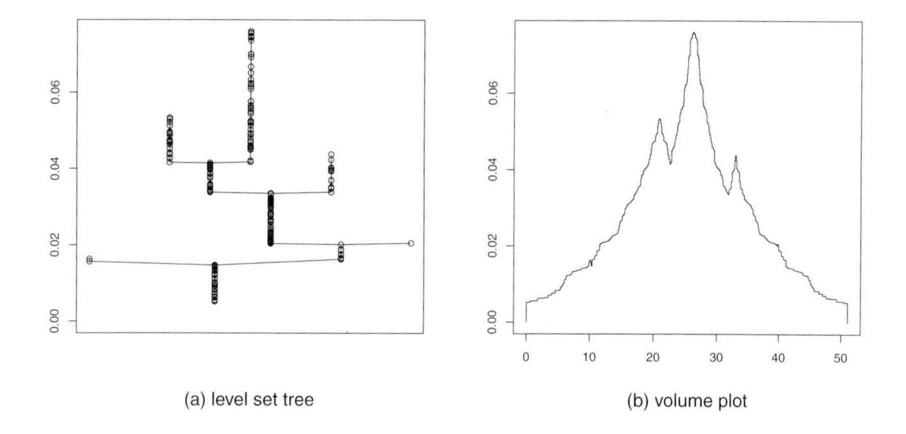

Figure 7.21 *Approximation using unions of balls.* (a) A level set tree of a kernel density estimate when the level sets are approximated using unions of balls. (b) A volume function when the volumes are calculated with Monte Carlo integration and interpolation.

the Monte Carlo sample. The level set tree shows that the approximation has two spurious modes. The spurious modes are not visible in the volume function. It can be seen that the estimated level sets are a blown-up version of the true level sets, by comparing the estimated level set tree and volume function to the true volume function in Figure 7.17(b). The level sets are separating in higher levels than the true level sets. The spurious modes can be made to disappear if we make the smoothing parameter h or the radius ρ larger, but then the level sets are even more blown up.

7.4.4 Barycenter Plot

We use a *barycenter plot* to visualize the locations of the barycenters of the separated components of the level sets of a function. In particular, a barycenter plot visualizes the locations of the modes.

The *barycenter* of set $A \subset \mathbf{R}^d$ is defined as

$$\text{barycenter}(A) = \frac{1}{\text{volume}(A)} \int_A x \, dx. \tag{7.12}$$

A barycenter is a d-dimensional vector which gives the center of mass of the set. It is the expectation of the random vector that is uniformly distributed on the set. The barycenter can lie outside set A

In a level set tree each node is associated with a set. We calculate the barycenter of the sets and use this information to plot the level set tree.

A barycenter plot consists of d windows when the function is defined in the d-dimensional Euclidean space, so that the barycenter of a set is a vector of d real numbers. We have a window for each coordinate. Each window shows the positions of one coordinate of the barycenters for different levels. The nodes of the tree are drawn as bullets.

1. The horizontal position of a node in the ith window is equal to the ith coordinate of the barycenter of the set associated with the node, where $i = 1, \ldots, d$.

2. The vertical position of a node is equal to the level of the the set associated with the node.

3. The parent–child relations are expressed by the line joining a child with the parent.

A barycenter plot visualizes the "skeleton" of the function. A barycenter plot shows the 1D curves that go through the barycenters of all separated components of the level sets.

To identify the nodes between different windows of a barycenter plot and between a volume function and a barycenter plot, we label the modes. The labeling of the modes will uniquely determine the correspondence of all nodes in different windows. To ease the identification of nodes and branches across different windows, we will also color the nodes. We will use the leafs-first coloring, That is, we first choose distinct colors for the leaf nodes and then travel toward the root nodes by changing the color always when two branches are merging. We also color the lines joining a child and a parent. The color of a line will be the same as the color of the child node; that is, the color of the line will be the same as the color of the node that is at the lower end of that line. The leafs-first coloring is appropriate because we want to highlight the modes of the function and thus we want to choose the colors for the modes in such a way that the modes are easy to distinguish from one another.

Figure 7.22 shows a barycenter plot of a 2D three-modal function shown in Figure 7.17. Panel (a) shows the first coordinate of the barycenter plot, and panel (b)

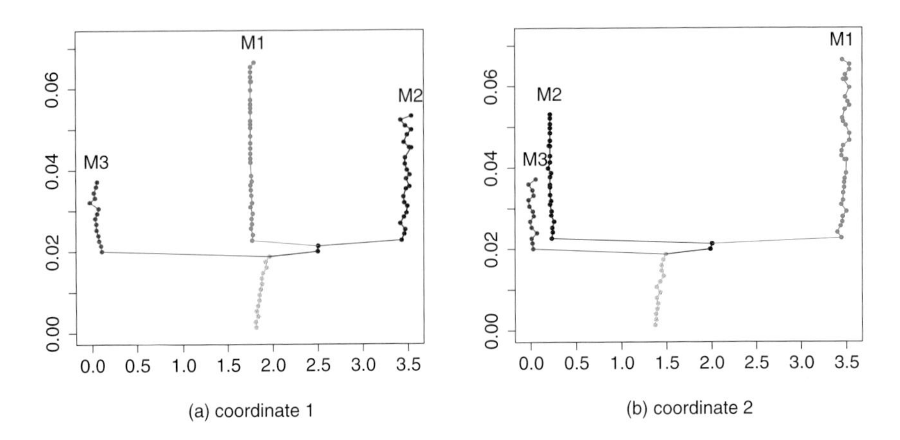

Figure 7.22 *Barycenter plot of a 2D density.* (a) A barycenter plot of the first coordinate. (b) A barycenter plot of the second coordinate.

shows the second coordinate of the barycenter plot. A level set tree of the density function was shown in Figure 7.9(b), and a volume function of the density function was shown in Figure 7.17(b).

Figure 7.23 shows that a barycenter plot can be accompanied with slices, as defined in Section 7.1. Alternatively, in the case of density estimation we can accompany a barycenter plot with marginal densities, and in the case of regression function estimation we can accompany a barycenter plot with a partial dependence function, as defined in Section 7.2. This was noted in Karttunen, Holmström & Klemelä (2014).

7.4.5 Level Set Trees in Regression Function Estimation

Let R_t be the S&P 500 return and

$$Y_t = R_t, \qquad X_t = (R_{t-1}, R_{t-2}).$$

We consider the estimation of the conditional variance with the kernel estimator, using the S&P 500 data described in Section 1.6.1. We define the news impact curve as

$$\sigma^2(x) = E(Y_t^2 \mid X_t = x).$$

We make first the copula transform to the standard Gaussian marginals, as defined in Section 1.7.2, and use the kernel estimator

$$\hat{\sigma}^2(x) = \sum_{t=1}^{T} p_t(x)\, Y_t^2,$$

where $p_t(x)$ are the kernel weights, defined in (3.7).

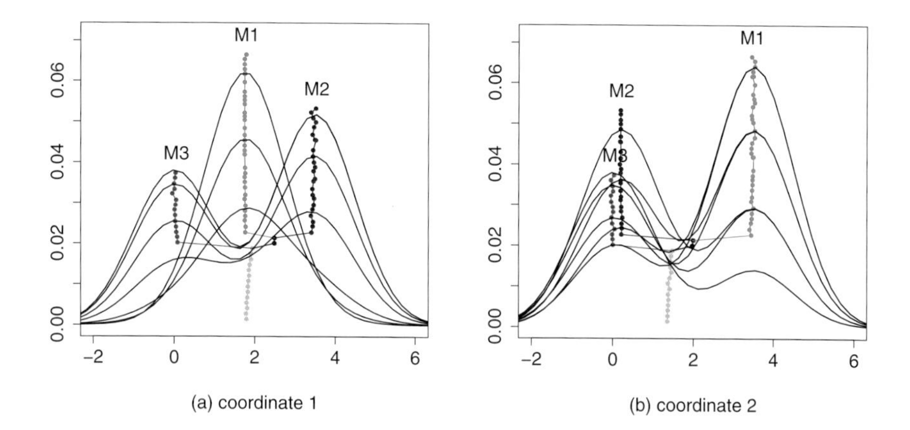

Figure 7.23 *Barycenter plot of a 2D density with slices.* (a) A barycenter plot of the first coordinate with slices. (b) The second coordinate.

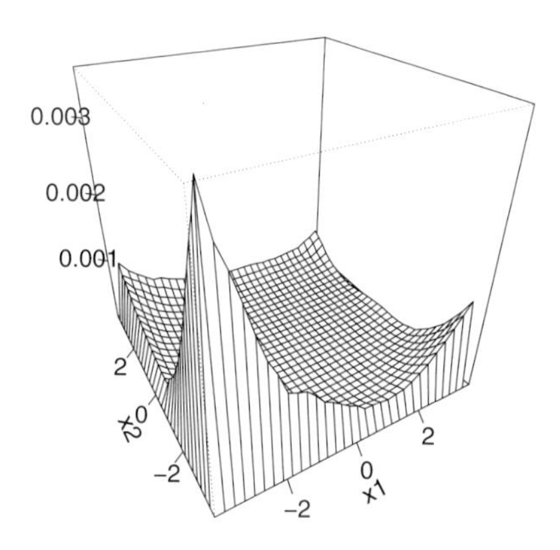

Figure 7.24 *News impact function.* A kernel estimate of the news impact function for the S&P 500 returns.

Figure 7.24 shows the estimate with smoothing parameter $h = 0.9$. Recall that a kernel estimate of the one-dimensional news impact curve is shown in Figure 3.28, and a local linear estimate is shown in Figure 5.7. An ARCH(∞) model for the estimation of the news impact curve is mentioned in (3.64).

Local maxima and minima do not reveal the partial effects of the explanatory variables. We have discussed partial effects in Section 1.1.3. To show the partial

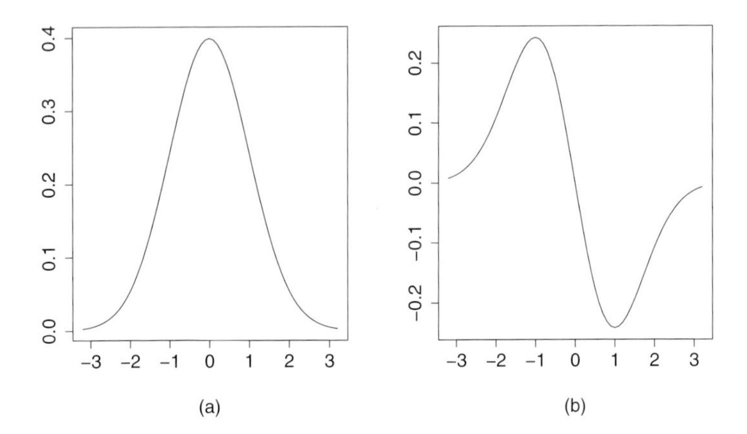

Figure 7.25 *A function and its derivative.* (a) The standard Gaussian density function. (b) The derivative of the standard Gaussian density function.

effects of the explanatory variables, we estimate the partial derivatives of the regression function. We have given two methods for the estimation of partial derivatives: (1) the partial derivatives of a kernel estimator in Section 3.2.9 and (2) a local linear estimator in Section 5.2.1. Before looking at the partial effects, let us recall the concept of derivative. Figure 7.25 illustrates the concept of derivative. Panel (a) shows the density function of the standard Gaussian distribution. The density function is

$$\phi(x) = (2\pi)^{-1/2} \exp\{-x^2/2\}, \qquad x \in \mathbf{R}.$$

Panel (b) shows the derivative of the standard Gaussian density function. The derivative is

$$\phi'(x) = -x\phi(x), \qquad x \in \mathbf{R}.$$

The derivative is positive when the function is increasing, it takes value zero at the maximum of the function, and then the derivative is negative when the function is decreasing. The maximum of the derivative is at the point where the increase of the argument gives the largest increase for the function. The minimum of the derivative is at the point where the increase of the argument gives the largest decrease for the function.

Figure 7.26 shows kernel estimators of the partial derivatives of the news impact function. Panel (a) shows an estimate of the first partial derivative, and panel (b) shows an estimate of the second partial derivative. We have used smoothing parameter $h = 0.9$ and the standard normal kernel.

We can visualize the partial effects with the level set tree-based methods. It is useful to decompose a partial derivative to the positive part and to the negative part. We define the positive and the negative part of a function $g : \mathbf{R}^d \to \mathbf{R}$ as

$$g_+(x) = \max\{g(x), 0\}, \qquad g_-(x) = -\min\{g(x), 0\}.$$

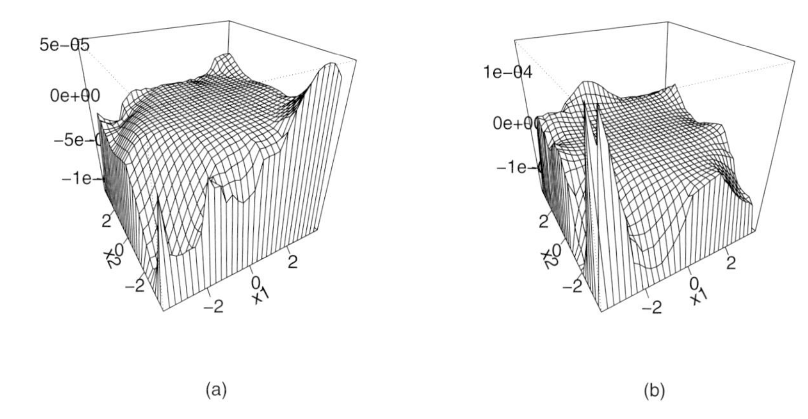

Figure 7.26 *Partial derivatives of the news impact function.* Kernel estimates of the partial derivatives of the news impact function for the S&P 500 returns.

Both the positive and the negative part of a function are nonnegative: $g_+(x) \geq 0$ and $g_-(x) \geq 0$.

Figure 7.27 visualizes the positive part of the kernel estimate of the first partial derivative, shown in Figure 7.26 (a). Panel (a) shows a volume function, panel (b) shows the first coordinate of the barycenter plot, and panel (c) shows the second coordinate of the barycenter plot. The volume function shows that there are two main areas where the partial effect is large. The barycenter plot identifies the areas to be in the two corners. The bigger effect is in the area where the first lag is large and the second lag is small (blue with label "M2"). The smaller effect is in the area where the first lag is large and the second lag is large (red with label "M1").

Figure 7.28 visualizes the negative part of the kernel estimate of the first partial derivative, shown in Figure 7.26 (a). Panel (a) shows a volume function, panel (b) shows the first coordinate of the barycenter plot, and panel (c) shows the second coordinate of the barycenter plot. The volume function shows that there are one main area where the partial effect is large negative. The barycenter plot identifies the area to be in the corner where the first lag is small and the second lag is small (green with label "M3").

7.5 UNIMODAL DENSITIES

Density visualization is useful also in the case of regression function estimation. In regression function estimation we estimate the regression function $f(x) = E(Y \mid X = x)$, but it is important to know how the explanatory variable $X \in \mathbf{R}^d$ is distributed. Having regression data (X_i, Y_i), $i = 1, \ldots, n$, we can use X_i, $i = 1, \ldots, n$, to estimate the density function of X. Level set trees are useful for visualizing multimodal densities; but to visualize unimodal densities, simpler methods can be used.

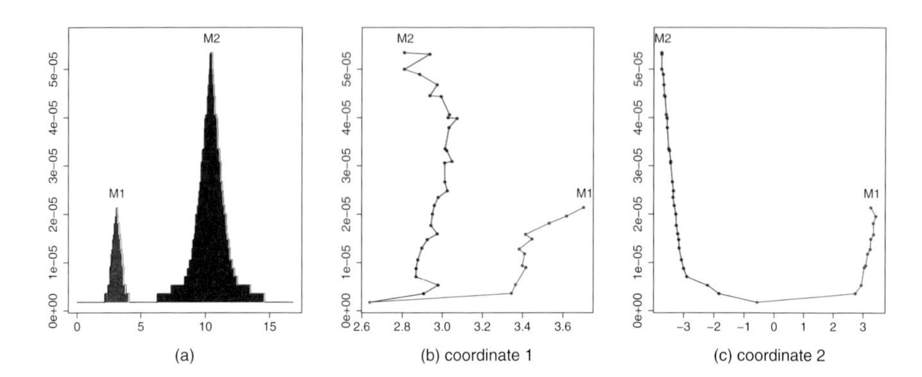

Figure 7.27 *The Positive part of the first partial derivative.* A kernel estimate of the positive part of the first partial derivative of the news impact function for the S&P 500 returns. (a) A volume function. (b) A barycenter plot of the first coordinate. (c) A barycenter plot of the second coordinate.

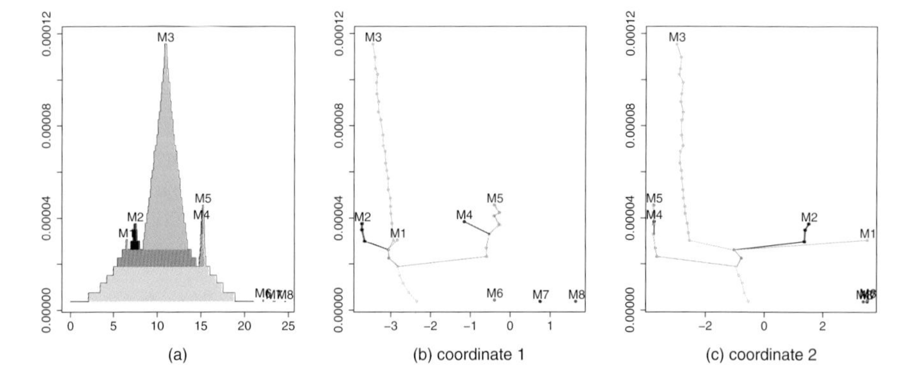

Figure 7.28 *The negative part of the first partial derivative.* A kernel estimate of the negative part of the first partial derivative of the news impact function for the S&P 500 returns. (a) A volume function. (b) A barycenter plot of the first coordinate. (c) A barycenter plot of the second coordinate.

7.5.1 Probability Content of Level Sets

We can describe the heaviness of the tails of the distribution with density function $f : \mathbf{R}^d \to \mathbf{R}$ by looking at the probability content of the level sets

$$\Lambda(f, \lambda) = \left\{x \in \mathbf{R}^d : f(x) \geq \lambda\right\}, \qquad \lambda \in \mathbf{R}.$$

Let us define the function

$$h(p) = P_f\left(\Lambda\left(f, p\,\|f\|_\infty\right)\right), \qquad p \in [0, 1].$$

For a uniform distribution, the function h is identically 1. For other distributions the function h is decreasing. We can estimate the function h with the help of density estimator $\hat{f} : \mathbf{R}^d \to \mathbf{R}$ by setting

$$\hat{h}(p) = \frac{1}{n}\,\#\left\{X_i : \hat{f}(X_i) \geq pM_n\right\},$$

where

$$M_n = \max\left\{\hat{f}(X_i) : i = 1, \ldots, n\right\}.$$

To calculate this estimate, it is enough to evaluate the density estimate $\hat{f}$ at the data points $X_1, \ldots, X_n$.

For a given point $x \in \mathbf{R}^d$, it is of interest to know whether the point is in the central area of the distribution or at the tails of the distribution. This can be found out by calculating

$$P(x) = \frac{1}{n}\,\#\left\{X_i : \hat{f}(X_i) \geq \hat{f}(x)\right\}.$$

If $P(x)$ is close to one, then the point x is in the tail area, and if $P(x)$ is close to zero, then the point x is in the central area. Note that

$$P(x) = \hat{h}\left(\hat{f}(x)/M_n\right).$$

7.5.2 Set Visualization

In addition to the visualization of the mode structure of a density, it is also important to characterize the tail behavior of the density. Some modifications of the level set tree based methods, such as shape trees and tail trees, introduced in Klemelä (2006) and Klemelä (2007), can be used to analyze the tail behavior. Shape trees visualize the shapes of the level sets of a function, and tail trees visualize the shapes of point clouds.

Set visualization can be reduced to the visualization of functions. Let $A \subset \mathbf{R}^d$ and let $f_A : A \to \mathbf{R}$ be a function defined on A. Now set A can be visualized by visualizing the function f_A. For example, we can choose f_A as a distance function, so that

$$f_A(x) = \|x - \mu\|,$$

where $\mu \in A$ is a center point. Alternatively, we can choose f_A as a height function, so that

$$f_A(x) = Px,$$

where $P : \mathbf{R}^d \to \mathbf{R}$ is a projection to one dimension.

APPENDIX A

R TUTORIAL

We start the tutorial with data visualization tools like QQ plots, tail plots, and smooth scatter plots. Then we give an introduction for calculating an estimator in the linear regression, kernel regression, local linear regression, additive model, single-index model, and forward stagewise modeling. Finally we introduce the calculation of linear and kernel quantile regression estimators. The functions are available in the R-package "regpro." This package can be used to learn and to implement regression methods. We introduce also programs of package "denpro," which contains tools for data visualization, function visualization, and density estimation.

A.1 DATA VISUALIZATION

A.1.1 QQ Plots

QQ plots are defined in Section 6.1.2. First we generate data from t-distribution. It is useful to set the seed of the random number generator using the function "set.seed." Then we can later reproduce the results. In the following sections of this tutorial we skip the application of function "set.seed" to save space.

Multivariate Nonparametric Regression and Visualization. By Jussi Klemelä

```
set.seed(1); dendat<-rt(1000,df=6)
```

We compare the sample to the normal distribution. The QQ plot is now the plot of the points

$$\left(x_{(i)}, F^{-1}(p_i)\right), \qquad i = 1, \ldots, n,$$

where $x_{(1)} \leq \cdots \leq x_{(n)}$ is the ordered sample, $p_i = (i - 1/2)/n$, and F is the distribution function of the standard normal distribution. We plot also the line $x = y$ with the function "segments."

```
x<-dendat[order(dendat)]; n<-length(x)
p<-(seq(1:n)-1/2)/n
y<-qnorm(p,mean=mean(x),sd=sd(x))
plot(x,y); segments(-6,-6,6,6)
```

A.1.2 Tail Plots

Tail plots are defined in Section 6.1.2. We show how a left tail plot can be made. The right tail plot can be made analogously. The left tail consists of the observations that are to the left from the median. The data are ordered before plotting. The left tail plot is a scatter plot of the observations in the left tail and the level, defined as the order statistics of the observations. The logarithmic scale is used for the y-axis.

```
split<-median(dendat)
left.tail<-dendat[(dendat<split)]
ord<-order(left.tail,decreasing=TRUE)
ordered.left.tail<-left.tail[ord]
level<-seq(length(left.tail),1)
plot(ordered.left.tail,level,log="y")
```

A.1.3 Two-Dimensional Scatter Plots

Function "plot" can bes used to plot two dimensional scatter plots.

```
n<-20000; dendat<-matrix(rnorm(2*n),n,2); plot(dendat)
```

When the sample size n is large, it is useful to plot binned data, as in Figure 6.1(b). Bin counts can be calculated with the function "pcf.histo" in package "denpro." Parameter `N` gives the number of bins for each direction.

```
N<-c(100,100); pcf<-pcf.histo(dendat,N)
```

Then we transform the bin counts to interval $[0, 1]$, make a scale of gray values, and use the function "plot.histo" from the package "denpro."

```
f<-sqrt(pcf$value)
colo<-1-(f-min(f)+0.5)/(max(f)-min(f)+0.5)
col<-gray(colo); plot.histo(pcf,col=col)
```

Available are also (a) the function "hexbin" in the R package "hexbin," which uses hexagonal binning, and (b) the function "hist2d" in the R package "gplots."

A.1.4 Three-Dimensional Scatter Plots

Let us simulate three-dimensional data.

```
dendat<-matrix(rnorm(3*20000),20000,3)
```

Then we calculate a three-dimensional histogram.

```
N<-c(100,100,100); pcf<-pcf.histo(dendat,N)
```

We use the function "histo2data" to make data points from the histogram. The centers of the bins are the new data points. We rotate the new data and project it to two dimension. Before plotting the data, we order it in such a way that the observations with the largest value for the third coordinate are plotted last.

```
hd<-histo2data(pcf)
alpha<-pi; beta<-pi; gamma<-0
rotdat<-rotation3d(hd$dendat,alpha,beta,gamma)
i1<-1; i2<-2; i3<-3
ord<-order(rotdat[,i3]); plotdat<-rotdat[ord,c(i1,i2)]
plot(plotdat,col=hd$col[ord])
```

A.2 LINEAR REGRESSION

In a two-dimensional linear model the response variable Y satisfies

$$Y = \beta_0 + \beta_1 X_1 + \beta_2 X_2 + \epsilon,$$

where X_1 and X_2 are the explanatory variables and ϵ is an error term.

First we simulate data from a linear model. Let the regression function coefficients be $\beta_0 = 0$, $\beta_1 = 2$, and $\beta_2 = -2$. Let the explanatory variable $X = (X_1, X_2)$ be uniformly distributed on $[0, 1]^2$ and let the error term be $\epsilon \sim N(0, \sigma^2)$. Sample size is n and the number of explanatory variables is $d = 2$.

```
n<-100; d<-2; x<-matrix(runif(n*d),n,d)
y<-matrix(x[,1]-2*x[,2]+0.1*rnorm(n),n,1)
```

The least squares estimator of linear regression coefficients was defined in (2.10) as

$$\hat{\beta} = (\mathbf{X}'\mathbf{X})^{-1}\mathbf{X}'\mathbf{y},$$

where $\mathbf{X}$ is the $n \times (d + 1)$ matrix whose first columns consists of ones, the other columns are the observations from the d explanatory variables, and $\mathbf{y}$ is the $n \times 1$ vector of the observed values of the repsonse variable. In the code below, we first insert the column vector of ones to the original $n \times d$ matrix x. The matrix transpose is calculated with function t(), the matrix multiplication is calculated using the operator $\% * \%$, and the matrix inversion is calculated with function"solve."

```
X<-matrix(0,n,d+1); X[,1]<-1; X[,2:(d+1)]<-x
```

```
XtX<-t(X)%*%X; invXtX<-solve(XtX,diag(rep(1,d+1)))
beta<-invXtX%*%t(X)%*%y
```

The ridge regression estimator was defined in (2.32) as

$$\hat{\beta} = (\mathbf{X}'\mathbf{X} + \lambda I)^{-1}\mathbf{X}'\mathbf{y},$$

where I is the $d \times d$ identity matrix, and $\lambda \geq 0$. In ridge regression the data are normalized to have mean expectations and unit variances.

```
Y<-(y-mean(y))/sd(c(y))
X<-(x-colMeans(x))/sqrt(colMeans(x^2)-colMeans(x)^2)
lambda<-10; XtX<-t(X)%*%X+lambda*diag(rep(1,d))
invXtX<-solve(XtX,diag(rep(1,d)))
beta<-invXtX%*%t(X)%*%Y
```

The above code is included in the function "linear" of package "regpro."

A.3 KERNEL REGRESSION

Kernel regression estimator was defined in (3.6). The kernel estimator is

$$\hat{f}(x) = \sum_{i=1}^{n} p_i(x)\, Y_i,$$

where

$$p_i(x) = \frac{K_h(x - X_i)}{\sum_{i=1}^{n} K_h(x - X_i)}, \qquad i = 1, \ldots, n,$$

$K : \mathbf{R}^d \to \mathbf{R}$ is the kernel function, $K_h(x) = K(x/h)/h^d$, and $h > 0$ is the smoothing parameter.

A.3.1 One-Dimensional Kernel Regression

First we simulate data. Let the regression function be

$$f(x) = \phi(x) + \phi(x - 3),$$

where ϕ is the density function of the standard normal density. Let the distribution of the explanatory variable X be uniform on $[-1, 4]^2$ and let the error term have distribution $\epsilon \sim N(0, \sigma^2)$. Sample size is n.

```
n<-500; x<-5*matrix(runif(n),n,1)-1
phi1D<-function(x){ (2*pi)^(-1/2)*exp(-x^2/2) }
func<-function(t){ phi1D(t)+phi1D(t-3) }
y<-matrix(func(x)+0.1*rnorm(n),n,1)
```

We calculate $\hat{f}(x_0)$, where $x_0 = 0.5$. Let us choose the kernel K to be the Bartlett kernel function $K(x) = (1 - x^2)_+$. The code below is implemented in function "kernesti.regr" of package "regpro."

```
arg<-0.5; h<-0.1
ker<-function(x){ (1-x^2)*(x^2<=1) }
w<-ker((arg-x)/h); p<-w/sum(w)
hatf<-sum(p*y)   # the estimated value
```

Let us plot the estimate on a grid of points, together with the true regression function and the data. We choose N evaluation points.

```
N<-40; t<-5*seq(1,N)/(N+1)-1
hatf<-matrix(0,length(t),1); f<-hatf
for (i in 1:length(t)){
  hatf[i]<-kernesti.regr(t[i],x,y,h=0.2)
  f[i]<-func(t[i])
}
plot(x,y)                                    # data
matplot(t,hatf,type="l",add=TRUE)            # estimate
matplot(t,f,type="l",add=TRUE,col="red") # true function
```

We can also use functions "pcf.kernesti" and "draw.pcf," which allow a more automatic plotting of the function. Function "draw.pcf" is included in package "denpro."

```
pcf<-pcf.kernesti(x,y,h=0.2,N=N)
dp<-draw.pcf(pcf)
plot(x[order(x)],y[order(x)])                 # data
matplot(dp$x,dp$y,type="l",add=TRUE)          # estimate
matplot(dp$x,func(dp$x),type="l",add=TRUE) # true func.
```

A.3.2 Moving Averages

We have defined moving averages of a time series in Section 3.2.4. The two-sided moving average is defined in (3.12) and the one-sided moving average is defined in (3.14). These can be calculated with the function "kernesti.regr" in the following way. In the case of two-sided moving averages we use a two-sided symmetric kernel $K : \mathbf{R} \to \mathbf{R}$. In the case of one-sided moving averages we use a one-sided nonsymmetric kernel $K : [0, \infty) \to \mathbf{R}$.

Let us simulate data from a GARCH(1,1) model. The GARCH(1,1) model was defined in Section 3.9.2 [see (3.65)], as

$$Y_t = \sigma_t \epsilon_t, \qquad \sigma_t^2 = \alpha_0 + \alpha_1 Y_{t-1}^2 + \beta \sigma_{t-1}^2.$$

We take the parameter values from (3.66), where the estimates for S&P 500 returns were given.

```
a0<-7.2*10^(-7); a1<-0.0077; b<-0.92
n<-1000     # sample size
y<-matrix(0,n,1); sigma<-matrix(a0,n,1)
for (t in 2:n){
   sigma[t]<-sqrt( a0+a1*y[t-1]^2+b*sigma[t-1]^2 )
   y[t]<-sigma[t]*rnorm(1)
}
```

We estimate the conditional variance sequentially, using one-sided exponentially weighted moving averages.

```
ewma<-matrix(0,n,1)
for (i in 1:n){
   ycur<-matrix(y[1:i]^2,i,1)
   xcur<-matrix(seq(1:i),i,1)
   ewma[i]<-kernesti.regr(i,xcur,ycur,h=10,kernel="exp")
}
plot(ewma,type="l")
```

To calculate one-sided moving averages, we can use the faster program "ma," contained in the package "regpro."

```
ewma<-matrix(0,n,1)
for (i in 1:n) ewma[i]<-ma(y[1:i]^2,h=10)
plot(ewma,type="l")
```

A.3.3 Two-Dimensional Kernel Regression

Two-dimensional regression function estimates can be visualized with perspective plots and contour plots. In addition, level set tree-based methods can be used.

First we simulate data. The regression function is

$$f(x) = \sum_{i=1}^{3} \phi(x - m_i), \tag{A.1}$$

where $m_1 = (0, 0)$, $m_2 = D \times (0, 1)$, $m_3 = D \times (1/2, \sqrt{3}/2)$, $D = 3$, and ϕ is the density function of the standard normal density. The distribution of the explanatory variables $X = (X_1, X_2)$ is uniform on $[-3, 5]^2$ and the error term ϵ has distribution $N(0, \sigma^2)$.

```
n<-1000; d<-2
x<-8*matrix(runif(n*d),n,d)-3
C<-(2*pi)^(-d/2)
phi<-function(x){ return( C*exp(-sum(x^2)/2) ) }
D<-3; c1<-c(0,0); c2<-D*c(1,0); c3<-D*c(1/2,sqrt(3)/2)
func<-function(x){phi(x-c1)+phi(x-c2)+phi(x-c3)}
for (i in 1:n) y[i]<-func(x[i,])+0.01*rnorm(1)
```

We calculate $\hat{f}(x_0)$, where $x_0 = (0.5, 0.5)$. Let us choose the kernel K to be the standard normal density function. The code below is implemented in function "kernesti.regr" of package "regpro."

```
arg<-c(0.5,0.5); h<-0.5
ker<-function(x){ return( exp(-rowSums(x^2)/2) ) }
argu<-matrix(arg,dim(x)[1],d,byrow=TRUE)
w<-ker((argu-x)/h); p<-w/sum(w)
hatf<-sum(p*y)  # the estimate
```

Perspective Plots and Contour Plots Let us plot the estimate on a grid of points. A 2D function can be plotted with a perspective plot or with a contour plot. We plot below also the true regression function.

```
num<-30  # number of grid points in one direction
t<-8*seq(1,num)/(num+1)-3; u<-t
hatf<-matrix(0,num,num); f<-hatf
for (i in 1:num){
   for (j in 1:num){
      arg<-matrix(c(t[i],u[j]),1,2)
      hatf[i,j]<-kernesti.regr(arg,x,y,h=0.5)
      f[i,j]<-phi(arg-c1)+phi(arg-c2)+phi(arg-c3)
   }
}
contour(t,u,hatf)    # kernel estimate
persp(t,u,hatf,ticktype="detailed",phi=30,theta=-30)
contour(t,u,f)        # true function
persp(t,u,f,phi=30,theta=-30,ticktype="detailed")
```

We can use also functions "pcf.kernesti" and "draw.pcf" that allow a more automatic plotting of the function. Function "draw.pcf" is in package "denpro."

```
pcf<-pcf.kernesti(x,y,h=0.5,N=c(num,num))
dp<-draw.pcf(pcf,minval=min(y))
persp(dp$x,dp$y,dp$z,phi=30,theta=-30)
contour(dp$x,dp$y,dp$z,nlevels=30)
```

Level Set Trees Function"pcf.kernesti" gives an output that can be used to calculate level set trees, using package "denpro." Funnction "leafsfirst" calculates a level set tree using the Leafsfirst algorithm. Function "plotvolu" plots a volume plot. Parameter "cutlev" can be used to cut the lower part of the tree and to zoom into the higher levels. Function "plotbary" plots the barycenter plot of the level set tree. Parameter "coordi" is used to choose the coordinate of the barycenter plot. Function "plottree" plots the tree structure of the level set tree. Function "treedisc" can be used to prune the level set tree to make the plotting faster.

```
pcf<-pcf.kernesti(x,y,h=0.5,N=c(15,15))
```

```
lst<-leafsfirst(pcf,lowest="regr")

plotvolu(lst,lowest="regr")
plotvolu(lst,lowest="regr",cutlev=0.05)

plotbary(lst,lowest="regr") # barycenter plot,1st coord.
plotbary(lst,coordi=2,lowest="regr") # 2nd coord.

plottree(lst,lowest="regr") # plot level set tree

td<-treedisc(lst,pcf,ngrid=30,lowest="regr")
plotvolu(td,modelabel=FALSE,lowest="regr")
```

A.3.4 Three- and Higher-Dimensional Kernel Regression

Three- and higher-dimensional regression function estimates can be visualized by one- or two-dimensional slices, partial dependency plots, and level set tree-based visualizations.

First we simulate data. The regression function is a mixture

$$f(x) = \sum_{i=1}^{4} \phi(x - c_i),$$

where $c_1 = D \times (1/2, 0, 0)$, $c_2 = D \times (-1/2, 0, 0)$, $c_3 = D \times (0, \sqrt{3}/2, 0)$, $c_4 = D \times (0, 1/(2\sqrt{3}), \sqrt{2/3})$, $D = 3$, and ϕ is the density function of the standard normal density. The distribution of the explanatory variables $X = (X_1, X_2, X_3)$ is uniform on $[-3, 3]^3$ and the error term ϵ has distribution $N(0, \sigma^2)$. The sample size is denoted n and the number of explanatory variables is denoted d.

```
n<-1000; d<-3; x<-8*matrix(runif(n*d),n,d)-3
C<-(2*pi)^(-d/2)
phi<-function(x){ return( C*exp(-sum(x^2)/2) ) }
D<-3; c1<-D*c(1/2,0,0);c2<-D*c(-1/2,0,0)
c3<-D*c(0,sqrt(3)/2,0);c4<-D*c(0,1/(2*sqrt(3)),sqrt(2/3))
fun<-function(x){phi(x-c1)+phi(x-c2)+phi(x-c3)+phi(x-c4)}
y<-matrix(0,n,1)
for (i in 1:n) y[i]<-fun(x[i,])+0.01*rnorm(1)
```

Slices A one-dimensional slice of regression function estimate $\hat{f} : \mathbf{R}^d \to \mathbf{R}$ is

$$g(x_1) = \hat{f}(x_1, x_{0,2}, \ldots, x_{0,d}),$$

where $x_{0,2}, \ldots, x_{0,d}$ is fixed point. We can choose $x_{0,k}$ to be the sample median of the kth coordinate of the vector of explanatory variables: $x_{0,k} = \text{median}(X_{1,k}, \ldots, X_{n,k})$ for $k = 2, \ldots, d$.

One-dimensional slices can be calculated using function "pcf.kernesti.slice." The parameter $p = 0.5$ indicates that the fixed point is the median. The parameter N gives the number of evaluation points.

```
pcf<-pcf.kernesti.slice(x,y,h=0.5,N=50,coordi=1,p=0.5)
dp<-draw.pcf(pcf); plot(dp$x,dp$y,type="l")
```

We can compare the one-dimensional slice to the one dimensional slice of the linear regression estimate.

```
coordi<-1; notcoordi<-c(2,3)
li<-linear(x,y); a<-li$beta1[notcoordi]
x0<-c(median(x[,2]),median(x[,3]))
flin<-li$beta0+li$beta1[coordi]*dp$x+sum(a*x0)
plot(dp$x,flin,type="l")
```

We can now plot the slice of kernel regression estimate and the slice of linear regression in the same figure.

```
ylim<-c(min(flin,dp$y),max(flin,dp$y))
matplot(dp$x,flin,type="l",ylim=ylim)
matplot(dp$x,dp$y,type="l",add=TRUE)
```

Partial Dependence Functions A partial dependence function eas defined in Section 7.2. A one-dimensional partial dependence function is

$$g_{X_1}(x_1) = Ef(x_1, X_2, \ldots, X_d),$$

where $f(x) = E(Y \mid X = x)$ and $X = (X_1, \ldots, X_d)$. We can use function "pcf.kernesti.marg" to calculate a kernel estimate of a one dimensional partial dependence function.

```
pcf<-pcf.kernesti.marg(x,y,h=0.5,N=30,coordi=1)
dp<-draw.pcf(pcf); plot(dp$x,dp$y,type="l")
```

Level Set Trees Function"pcf.kernesti" gives an output that can be used to calculate level set trees, using package "denpro." We proceed as explained in Section A.3.3, where two-dimensional regression function was visualized.

```
pcf<-pcf.kernesti(x,y,h=0.5,N=c(15,15,15))
lst<-leafsfirst(pcf,lowest="regr")
td<-treedisc(lst,pcf,ngrid=30,lowest="regr")
plotvolu(td,modelabel=FALSE,lowest="regr")
plotbary(td,coordi=3,lowest="regr") # 3rd coord.
plottree(td,lowest="regr") # level set tree
```

A.3.5 Kernel Estimator of Derivatives

The kernel estimator of a partial derivative was defined in Section 3.2.9; see (3.28). The kernel estimator of the kth partial derivative, with the Gaussian kernel, is

$$\widehat{D_k f}(x) = \sum_{i=1}^{n} q_i(x)\, Y_i,$$

where

$$q_i(x) = \frac{1}{\sum_{i=1}^{n} K_h(x - X_i)} \left(\frac{\partial}{\partial x_k} K_h(x - X_i) - p_i(x) \sum_{i=1}^{n} \frac{\partial}{\partial x_k} K_h(x - X_i) \right),$$

where

$$\frac{\partial}{\partial x_k} K_h(x - X_i) = \frac{1}{h^{d+1}} (D_k K) \left(\frac{x - X_i}{h} \right),$$

$D_k K(x) = -x_k K(x)$, and $p_i(x)$ are the weights of the kernel estimator.

One-Dimensional Estimator of Derivatives First we simulate data. Let the regression function be

$$f(x) = \int_{-1}^{x} (\phi(t) + \phi(t-3))\, dt, \qquad x \in [-1, 4],$$

where ϕ is the density function of the standard normal density. Let the distribution of the explanatory variable X be uniform on $[-1, 4]^2$ and let the error term ϵ has distribution $N(0, \sigma^2)$.

```
n<-1000; x<-5*matrix(runif(n),n,1)-1
phi1D<-function(x){ (2*pi)^(-1/2)*exp(-x^2/2) }
func0<-function(t){ phi1D(t)+phi1D(t-3) }
func<-function(t){
    ngrid<-1000; step<-5/ngrid; grid<-seq(-1,4,step)
    i0<-floor((t+1)/step)
    return( step*sum( func0(grid[1:i0]) ) )
}
y<-matrix(0,n,1)
for (i in 1:n) y[i]<-func(x[i])+0.001*rnorm(1)
```

The code below calculates $\widehat{f}'(x_0)$, where $x_0 = 0$. This code is implemented in function "kernesti.der" of package "regpro."

```
arg<-0; h<-0.5; ker<-phi1D
dker<-function(t){ return( -t*exp(-t^2/2) ) }
w<-ker((arg-x)/h); p<-w/sum(w)
u<-dker((arg-x)/h)/h^2; q<-1/sum(w)*(u-p*sum(u))
hatf<-sum(y*q) # the estimated value
```

Let us plot the estimate on a grid of points.

```
t<-5*seq(1,n)/(n+1)-1
hatf<-matrix(0,length(t),1); f<-hatf
for (i in 1:length(t)){
  hatf[i]<-kernesti.der(t[i],x,y,h=0.5)
  f[i]<-func0(t[i])
}
ylim<-c(min(f,hatf),max(f,hatf))
matplot(t,hatf,type="l",ylim=ylim)        # estimate
matplot(t,f,type="l",add=TRUE,col="red") # true func.
```

We can use also functions "pcf.kernesti.der" and "draw.pcf," which allow a more automatic plotting of the function. Function "draw.pcf" is included in package "denpro."

```
pcf<-pcf.kernesti.der(x,y,h=0.5,N=n)
dp<-draw.pcf(pcf)
plot(x,y); matplot(dp$x,dp$y,type="l",add=TRUE)
```

Two- and Higher-Dimensional Estimator of Derivatives First we simulate data. Let the regression function be

$$f(x_1, x_2) = f_1(x_1) f_1(x_2),$$

where f_1 is the function defined in (A.1) and $x_1, x_2 \in [-1, 4]$. The distribution of the explanatory variable X is uniform on $[-1, 4]^2$. The error term ϵ has distribution $N(0, \sigma^2)$. The sample size is n and the number of explanatory variables is $d = 2$.

```
n<-1000; d<-2; x<-5*matrix(runif(n*d),n,d)-1
phi1D<-function(x){ (2*pi)^(-1/2)*exp(-x^2/2) }
func0<-function(t){ phi1D(t)+phi1D(t-3) }
func1<-function(t){
    ngrid<-1000; step<-5/ngrid; grid<-seq(-1,4,step)
    i0<-floor((t+1)/step)
    return( step*sum( func0(grid[1:i0]) ) )
}
func<-function(t){ func1(t[1])*func1(t[2]) }
y<-matrix(0,n,1)
for (i in 1:n) y[i]<-func(x[i,])+0.001*rnorm(1)
```

The code below calculates the first partial derivative $\widehat{D_1 f}(0)$, where $D_1 f(0) = \partial f(x)/\partial x_1|_{x=0}$. This code below is implemented in function "kernesti.der" of package "regpro." Parameter `direc` is set equal to one to estimate the first partial derivative.

```
direc<-1; arg<-c(0,0); h<-0.5
ker<-function(xx){ exp(-rowSums(xx^2)/2) }
```

```
C<-(2*pi)^(-1)
dker<-function(xx){-C*xx[,direc]*exp(-rowSums(xx^2)/2)}
argu<-matrix(arg,n,d,byrow=TRUE)
w<-ker((argu-x)/h); p<-w/sum(w)
u<-dker((argu-x)/h)/h^(d+1); q<-1/sum(w)*(u-p*sum(u))
value<-q%*%y  # the estimated value
```

Let us plot a perspective plot and contour plot of the estimate of the first partial derivative and of the true first partial derivative. Parameter `num` gives the number of grid points in one direction.

```
num<-30; t<-5*seq(1,num)/(num+1)-1; u<-t
hatf<-matrix(0,num,num); df<-hatf
for (i in 1:num){ for (j in 1:num){
      arg<-matrix(c(t[i],u[j]),1,2)
      hatf[i,j]<-kernesti.der(arg,x,y,h=0.5)
      df[i,j]<-func0(arg[1])*func1(arg[2])
} }
contour(t,u,hatf)   # kernel estimate
persp(t,u,hatf,ticktype="detailed",phi=3,theta=-30)
contour(t,u,df)     # true function
persp(t,u,df,phi=30,theta=-30,ticktype="detailed")
```

We can use functions "pcf.kernesti.der" and "draw.pcf" to make perspective plots and contour plots. Function "draw.pcf" is in package "denpro."

```
pcf<-pcf.kernesti.der(x,y,h=0.5,N=c(50,50),direc=1)
dp<-draw.pcf(pcf)
persp(dp$x,dp$y,dp$z,phi=3,theta=-30)
contour(dp$x,dp$y,dp$z)
```

We can also use level set tree-based methods, as in Section A.3.3.

```
pcf<-pcf.kernesti.der(x,y,h=0.5,N=c(15,15))
lst<-leafsfirst(pcf,lowest="regr")
plotvolu(lst,lowest="regr")           # volume plot
plotbary(lst,coordi=1,lowest="regr") # barycenter plot
plottree(lst,lowest="regr")           # level set tree
```

A.3.6 Combined State– and Time–Space Smoothing

Locally stationary data were discussed in Section 3.2.5. To estimate a regression function with locally stationary data, it can be useful to combine state–space kernel estimator with moving averages. The kernel estimator combining time– and state–space smoothing, as defined in (3.20), is

$$\hat{f}_t(x) = \sum_{i=1}^{t} w_i(x,t)\, Y_i,$$

where the weights have the form

$$w_i(x,t) = \frac{K((x - X_i)/h)\, L((t-i)/g)}{\sum_{j=1}^{n} K((x - X_j)/h)\, L((t-j)/g)}, \qquad i = 1, \ldots, t,$$

where $K : \mathbf{R}^d \to \mathbf{R}$, $L : \mathbf{R} \to \mathbf{R}$ are kernel functions and $h > 0$, $g > 0$ are smoothing parameters.

Let us simulate locally stationary data. Let the regression functions be

$$f_t(x) = 0.5\,\phi\left(x - \mu_t^{(1)}\right) + 0.5\,\phi\left(x - \mu_t^{(2)}\right),$$

where $\mu_t^{(1)} = -2t/T$, $\mu_t^{(2)} = 2t/T$, and ϕ is the density function of the standard normal distribution. The design variables X_t are i.i.d. $N(0,1)$ and the errors ϵ_t are i.i.d. $N(0, 0.1^2)$.

```
n<-1000; x<-matrix(rnorm(n),n,1); y<-matrix(0,n,1)
for (i in 1:n){
   mu1<--i/n*2; mu2<-i/n*2
   func<-function(t){
        return( 0.5*dnorm(t-mu1)+0.5*dnorm(t-mu2) )
   }
   y[i]<-func(x[i])+0.1*rnorm(1)
}
```

Now we apply function “kernesti.regr.” The smoothing parameter h is the state–space smoothing parameter and the smoothing parameter g is the time–space smoothing parameter.

```
arg<-0; kernesti.regr(arg,x,y,h=1,g=10,gernel="exp")
```

A.4 LOCAL LINEAR REGRESSION

Local linear estimator was discussed in Section 5.2.1.

A.4.1 One-Dimensional Local Linear Regression

We use the same simulated data as was used in Section A.3.1 to illustrate one–dimensional kernel regression.

Let us choose the kernel K to be the Bartlett function $K(x) = (1 - x^2)_+$. We want to calculate $\hat{f}(x_0)$, where $x_0 = 0.5$. The weights of the one–dimensional local linear regression are given in (5.9).

```
arg<-0.5; h<-0.5
ker<-function(x){ (1-x^2)*(x^2<=1) }
w<-ker((arg-x)/h); p<-w/sum(w)
```

```
barx<-sum(p*x); bary<-sum(p*y)
q<-p*(1-((x-barx)*(barx-arg))/sum(p*(x-barx)^2))
hatf<-sum(q*y)
```

The above code is implemented in function "loclin" of package "regpro." Let us plot the estimate on a grid of points.

```
N<-40; t<-5*seq(1,N)/(N+1)-1
hatf<-matrix(0,length(t),1); f<-hatf
for (i in 1:length(t)){
   hatf[i]<-loclin(t[i],x,y,h=0.5,kernel="bart")
   f[i]<-phi1D(t[i])+phi1D(t[i]-3)
}
plot(x,y)                              # data
matplot(t,hatf,type="l",add=TRUE) # estimate
matplot(t,f,type="l",add=TRUE)    # true function
```

We can use also functions "pcf.loclin" and "draw.pcf" that allow a more automatic plotting of the function. Function "draw.pcf" is in package "denpro."

```
pcf<-pcf.loclin(x,y,h=0.5,N=40,kernel="bart")
dp<-draw.pcf(pcf)
plot(x,y)
matplot(dp$x,dp$y,type="l",add=TRUE)
matplot(dp$x,func(dp$x),type="l",add=TRUE)
```

A.4.2 Two-Dimensional Local Linear Regression

We use the same two-dimensional simulated data as was used in Section A.3.3 to illustrate two-dimensional kernel regression.

The local linear regression estimator was defined in (5.6) as a solution of weighted linear least squares estimator. The weighted least squares estimator is

$$\hat{\beta} = (\mathbf{X}'W(x)\mathbf{X})^{-1}\mathbf{X}'W(x)\mathbf{y},$$

where $\mathbf{X}$ is the $n \times (d+1)$ matrix whose first column consists of ones, the other columns are the observations from the d explanatory variables, and $\mathbf{y}$ is the $n \times 1$ vector of the observed values of the response variable.

First we calculate the matrix `W` of the kernel weights. Object `x` is the $n \times d$ matrix of the observed values of the explanatory variables. In this example we have $d = 2$.

```
arg<-c(0,0); argu<-matrix(arg,n,d,byrow=TRUE)
ker<-function(x){ return( exp(-rowSums(x^2)/2) ) }
w<-ker((x-argu)/h); weights<-w/sum(w); W<-diag(weights)
```

In the code below we first insert the column vector of ones to the original $n \times d$ matrix `x`. The matrix transpose is calculated with function `t()`, and the matrix multiplication is calculated using the operator $\% * \%$. The matrix inversion can be calculated with function"solve." We obtain the vector `esti` of three elements. The first element is the estimate of the regression function at one point. The second element is the estimate of the first partial derivative, and the third element is the estimate of the second partial derivative.

```
X<-cbind(matrix(1,n,1),x-argu)
A<-t(X)%*%W%*%X; invA<-solve(A,diag(rep(1,d+1)))
esti<-invA%*%t(X)%*%W%*%y; estimate<-esti[1]
```

We can use also functions "pcf.loclin" and "draw.pcf," which allow a more automatic plotting of the function. Function "draw.pcf" is included in package "denpro."

```
pcf<-pcf.loclin(x,y,h=0.5,N=c(20,20))
dp<-draw.pcf(pcf)
persp(dp$x,dp$y,dp$z,ticktype="detailed",phi=30,theta=3)
contour(dp$x,dp$y,dp$z,nlevels=30)
```

A.4.3 Three- and Higher-Dimensional Local Linear Regression

When the functions are three- and higher-dimensional, we cannot use perspective plots and contour plots. However, we can use level set tree-based methods. Let us use the same three-dimensional data which was used to illustrate three-dimensional kernel estimation in Section A.3.4.

Function"pcf.loclin" gives an output that can be used to calculate level set trees, using package "denpro." We proceed as explained in Section A.3.3, where two-dimensional regression function was visualized. After calculating a level set tree, we plot volume plots and a barycenter plot.

```
pcf<-pcf.loclin(x,y,h=0.5,N=c(15,15,15))
lst<-leafsfirst(pcf,lowest="regr")

td<-treedisc(lst,pcf,ngrid=30,lowest="regr")
plotvolu(td,modelabel=FALSE,lowest="regr")
plotvolu(td,modelabel=FALSE,lowest="regr",cutlev=0.03)

plotbary(td,coordi=1,lowest="regr")
```

A.4.4 Local Linear Derivative Estimation

We can use the same two-dimensional data to illustrate local linear partial derivative estimation as was used to illustrate kernel estimation of partial derivatives in Section A.3.5. We use function "pcf.loclin" with the argument `type`. The argument `type` is set to 1 to estimate the first partial derivative, it is set to 2 to estimate the second partial derivative, and similarily in the higher- than two-dimensional cases. Function "pcf.loclin" is included in package "regpro," and function "draw.pcf" is included in package "denpro."

```
pcf<-pcf.loclin(x,y,h=0.5,N=c(20,20),type=1)
dp<-draw.pcf(pcf)
persp(dp$x,dp$y,dp$z,ticktype="detailed",phi=3,theta=30)
contour(dp$x,dp$y,dp$z,nlevels=30)
```

A.5 ADDITIVE MODELS: BACKFITTING

The additive models were covered in Section 4.2, where the backfitting algorithm was presented. In the two-dimensional additive model the response variable Y can be written

$$Y = f_1(X_1) + f_2(X_2) + \epsilon,$$

where X_1 and X_2 are the explanatory variables, $f_k : \mathbf{R} \to \mathbf{R}$ are the unknown components, and ϵ is an error term.

First we simulate data from an additive model. The distribution of the explanatory variable $X = (X_1, X_2)$ is uniform on $[0, 1]^2$. The regression function is $f(x_1, x_2) = f_1(x_1) + f_2(x_2)$, where $f_1(x_1) = x_1^2 - EX_1^2$ and $f_2(x_2) = \log(x_2) - E\log(X_2)$. The response variable is $Y = f(x_1, x_2) + \epsilon$, where $\epsilon \sim N(0, \sigma^2)$.

```
n<-100; d<-2; x<-matrix(runif(n*d),n,d)
fun1<-function(t){t^2}; fun2<-function(t){log(t)}
f<-matrix(0,n,d)
f[,1]<-fun1(x[,1])-mean(fun1(x[,1]))
f[,2]<-fun2(x[,2])-mean(fun2(x[,2]))
y<-f[,1]+f[,2]+0.1*rnorm(n)
```

We estimate the additive model using function "additive." We need to give as arguments the $n \times d$-matrix x of the values of the explanatory variables, the n vector y of the values of the response variable, the smoothing parameter h > 0, and the number of iterations M ≥ 1.

```
h<-0.1; M<-5; est<-additive(x,y,h=h,M=M)
```

The output est$eval is an $n \times d$ matrix that contains the evaluations $\hat{f}_k(X_{i,k})$, $i = 1, \ldots, n$, $k = 1, \ldots, d$, where X_{ik} is the kth component of the ith observation. Next we plot the components of the estimate. The code below plots the estimate of the first component and the true first component, The functions are plotted at the observations X_{i1}, $i = 1, \ldots, n$.

```
or<-order(x[,1]); t<-x[or,1]
hatf1<-est$eval[or,1]; f1<-f[or,1]
plot(t,y[or])                        # data
matplot(t,hatf1,type="l",add=TRUE)   # estimate
matplot(t,f1,type="l",add=TRUE)      # true function
```

We can evaluate the estimate on a regular grid. We need to give the matrix est$eval as an argument. The matrix was calculated at a previous step. The code below calculates the estimates $\hat{f}_1$ and $\hat{f}_2$ and the estimate $\hat{f}(x_1, x_2) = \bar{y} + \hat{f}_1(x_1) + \hat{f}_2(x_2)$ on a grid, where $\bar{y}$ is the mean of the values of the response variable. Parameter `num` gives the number of grid points in one direction.

```
num<-50; t<-seq(1,num)/(num+1); u<-t
hatf<-matrix(0,num,num)
hatf1<-matrix(0,num,1); hatf2<-matrix(0,num,1)
func<-function(arg){
  additive(x,y,arg,h=h,M=M,eval=est$eval)$valvec
}
for (i in 1:num){ for (j in 1:num){
      valvec<-func(c(t[i],u[j]))
      hatf1[i]<-valvec[1]; hatf2[j]<-valvec[2]
      hatf[i,j]<-mean(y)+sum(valvec)
} }
plot(t,hatf1,type="l")
persp(t,u,hatf,ticktype="detailed",phi=30,theta=-30)
```

Function "pcf.additive" can be used for perspective plots, contour plots, and level set trees.

```
N<-c(50,50)
pcf<-pcf.additive(x,y,N=N,h=h,eval=est$eval,M=M)

dp<-draw.pcf(pcf,minval=min(pcf$value))
persp(dp$x,dp$y,dp$z,phi=30,theta=-30)
contour(dp$x,dp$y,dp$z,nlevels=30)

lst<-leafsfirst(pcf)
plotvolu(lst,lowest="regr")
```

A.6 SINGLE-INDEX REGRESSION

The single-index model was defined in Section 4.1. In the single-index model the response variable can be written

$$Y = g(\theta' X) + \epsilon,$$

where $X \in \mathbf{R}^d$, $\theta \in \mathbf{R}^d$ is an unknown direction vector with $\|\theta\| = 1$, and $g : \mathbf{R} \to \mathbf{R}$ is an unknown link function.

First we simulate data from a single-index model. The distribution of the vector $X = (X_1, X_2)$ of the explanatory variables is the standard normal 2D distribution. The error term is $\epsilon \sim N(0, \sigma^2)$. The index vector is $\theta = (0, 1)$ and the link function g is the distribution function Φ of the standard normal distribution.

```
n<-100; x<-matrix(rnorm(n*2),n,2)
theta<-matrix(c(0,1),2,1); x1d<-x%*%theta
y<-pnorm(x1d)+0.1*rnorm(n)
```

A.6.1 Estimating the Index

We cover four estimators discussed in Section 4.1.2: derivative method, average derivative method, numerical minimization to find the least squares solution, and a stagewise algorithm to find the least squares solution.

The average derivative method was defined in (4.9). The following commands can be used to to estimate the direction vector θ with the average derivative method.

```
method<-"aved"; h<-1.5
hat.theta<-single.index(x,y,h=h,method=method)
```

The derivative method was defined in (4.8). In the derivative method we have to specify additionally the point at which the gradient is estimated. We choose the point $(0, 0)$.

```
method<-"poid"; h<-1.5; argd<-c(0,0)
hat.theta<-single.index(x,y,h=h,method=method,argd=argd)
```

The direction vector can be estimated using numerical minimization to find the solution in the least squares problem (4.2). The numerical minimization can be performed with the following commands.

```
method<-"nume"; h<-1.5
hat.theta<-single.index(x,y,h=h,method=method)
```

The least squares problem (4.2) can be solved by using an iterative method, as explained in Section 4.1.2. The following commands can be used to apply the iterative method. Argument `M` gives the number of iterations.

```
method<-"iter"; h<-1.5; M<-10
hat.theta<-single.index(x,y,h=h,method=method,M=M)
```

A.6.2 Estimating the Link Function

After estimating the direction vector θ, we have to estimate the link function $g : \mathbf{R} \to \mathbf{R}$. We do this by the kernel estimator using the following commands, which plot also the true link function.

```
x1d<-x%*%hat.theta  # project data to 1D
pcf<-pcf.kernesti(x1d,y,h=0.3,N=20)
dp<-draw.pcf(pcf)
matplot(dp$x,dp$y,type="l",ylim=c(min(y,0),max(y,1)))
matplot(dp$x,pnorm(dp$x),type="l",add=TRUE)
```

A.6.3 Plotting the Single-Index Regression Function

We can estimate the regression function $f(x) = g(\theta' x)$ on a grid with the function "pcf.single.index."

```
h<-0.3; N<-c(40,40); method<-"poid"
pcf<-pcf.single.index(x,y,h=h,N=N,method=method)
dp<-draw.pcf(pcf)
persp(dp$x,dp$y,dp$z,ticktype="detailed",phi=3,theta=30)
```

If the data do not come from a single-index model, it can be better to use the derivative method, and estimate the gradient at the point of evaluation. We give an example where the regression function is the standard normal density function. First we simulate data.

```
n<-1000; x<-matrix(6*runif(n*2)-3,n,2); C<-(2*pi)^(-1)
phi<-function(x){ return( C*exp(-rowSums(x^2)/2) ) }
y<-phi(x)+0.1*rnorm(n)
```

Then we use the derivative method and estimate the direction so that the gradient is estimated separately at each the point of evaluation.

```
method<-"poid"; h<-1.5; num<-50
t<-6*seq(1,num)/(num+1)-3; u<-t
hatf<-matrix(0,num,num)
for (i in 1:num){  for (j in 1:num){
     arg<-c(t[i],u[j])
     hatf[i,j]<-
     single.index(x,y,arg=arg,h=h,method=method,argd=arg)
} }
persp(t,u,hatf,ticktype="detailed",phi=30,theta=-30)
```

A.7 FORWARD STAGEWISE MODELING

Algorithms for forward stagewise modeling were given in Section 5.4. We include the stagewise fitting of additive models of Section 5.4.2 and projection pursuit regression of Section 5.4.3 to this tutorial.

A.7.1 Stagewise Fitting of Additive Models

An algorithm for stagewise fitting of additive models was given in Section 5.4.2. The stagewise fitting is an alternative method to backfitting for finding an estimate with the additive structure. We use the same simulated data as in the case of backfitting the additive model in Section A.5.

We estimate the additive model using function "additive.stage." The arguments are the $n \times d$ matrix x of the values of the explanatory variables, the n vector y of the values of the response variable, the smoothing parameter $h > 0$, and the number of iterations M.

```
h<-0.1; M<-5; est<-additive.stage(x,y,h=h,M=M)
```

The output "est$eval" is the $n \times d$ matrix that contains the evaluations $\hat{f}_k(X_{i,k})$, $i = 1, \ldots, n$, $k = 1, \ldots, d$. We can use the same code to plot the components as in the case of the backfitting algorithm.

We can evaluate the estimate on a regular grid. The evaluation proceeds differently than in the case of backfitting, because now we need to give as arguments the $n \times M$ matrix of residuals and the M vector, which indicates which variable is chosen at each step. These are given as "est$residu" and "est$deet." Note that in the case of backfitting it is enough to calculate the $n \times d$ matrix of final y-values, whereas to evaluate the stagewise estimator at an arbitrary point, we need to calculate and save the complete $n \times M$ matrix of residuals and the indicator vector which contains the estimated direction for each step.

```
num<-50; t<-seq(1,num)/(num+1); u<-t
hatf<-matrix(0,num,num)
funi<-function(t,u){
   additive.stage(x,y,c(t,u),h=h,M=M,residu=est$residu,
                  deet=est$deet)$value
}
for (i in 1:num){ for (j in 1:num){
      hatf[i,j]<-funi(t[i],u[j])
} }
persp(t,u,hatf,ticktype="detailed",phi=30,theta=3)
```

A.7.2 Projection Pursuit Regression

An algorithm for projection pursuit regression was given in Section 5.4.3. We simulate data where $X = (X_1, X_2)$ is uniformly distributed on $[-3, 3]^2$, $\epsilon \sim N(0, \sigma^2)$, and the regression function is the density function of the standard 2D normal distribution.

```
n<-1000; x<-matrix(6*runif(n*2)-3,n,2)
phi<-function(x){ (2*pi)^(-1)*exp(-rowSums(x^2)/2) }
y<-phi(x)+0.1*rnorm(n)
```

We calculate the projection pursuit regression function estimate using function "pp.regression." The arguments are the $n \times d$-matrix x of the values of the explanatory variables, the n vector y of the values of the response variable, the smoothing parameter $h > 0$, and the number of iterations M.

```
h<-0.5; M<-3; est<-pp.regression(x,y,h=h,M=M)
```

The output "est$eval" is a vector of length n that contains the evaluations $\hat{f}(X_i)$, $i = 1, \ldots, n$. We can evaluate the estimate on a regular grid. We need to give as arguments the $n \times M$ matrix of residuals and the M vector of directions $\hat{\theta}_m$. These are given as "est$residu" and "est$teet."

```
num<-30; t<-6*seq(1,num)/(num+1)-3; u<-t
```

```
hatf<-matrix(0,num,num)
funi<-function(t,u){
    pp.regression(x,y,c(t,u),h=h,M=M,residu=est$residu,
    teet=est$teet)$value
}
for (i in 1:num){  for (j in 1:num){
      hatf[i,j]<-funi(t[i],u[j])
} }
persp(t,u,hatf,ticktype="detailed",phi=20,theta=-30)
```

A.8 QUANTILE REGRESSION

Quantile regression with the kernel method was introduced in Section 3.8. Linear quantile regression was introduced in Section 5.1.2. Let us use the same data as was applied in Section A.3.1 to illustrate one-dimensional kernel regression.

A.8.1 Linear Quantile Regression

Function "linear.quan" implements linear quantile regression and it is included in package "regpro."

```
li<-linear.quan(x,y,p=0.1)
N<-50; t<-5*seq(1,N)/(N+1)-1; qhat<-matrix(0,N,1)
for (i in 1:N) qhat[i]<-li$beta0+li$beta1*t[i]
plot(x,y); lines(t,qhat)
```

Let us look at the code of function "linear.quan." First we define function "fn," which calculates the quantile loss, when the argument "b" is the $d+1$ vector of the intercept and the coefficients of a linear function.

```
p<-0.1; n<-dim(x)[1]; d<-dim(x)[2]
rho<-function(t){ t*(p-(t<0)) }
fn<-function(b) {
    b2<-matrix(b[2:(d+1)],d,1); gx<-b[1]+x%*%b2
    ro<-rho(y-gx); return(sum(ro)/n)
}
```

Then we use function "optim," which performs the numerical optimization. As the initial value for the optimization we give the solution of the least squares regression.

```
li<-linear(x,y); par<-c(li$beta0,li$beta1)#initial value
op<-optim(par=par,fn=fn,method="L-BFGS-B")
beta0<-op$par[1]        # the intercept
beta1<-op$par[2:(d+1)]  # the coefficients
```

A.8.2 Kernel Quantile Regression

Function "pcf.kern.quan" implements kernel quantile regression.

```
pcf<-pcf.kern.quan(x,y,h=0.5,N=50,p=0.1)
dp<-draw.pcf(pcf); plot(x,y); lines(dp$x,dp$y,type="l")
```

Let us look at the code of function "pcf.kern.quan." We want to calculate the estimate at point `arg`. First we calculate the weights, similarily as in kernel mean regression.

```
arg<-1; h<-0.5; p<-0.1
ker<-function(x){ (1-x^2)*(x^2<=1) }
w<-ker((arg-x)/h); ps<-w/sum(w)
```

Then we implement the rule given in (3.55). The estimate of p-quantile is "hatq."

```
or<-order(y); ps.ord<-ps[or]; i<-1; zum<-0
while ((i<=n) && (zum<p)){zum<-zum+ps.ord[i]; i<-i+1}
if (i>n) hatq<-max(y) else hatq<-y[or[i]]
```

REFERENCES

Aaron, C. (2013). Estimation of the support of the density and its boundary using random polyhedrons, *Technical report*, Université Blaise Pascal.

Abegaz, F., Gijbels, I. & Veraverbeke, N. (2012). Semiparametric estimation of conditional copulas, *J. Multivariate Anal.* **110**: 43–73.

Ai, C. (1997). A semiparametric maximum likelihood estimator, *Econometrica* **65**: 933–963.

Aït-Sahalia, Y. & Brandt, M. W. (2001). Variable selection for portfolio choice, *J. Finance* **56**(4): 1297–1351.

Akaike, H. (1973). Maximum likelihood identification of Gaussian autoregressive moving average models, *Biometrika* **60**: 255–265.

Andrews, D. (1972). Plots of high-dimensional data, *Biometrika* **28**: 125–136.

Andriyashin, A., Härdle, W. & Timofeev, R. (2008). Recursive portfolio selection with decision trees, *SFB 649 Discussion paper 009*, Humboldt-Universität zu Berlin.

Bauwens, L., Laurent, S. & Rombouts, V. K. (2006). Multivariate GARCH models: A survey, *J. Appl. Econ.* **21**: 79–109.

Bellman, R. E. (1961). *Adaptive Control Processes*, Princeton University Press, Princeton, NJ.

Bertin, J. (1967). *Semiologie Graphique*, Gauthier Villars, Paris.

Bertin, J. (1981). *Graphics and Graphic Information-Processing*, de Gruyter, Berlin.

Besbeas, P., de Feis, I. & Sapatinas, T. (2004). A comparative simulation study of wavelet shrinkage estimators for Poisson counts, *Int. Statist. Rev.* **72**: 209–237.

Billingsley, P. (2005). *Probability and Measure*, Wiley, New York.

Bollerslev, T. (1986). Generalized autoregressive conditional heteroscedasticity, *J. Econometrics.* **31**: 307–327.

Bollerslev, T. (1990). Modeling the coherence in short-run nominal exchange rates: A multivariate generalized ARCH model, *Rev. Econ. Statist.* **31**: 307–327.

Bollerslev, T., Engle, R. F. & Wooldridge, J. M. (1988). A capital asset pricing model with time-varying covariances, *J. Political Econ.* **96**: 116–131.

Bouchaud, J.-P. & Potters, M. (2003). *Theory of Financial Risks*, Cambridge University Press, Cambridge.

Bougerol, P. & Picard, N. (1992). Stationarity of GARCH processes and some nonnegative time series, *J. Econometrics* **52**: 115–127.

Box, G. E. P. & Cox, D. R. (1962). An analysis of transformations, *J. Roy. Statist. Soc. Ser. B* **26**: 211–252.

Brandt, M. W. (1999). Estimating portfolio and consumption choice: A conditional Euler approach, *J. Finance* **54**: 1609–1646.

Breiman, L. (1996). Bagging predictors, *Machine Learning* **24**: 123–140.

Breiman, L. (2001). Random forests, *Machine Learning* **45**: 5–32.

Breiman, L., Friedman, J., Olshen, R. & Stone, C. J. (1984). *Classification and Regression Trees*, Chapman and Hall, New York.

Brockwell, P. J. & Davis, R. A. (1991). *Time Series: Theory and Methods*, 2nd edn, Springer, Berlin.

Brown, L., Cai, T. T. & Zhou, H. H. (2010). Nonparametric regression in exponential families, *Ann. Statist.* **38**: 2005–2046.

Brown, L. D. (1986). *Fundamentals of Statistical Exponential Families with Applications in Statistical Decision Theory*, IMS, Hayward, CA.

Bühlmann, P. & Yu, B. (2003). Boosting with the L2 loss: regression and classification, *J. Am. Statist. Assoc.* **98**: 324–339.

Buja, A., Hastie, T. & Tibshirani, R. (1989). Linear smoothers and additive models, *Ann. Statist.* **17**(2): 453–510.

Carlsson, G. (2009). Topology and data, *Bulletin Am. Math. Soc.* **46**(2): 255–308.

Carr, D. B., Littlefield, R. J., Nicholson, W. L. & Littlefield, J. S. (1987). Scatterplot matrix techniques for large N, *J. Am. Statist. Assoc.* **82**: 424–436.

Carroll, R. J., Fan, J. Q., Gijbels, I. & Wand, M. P. (1997). Generalized partially linear single-index models, *J. Am. Statist. Assoc.* **92**: 477–489.

Chaudhuri, P. & Marron, J. S. (1999). Sizer for exploration of structures in curves, *J. Am. Statist. Assoc.* **94**: 807–823.

Chernoff, H. (1973). Using faces to represent points in k-dimensional space graphically, *J. Am. Statist. Assoc.* **68**: 361–368.

Cook, D., Buja, A. & Cabrera, J. (1993). Projection pursuit indexes based on orthonormal function expansions, *J. Comput. Graph. Statist.* **2**(3): 225–250.

Cook, D., Buja, A., Cabrera, J. & Hurley, C. (1995). Grand tour and projection pursuit, *J. Comput. Graph. Statist.* **4**(3): 155–172.

Dahlhaus, R. (1997). Fitting time series models to nonstationary processes, *Ann. Statist.* **25**: 1–37.

Delecroix, M., Härdle, W. & Hristache, M. (2003). Efficient estimation in conditional single-index regression, *J. Multivariate Anal.* **86**(2): 213–226.

Delecroix, M. & Hristache, M. (1999). M-estimateurs semi-paramétriques dans les modéles à direction révélatrice unique, *Bull. Belg. Math. Soc.* **6**: 161–185.

Diebold, F. X. & Mariano, R. S. (1995). Comparing predictive accuracy, *J. Bus. Econ. Statist.* **13**: 225–263.

Donoho, D. L. (1997). Cart and best-ortho-basis: A connection, *Ann. Statist.* **25**: 1870–1911.

Duan, N. & Li, K.-C. (1991). Slicing regression: a link-free regression method, *Ann. Statist.* **19**: 505–530.

Efron, B. (1982). Transformation theory: How normal is a family of a distributions?, *Ann. Statist.* **10**: 323–339.

Engle, R. F. (1982). Autoregressive conditional heteroscedasticity with estimates of the variance of U.K. inflation, *Econometrica* **50**: 987–1008.

Engle, R. F. (2002). Dynamic conditional correlation: A simple class of multivariate generalized autoregressive conditional heteroskedasticity models, *J. Bus. Econ. Statist.* **20**: 339–350.

Engle, R. F. & Kroner, K. F. (1995). Multivariate simultaneous generalized ARCH, *Econometric Theory* **11**: 122–150.

Fan, J. & Gu, J. (2003). Semiparametric estimation of Value at Risk, *Econometrics J.* **6**: 261–290.

Fan, J. & Yao, Q. (1998). Efficient estimation of conditional variance functions in stochastic regression, *Biometrika* **85**: 645–660.

Fan, J. & Yao, Q. (2005). *Nonlinear Time Series*, Springer, Berlin.

Flury, B. & Riedwyl, H. (1981). Graphical representation of multivariate data by means of asymmetrical faces, *J. Am. Statist. Assoc.* **76**: 757–765.

Flury, B. & Riedwyl, H. (1988). *Multivariate Statistics: A Practical Approach*, Cambridge University Press, Cambridge.

Franke, J., Härdle, W. & Hafner, C. M. (2004). *Statistics of Financial Markets: An Introduction*, Springer, Berlin.

Friedman, J. H. & Stuetzle, W. (1981). Projection pursuit regression, *J. Am. Statist. Assoc.* **76**: 817–823.

Friedman, J. H. & Tukey, J. (1974). A projection pursuit algorithm for exploratory data analysis, *IEEE Trans. Comput.* **C-23**: 881–889.

Fung, W. & Hsieh, D. A. (2004). Hedge fund benchmarks: A risk based approach, *Financial Analyst Journal* **60**: 65–80.

Gasser, T. & Müller, H.-G. (1979). Kernel estimation of regression functions, *Smoothing Techniques for Curve Estimation*, Vol. 757 of *Lecture Notes in Mathematics*, Springer, New York, pp. 23–68.

Gasser, T. & Müller, H.-G. (1984). Estimating regression functions and their derivatives by the kernel method, *Scand. J. Statist.* **11**: 171–185.

Gasser, T., Sroka, L. & Jennen-Steinmetz, C. (1986). Residual variance and residual pattern in nonlinear regression, *Biometrika* **73**: 625–633.

Gijbels, I., Pope, A. & Wand, M. P. (1999). Understanding exponential smoothing via kernel regression, *J. R. Statist. Soc. B* **61**: 39–50.

Giraitis, L., Kokoszka, P. & Leipus, R. (2000). Stationary ARCH models: Dependence structure and central limit theorem, *Econometric Theory* **16**: 3–22.

Györfi, G., Lugosi, G. & Udina, F. (2006). Nonparametric kernel-based sequential investment strategies, *Mathematical Finance* **16**(2): 337–357.

Györfi, L., Kohler, M., Krzyzak, A. & Walk, H. (2002). *A Distribution-Free Theory of Nonparametric Regression*, Springer, Berlin.

Györfi, L., Ottucsác, G. & Walk, H. (2012). *Machine Learning for Financial Engineering*, Imperial College Press, London.

Györfi, L. & Schäfer, D. (2003). Nonparametric prediction, *in* J. A. K. Suykens, G. Horváth, S. Basu, C. Micchelli & J. Vandevalle (eds), *Advances in Learning Theory: Methods, Models and Applications*, IOS Press, NATO Science Series, pp. 339–354.

Györfi, L., Udina, F. & Walk, H. (2008). Nonparametric nearest neighbor based empirical portfolio selection strategies, *Statistics and Decisions* **22**: 145–157.

Györfi, L., Urbán, A. & Vajda, I. (2007). Kernel-based semi-log-optimal empirical portfolio selection strategies, *Int. J. Theor. Appl. Finance* **10**(5): 505–516.

Hall, P. & Carroll, R. (1989). Variance function estimation in regression: The effect of estimating the mean, *J. R. Statist. Soc. B* **51**: 3–14.

Hall, P. & Horowitz, J. L. (2005). Nonparametric methods for inference in the presence of instrumental variables, *Ann. Statist.* **33**(6): 2904–2929.

Hall, P., Kay, J. & Titterington, D. (1990). Asymptotically optimal difference-based estimation of variance in nonparametric regression, *Biometrika* **77**: 521–528.

Hall, P., Kay, J. & Titterington, D. (1991). On estimation of noise variance in two-dimensional signal processing, *Adv. Appl. Probab.* **23**: 476–495.

Hall, P. & Marron, J. (1990). On variance estimation in nonparametric regression, *Biometrika* **77**: 415–419.

Hansen, L. P. (1982). Large sample properties of generalized method of moments estimators, *Econometrica* **50**: 1029–1054.

Härdle, W. (1990). *Applied Nonparametric Regression*, Cambridge University Press, Cambridge.

Härdle, W. & Mammen, E. (1993). Testing parametric versus nonparametric regression, *Ann. Statist.* **21**: 1926–1947.

Härdle, W., Müller, M., Sperlich, S. & Werwatz, A. (2004). *Nonparametric and Semiparametric Models*, Springer, Berlin.

Härdle, W. & Simar, L. (2003). *Applied Multivariate Statistical Analysis*, Cambridge University Press, Cambridge.

Härdle, W. & Tsybakov, A. B. (1993). How sensitive are average derivatives?, *J. Econometr.* **58**: 31–48.

Härdle, W. & Tsybakov, A. B. (1997). Local polynomial estimators of the volatility function in nonparametric autoregression, *J. Econometr.* **81**: 223–242.

Hastie, T. J. & Tibshirani, R. J. (1990). *Generalized Additive Models*, Chapman and Hall, London.

Hastie, T. & Tibshirani, R. (1993). Varying-coefficient models, *J. Roy. Statist. Soc. Ser. B* **55**: 757–796.

Hastie, T., Tibshirani, R. & Friedman, J. (2001). *The Elements of Statistical Learning: Data Mining, Inference, and Prediction*, Springer, Berlin.

Hoerl, A. E. & Kennard, R. W. (1970). Ridge-regression: Biased estimation for nonorthogonal problems, *Technometrics* **8**: 27–51.

Horowitz, J. L. (2009). *Semiparametric and Nonparametric Methods in Econometrics*, Springer, Berlin.

Hristache, M., Juditsky, A., Polzehl, J. & Spokoiny, V. (2001). Structure adaptive approach for dimension reduction, *Ann. Statist.* **29**: 1537–1566.

Hristache, M., Juditsky, A. & Spokoiny, V. (2001). Direct estimation of the index coefficient in a single-index model, *Ann. Statist.* **29**(3): 595–623.

Huber, P. J. (1964). Robust estimation of a location parameter, *Ann. Math. Statist.* **53**: 73–101.

Huber, P. J. (1985). Projection pursuit, *Ann. Statist.* **13**(2): 435–475.

Hull, J. C. (2010). *Risk Management and Financial Institutions*, 2nd edn, Pearson Education, Boston.

Ibragimov, I. A. & Linnik, Y. V. (1971). *Independent and Stationary Sequencies of Random Variables*, Walters-Noordhoff, Gröningen.

Ichimura, H. (1993). Semiparametric least squares (SLS) and weighted SLS estimation of single-index models, *J. Econometrics* **58**: 71–120.

Ichimura, H. & Lee, L.-F. (1991). Semiparametric least squares estimation of multiple index models: Single equation estimation, *in* W. A. Barnett, J. Powell & G. Tauchen (eds), *Nonparametric and Semiparametric Methods in Econometrics and Statistics*, Cambridge University Press, Cambridge, pp. 3–50.

Inselberg, A. (1985). The plane with parallel coordinates, *Visual Computer* **1**: 69–91.

Inselberg, A. (1997). Multidimensional detective, *Proceedings IEEE Information Visualization'97*, IEEE, Washington, DC, pp. 100–107.

Inselberg, A. & Dimsdale, B. (1990). Parallel coordinates: A tool for visualizing multi-dimensional geometry, *Proceedings IEEE Information Visualization'90*, IEEE, Washington, DC, pp. 361–378.

James, W. & Stein, C. (1961). Estimation with quadratic loss, *Proceedings Fourth Berkeley Symposium on Mathematical Statistics and Probability*, Vol. 1, pp. 361–379.

JPMorgan (1996). *RiskMetrics–Technical Document*, 4th edn, JPMorgan, New York.

Karttunen, K., Holmström, L. & Klemelä, J. (2014). Level set trees with enhanced marginal density visualization: Application to flow cytometry, *Proceedings 5th International Conference on Information Visualization Theory and Applications*.

Klein, R. W. & Spady, R. H. (1993). An efficient semiparametric estimator for binary response, *Econometrica* **61**: 387–421.

Klemelä, J. (2004). Visualization of multivariate density estimates with level set trees, *J. Comput. Graph. Statist.* **13**(3): 599–620.

Klemelä, J. (2006). Visualization of multivariate density estimates with shape trees, *J. Comput. Graph. Statist.* **15**(2): 372–397.

Klemelä, J. (2007). Visualization of multivariate data with tail trees, *Inform. Visualization* **6**: 109–122.

Klemelä, J. (2009). *Smoothing of Multivariate Data: Density Estimation and Visualization*, Wiley, New York.

Koenker, R. (2005). *Quantile Regression*, Cambridge University Press, Cambridge.

Koenker, R. & Bassett, G. (1978). Regression quantiles, *Econometrica* **46**: 33–50.

Kohler, M. (1999). Nonparametric estimation of piecewise smooth regression functions, *Statist. Probab. Lett.* **43**: 49–55.

Koltchinskii, V. I. (2008). *Oracle Inequalities in Empirical Risk Minimization and Sparse Recovery Problems*, Vol. 2033 of *Lecture Notes in Mathematics*, Springer, Berlin.

Korostelev, A. P. & Korosteleva, O. (2010). *Mathematical Statistics: Asymptotic Minimax Theory*, Vol. 119 of *Graduate Studies in Mathematics*, AMS, Providence, RI.

Li, K.-C. (1991). Sliced inverse regression for dimension reduction (with discussion), *J. Am. Statist. Assoc.* **86**: 316–342.

Li, K.-C. & Duan, N. (1989). Regression analysis under link violation, *Ann. Statist.* **17**: 1009–1052.

Li, Q. & Racine, S. (2007). *Nonparametric Econometrics: Theory and Practice*, Princeton University Press, Princeton, NJ.

Linton, O. B. (2009). Semiparametric and nonparametric ARCH modeling, *in* T. G. Andersen, R. A. Davis, J.-P. Kreiss & T. Mikosch (eds), *Handbook of Financial Time Series*, Springer, New York, pp. 157–167.

Linton, O. B. & Mammen, E. (2005). Estimating semiparametric ARCH(∞) models by kernel smoothing methods, *Econometrica* **73**: 771–836.

Linton, O. & Nielsen, J. P. (1995). A kernel method of estimating structured nonparametric regression based on marginal integration, *Biometrika* **82**: 93–100.

Malevergne, Y. & Sornette, D. (2005). *Extreme Financial Risks: From Dependence to Risk Management*, Springer, Berlin.

Mallows, C. L. (1973). Some comments on C_p, *Technometrics* **15**: 661–675.

Mammen, E., Linton, O. & Nielsen, J. (1999). The existence and asymptotic properties of a backfitting projection algorithm under weak conditions, *Ann. Statist.* **27**: 1443–1490.

Mammen, E. & Tsybakov, A. B. (1999). Smooth discrimination analysis, *Ann. Statist.* **27**: 1808–1829.

Mandelbrot, B. (1963). The variation of certain speculative prices, *J. Business* **36**(4): 394–419.

Markowitz, H. (1952). Portfolio selection, *J. Finance* **7**: 77–91.

Markowitz, H. (1959). *Portfolio Selection*, Wiley, New York.

McClellan, M., McNeil, B. & Newhouse, J. P. (1994). Does more intensive treatment of acute myocardial infarction in the elderly reduce mortality, *J. Am. Med. Assoc.* **272**(11): 859–866.

McCullagh, P. & Nelder, J. A. (1989). *Generalized Linear Models*, Vol. 37 of *Monographs on Statistics and Applied Probability*, 2nd edn, Chapman and Hall, London.

McNeil, A. J., Frey, R. & Embrechts, P. (2005). *Quantitative Risk Management: Concepts, Techniques, and Tools*, Princeton University Press, Princeton, NJ.

Minnotte, M. C. & West, B. W. (1999). The data image: A tool for exploring high dimensional data sets, *1998 Proceedings ASA Section on Statistical Graphics*, pp. 25–33.

Morgan, J. N. & Sonquist, J. A. (1963). Problems in the analysis of survey data, and a proposal, *J. Am. Statist. Assoc.* **58**: 415–434.

Müller, H. & Stadtmüller, U. (1987). Estimation of heteroscedasticity in regression analysis, *Ann. Statist.* **15**: 610–625.

Munk, A., Bissantz, N., Wagner, T. & Freitag, G. (2005). On difference-based variance estimation in nonparametric regression when the covariate is high dimensional, *J. R. Statist. Soc. B* **67**: 19–41.

Nadaraya, E. A. (1964). On estimating regression, *Theory Probab. Appl.* **10**: 186–190.

Nelder, J. A. & Wedderburn, R. W. M. (1972). Generalized linear models, *J. Roy. Statist. Soc., Ser. A* **135**(3): 370–384.

Nelsen, R. B. (1999). *An Introduction to Copulas*, Springer, Berlin.

Neumann, M. H. (1994). Fully data-driven nonparametric variance estimation, *Statistics* **25**: 189–212.

Newey, W. K. & West, K. D. (1987). A simple, positive semi-definite, heteroskedasticity and autocorrelation consistent covariance matrix, *Econometrica* **55**(3): 703–708.

Nielsen, J. P. & Sperlich, S. (2005). Smooth backfitting in practice, *J. Roy. Statist. Soc. B* **67**: 43–61.

Peligrad, M. (1986). Recent advances in the central limit theorems and its weak invariance principle for mixing sequencies of random variables (a survey), *Dependence in Probability and Statistics*, Birkhäuser, Boston, pp. 193–223.

Powell, J. L., Stock, J. H. & Stoker, T. M. (1989). Semiparametric estimation of index coefficients, *Econometrica* **51**: 1403–1430.

Priestley, M. B. & Chao, M. T. (1972). Non-parametric function fitting, *J. Roy. Statist. Soc. Ser. B* **34**: 385–392.

Rebonato, R. (2007). *Plight of the Fortune Tellers; Why We Need to Manage Financial Risk Differently*, Princeton University Press, Princeton, NJ.

Reeb, G. (1946). Sur les points singuliers d'une forme de pfaff completement integrable ou d'une fonction numerique, *Comptes Rend. Acad. Sci. Paris* **222**: 847–849.

Rice, J. (1984). Bandwidth choice for nonparametric regression, *Ann. Statist.* **12**: 1215–1230.

Ruppert, D. (2004). *Statistics and Finance*, Springer, New York.

Ruppert, D., Wand, M., Holst, U. & Hössjer, O. (1997). Local polynomial variance-function estimation, *Technometrics* **39**: 262–272.

Ruppert, D., Wand, M. P. & Carroll, R. J. (2003). *Semiparametric Regression*, Cambridge University Press, Cambridge.

Samarov, A. M. (1991). On asymptotic efficiency of average derivative estimates, *Nonparametric Functional Estimation and Related Topics*, Vol. 335, NATO Advanced Science Institutes Series C, pp. 167–172.

Samarov, A. M. (1993). Exploring regression structure using nonparametric functional estimation, *J. Am. Statist. Assoc.* **88**: 836–847.

Scheffé, H. (1959). *The Analysis of Variance*, Wiley, New York.

Seber, G. A. F. (1977). *Linear Regression Analysis*, Wiley, New York.

Silvennoinen, A. & Teräsvirta, T. (2009). Multivariate GARCH models, *in* T. G. Andersen, R. A. Davis, J.-P. Kreiss & T. Mikosch (eds), *Handbook of Financial Time Series*, Springer, New York, pp. 201–232.

Simonoff, J. S. (1996). *Smoothing Methods in Statistics*, Springer, Berlin.

Sklar, A. (1959). Fonctions de répartition à n dimensions et leurs marges, *Publ. Inst. Statist. Univ. Paris* **8**: 229–231.

Sornette, D. (2003). *Why Stock Markets Crash*, Princeton University Press, Princeton, NJ.

Spokoiny, V. (2000). Multiscale local change point detection with applications to value-at-risk, *Ann. Statist.* **37**(3): 1405–1436.

Spokoiny, V. (2002). Variance estimation for high-dimensional regression models, *J. Multiv. Anal.* **82**: 111–133.

Spokoiny, V. (2010). *Local Parametric Methods in Nonparametric Estimation*, Springer, Berlin.

Stone, C. J. (1985). Additive regression and other nonparametric models, *Ann. Statist.* **13**: 689–705.

Sun, J. & Loader, C. R. (1994). Simultaneous confidence bands for linear regression and smoothing, *Ann. Statist.* **22**: 1328–1345.

Thompson, A., Kay, J. & Titterington, D. (1991). Noise estimation in signal restoration using regularization, *Biometrika* **78**: 475–488.

Tibshirani, R. (1996). Regression shrinkage and selection via the LASSO, *J. Roy. Statist. Soc. B* **58**(1): 267–288.

Tjøstheim, D. & Auestadt, B. (1994). Nonparametric identification of nonlinear time series: Projections, *J. Am. Statist. Assoc.* **89**: 1398–1409.

Tukey, J. (1957). The comparative anatomy of transformations, *Ann. Math. Statist.* **28**: 602–632.

Tukey, J. (1961). Curves as parameters, and touch estimation, *Proceedings 4th Berkeley Symposium*, pp. 681–694.

Tukey, J. (1977). *Exploratory Data Analysis*, Addison-Wesley, Reading, MA.

Vapnik, V. V. (1995). *The Nature of Statistical Learning Theory*, Springer, Berlin.

Vapnik, V. V., Golowich, S. E. & Smola, A. J. (1997). Support vector method for function approximation, regression estimation and signal processing, *in* M. Mozer, M. I. Jordan & T. Petsche (eds), *Advances in Neural Information Processing Systems 9*, MIT Press, Cambridge, MA, pp. 281–287.

von Neumann, J. (1941). Distribution of the ratio of the mean squared successive difference to the variance, *Ann. Math. Statist.* **12**: 367–395.

Wahba, G. (1990). *Spline Models for Observational Data*, Society for Industrial and Applied Mathematics, Philadelphia.

Wang, L., Brown, L. D., Cai, T. & Levine, M. (2008). Effect of mean on variance function estimation on nonparametric regression, *Ann. Statist.* **36**: 646–664.

Wasserman, L. (2005). *All of Nonparametric Statistics*, Springer, New York.

Watson, G. S. (1964). Smooth regression analysis, *Sankhya Ser. A* **26**: 359–372.

Wegman, E. J. (1990). Hyperdimensional data analysis using parallel coordinates, *J. Am. Statist. Assoc.* **85**(411): 664–675.

Weisberg, S. & Welsh, A. H. (1994). Adapting for the missing link, *Ann. Statist.* **22**: 1674–1700.

White, H. (1980). Heterscedasticity-consistent covariance matrix estimator and a direct test for heteroscedasticity, *Econometrica* **48**: 817–838.

White, H. (1982). Instrumental variables regression with independent observations, *Econometrica* **50**: 483–499.

Wong, H., Ip, W.-c. & Zhang, R. (2008). Varying-coefficient single-index model, *Comput. Statist. Data Anal.* **52**: 1458–1476.

Wooldridge, J. M. (2005). *Econometric Analysis of Cross Section and Panel Data*, MIT Press, Cambridge, MA.

Yu, K. & Jones, M. (2004). Likelihood-based local linear estimation of the conditional variance, *J. Am. Statist. Assoc.* **99**: 139–144.

Zhang, W., Lee, S. & Song, X. (2002). Local polynomial fitting in semivarying coefficient model, *J. Multivariate Anal.* **82**: 166–188.

Ziegelmann, F. A. (2002). Nonparametric estimation of volatility functions: The local exponential estimator, *Econometric Theory* **18**: 985–991.

Zomorodian, A. (2010). Fast construction of the Vietoris–Rips complex, *Computer and Graphics* **34**: 263–271.

Zomorodian, A. (2012). Topologocal data analysis, *in* A. Zomorodian (ed.), *Advances in Applied and Computational Topology*, Vol. 70, American Mathematical Society, pp. 1–40.

AUTHOR INDEX

Multivariate Nonparametric Regression and Visualization. By Jussi Klemelä

TOPIC INDEX